T0251225

THEORIES AND PRACTICE
IN INTERACTION DESIGN

HUMAN FACTORS AND ERGONOMICS
Gavriel Salvendy, Series Editor

Aykin, N. (Ed.): *Usability and Internationalization of Information Technology*

Bagnara, S., Crampton Smith, G., (Eds.): *Theories and Practice in Interaction Design*

Carayon, P. (Ed): *Handbook of Human Factors and Ergonomics in Health Care and Patient Safety*

Hendrick, H., & Kleiner, B. (Eds.): *Macroergonomics: Theory, Methods, and Applications*

Hollnagel, E. (Ed.): *Handbook of Cognitive Task Design*

Jacko, J. A., & Sears, A. (Eds.): *The Human–Computer Interaction Handbook: Fundamentals, Evolving Technologies, and Emerging Applications*

Karwowski, W. (Ed.): *Handbook of Standards and Guidelines in Ergonomics and Human Factors*

Meister, D. (Au.): *Conceptual Foundations of Human Factors Measurement*

Meister, D., & Enderwick, T. (Eds.): *Human Factors in System Design, Development, and Testing*

Proctor, R., & Vu, K. (Eds.): *Handbook of Human Factors in Web Design*

Schmorrow, D. (Ed.): *Foundations of Augmented Cognition*

Stanney, K. (Ed.): *Handbook of Virtual Environments: Design, Implementation, and Applications*

Stephanidis, C. (Ed.): *User Interfaces for All: Concepts, Methods, and Tools.*

Wogalter, M. (Ed.): *Handbook of Warnings*

Ye, N. (Ed.): *The Handbook of Data Mining*

Also in this Series

HCI International 1999 Proceedings 2 Volume Set
HCI International 2001 Proceedings 3 Volume Set
HCI International 2003 Proceedings 4 Volume Set
HCI International 2005 Proceedings 11 Volume CD Rom Set
ISBN# 0-8058-5807-5

THEORIES AND PRACTICE IN INTERACTION DESIGN

Edited by

Sebastiano Bagnara
Politecnico di Milano

Gillian Crampton Smith
Interaction Design Institute Ivrea

INTERACTION DESIGN INSTITUTE IVREA
Ivrea, Italy

CRC Press is an imprint of the
Taylor & Francis Group, an **informa** business

First published 2006 by Lawrence Erlbaum Associates, Inc.

Published 2019 by CRC Press
Taylor & Francis Group
6000 Broken Sound Parkway NW, Suite 300
Boca Raton, FL 33487-2742

© 2006 by Taylor & Francis Group, LLC
CRC Press is an imprint of Taylor & Francis Group, an informa business

No claim to original U.S. Government works

ISBN 13: 978-0-8058-5618-7 (hbk)

This book contains information obtained from authentic and highly regarded sources. Reasonable efforts have been made to publish reliable data and information, but the author and publisher cannot assume responsibility for the validity of all materials or the consequences of their use. The authors and publishers have attempted to trace the copyright holders of all material reproduced in this publication and apologize to copyright holders if permission to publish in this form has not been obtained. If any copyright material has not been acknowledged please write and let us know so we may rectify in any future reprint.

Except as permitted under U.S. Copyright Law, no part of this book may be reprinted, reproduced, transmitted, or utilized in any form by any electronic, mechanical, or other means, now known or hereafter invented, including photocopying, microfilming, and recording, or in any information storage or retrieval system, without written permission from the publishers.

For permission to photocopy or use material electronically from this work, please access www. copyright.com (http://www.copyright.com/) or contact the Copyright Clearance Center, Inc. (CCC), 222 Rosewood Drive, Danvers, MA 01923, 978-750-8400. CCC is a not-for-profit organization that provides licenses and registration for a variety of users. For organizations that have been granted a photocopy license by the CCC, a separate system of payment has been arranged.

Trademark Notice: Product or corporate names may be trademarks or registered trademarks, and are used only for identification and explanation without intent to infringe.

**Visit the Taylor & Francis Web site at
http://www.taylorandfrancis.com**

**and the CRC Press Web site at
http://www.crcpress.com**

Cover design by Kathryn Houghtaling-Lacey

Library of Congress Cataloging-in-Publication Data

Theories and practice in interaction design / edited by Sebastiano
  Bagnara, Gillian Crampton Smith.
      p.   cm.
  Includes bibliographical references and index.
ISBN 0-8058-5618-8 (alk. paper)
1. Human-machine systems. 2. Human-computer interaction.
  I. Bagnara, Sebastiano.   II. Crampton Smith, Gillian.

TA167.T54   2005
620.8'2—dc22   2005052979
                                                         CIP

The graceful English of this volume
is due to the dedication of our editor, Philip Tabor,
who strove to maintain the clarity and sparkle
of every contribution.

Contents

Contributors

Sebastiano Bagnara is Professor of Cognitive Ergonomics at Milan Polytechnic. Previously head of the Communication Sciences department, Siena University, and President of the European Association of Cognitive Ergonomics, he is currently General Secretary of the International Ergonomics Association. He is a member of the editorial board of several international scientific journals, and associate editor of *Theoretical Issues in Ergonomic Sciences.*

Massimo Bergamasco is Associate Professor of Applied Mechanics at the Scuola Superiore Sant'Anna, Pisa, where he founded and now directs its PERCRO laboratory and is now establishing a new center of applied research, CODE (Collaborative Design). At Pisa University, he teaches Mechanics of Robots in the Mechanical Engineering faculty and Virtual Environments in the Computer Science faculty.

Maria A. Brandimonte is Professor of Learning and Memory at Suor Orsola Benincasa University, Naples, where she leads the Laboratory of Experimental Psychology. Her research interests include visual memory, imagery, and prospective memory. In 1992 she received a Carnegie Mellon Fellowship Award and in 1999, an Early Career Achievement Award from the European Society for Cognitive Psychology (ESCoP). She is associate editor of the *European Journal of Cognitive Psychology* and *Cognitive Technology,* and consulting editor of *Psicologica.*

Cristiano Castelfranchi is Professor of Cognitive Sciences at Siena University, and Director of the Institute of Cognitive Sciences and Technologies of the National Research Council, Rome. A cognitive scientist with a background in linguistics and psychology, he is active in the artificial intelligence and the social simulation communities. He is an advisory member of several international conferences, societies, and journals like *Cognitive Science, Autonomous Agents and MAS, Cognitive Science Quarterly,* and the MIT CogNet, and is a fellow of the European Coordinating Committee for Artificial Intelligence.

Claudio Ciborra was Professor of Information Systems at the London School of Economics, Professor at the University of Languages and Communication (IULM), Milan, and Visiting Professor at the HEC School of Management, Paris, and universities including those of Oslo, Gothenburg, and New York. He wrote many articles, books, and reports on new technologies and organization, consulted for European and U.S. public and private organizations, and was keynote speaker at international conferences on information systems. Claudio Ciborra died on February 13, 2005.

Gillian Crampton Smith is Director of Interaction Design Institute Ivrea, in Ivrea, Italy. After studying philosophy and history of art at Cambridge University, she spent the 1970s as a designer, first in book publishing, then on the *Sunday Times* and *Times Literary Supplement.* In 1983 she joined Saint Martin's School of Art, London, where she set up a new postgraduate course, Graphic Design and Computers for Practicing Designers. In 1989 she moved to the Royal College of Art to establish the Computer Related Design (CRD, now Interaction Design) department, where artists and designers apply their traditional skills to interactive products and systems. Under her guidance, the CRD Research Studio achieved an international reputation as a leading center for interaction design, supported by a wide range of industrial and government sponsors.

Giorgio De Michelis is Director of the Informatics, Systems, Communications department (DISCo) at Milano-Bicocca University, Milan, where he teaches Theoretical Computer Science and Information Systems. His research focuses on models of concurrent systems (Petri nets), computer-supported cooperative work, community-ware, knowledge management, and interaction design. His research group is developing prototype support systems for cooperative processes and knowledge management systems. He is on the editorial board of *Computer Supported Cooperative Work: The Journal of Collaborative Computing* and *Studi Organizzativi* and has written four volumes and more than a hundred articles.

Roberta De Monticelli is Professor of the Philosophy of the Person at the San Raffaele University, Milan, and former Professor of Modern and Contemporary Philosophy at Geneva University. Her books include *L'Avenir de la phenomenologie, L'ordine del cuore,* and *L'allegria della mente.*

Pelle Ehn is Professor at the School of Arts and Communication at Malmö University, Sweden, where he directs the Interaction Design PhD program. For three decades he has played a major role within participatory design research. During the last decade he has also been engaged in shaping a true design discipline in information technology. In 1998, he cofounded the Digital Bauhaus. His many volumes include *Work-Oriented Design of Computer Artifacts, Manifesto for a Digital Bauhaus,* and *Participatory Design and the Collective Designer.*

Yrjö Engeström is Professor of Communication at the University of California, San Diego, Professor of Adult Education at Helsinki University, and Director of the Center for Activity Theory and Developmental Work Research, a Finnish Centre of Excellence. He applies and develops cultural-historical activity theory to studies of work activities and learning and development processes associated with transformations in organizations, communities, and technologies. His current research focuses on the emergence of coconfiguration, "knotworking," and other new forms of collaborative work, expansive learning and horizontal expertise in workplaces and organizational fields, particularly health care and education. He has just finished the book *Collaborative Expertise: Expansive Learning in Medical Work.*

Thomas Erickson practices interaction design and research at the IBM T. J. Watson Research Center. His work focuses on designing computer-mediated communication systems, and his approach to systems design is shaped by theories, methods, and analytical techniques drawn from architecture, psychology, rhetoric, sociology, and urban design. Over the last two decades, he has published about 50 articles and been involved in the design of over a dozen systems. Before joining IBM in 1997 he spent nine years at Apple Research, 5 years at a small start-up company, and 5 years studying Cognitive Psychology at the University of California, San Diego.

Maurizio Franzini is Professor of Economic Policy and Director of the Inter-University Research Center on the Welfare State (CRISS) and of the PhD program in Economics at the University of Rome "La Sapienza." He coordinates the Economy-Environment Interaction research area of the European Association for Evolutionary Political Economy (EAEPE) and has coedited the volumes *The Politics and Economics of Power* and *Environment, Inequality and Collective Action.*

Charles Goodwin is Professor of Applied Linguistics at the University of California, Los Angeles, and on the core faculty of the Sloan Center on Everyday Lives of Families. His interests and teaching include video analysis of talk-in-interaction, grammar in context, cognition in the lived social world, gesture, gaze, and embodiment as interactively organized social practices, aphasia in discourse, language in the professions, and the ethnography of science. As part of the Workplace Project at Xerox PARC he investigated cognition and talk-in-interaction in a complex work setting: ground operations at a major airport. Recent publications include *Conversation and Brain Damage, Action and Embodiment within Situated Human Interaction, Practices of Seeing, Time in Action,* and *The Body in Action.*

Marco Guerini is a PhD student at the Institute for Scientific and Technological Research (IRST), Trento, Italy. He is currently working on mechanisms for flexible automatic persuasion.

Isao Hosoe received a BSc and a Masters in Aerospace Engineering at Nihon University, Tokyo. In 1985 he founded Isao Hosoe Design in Milan. He has won numerous international awards for his design contributions to transportation, office furniture, products, telecommunications, electronics, and domestic furniture, has had numerous personal exhibitions, and taken part in collective exhibitions in Italy, Europe, and the United States. He has also been a visiting professor in many European, U.S., and Japanese universities.

Giovanna Leone is Professor of Social Psychology at Bari University, Italy. Her research focuses on social and collective dimensions of autobiographical memory, and the psychological costs of seeking and receiving help.

Patrizia Marti is Professor of Educational Technologies at the Communication Science department of Siena University. She has been a principal researcher on European- and Italian-funded projects in technologies for learning, robotics, and health care. Her current research includes the design of human activities in context (situated interaction), taking into account the multisensorial, emotional, and sociocultural aspects of interaction with artifacts. She has edited special issues of international journals like *Cognition, Technology and Work* and *Travail Humain.*

Mario Mattioda, an interaction designer with a background in psychology and psychotherapy, is concerned with the role of hermeneutics in design. His projects have been mainly related to health systems, focusing on the design and implementation of interfaces for managing and comparing scientific data and patients' files. He is affiliated to Interaction Design Institute Ivrea.

Bill Moggridge founded Moggridge Associates in 1969, expanding it throughout the 1970s with clients worldwide. In 1979 he added a second office in San Francisco and in 1980 designed the first laptop computer, the GRiD Compass. Over the next few years he pioneered user interface design as a discipline to be an integrated part of product development, and coined the phrase "interaction design." In 1991 he merged his company with David Kelley and Mike Nuttall to form IDEO. Throughout his career he has been active in design education at London's Royal College of Art, the London Business School, and Stanford University.

Thomas Moran was a pioneer in establishing the field of human–computer interaction (HCI) within computer science in the early 1980s. He was Principal Scientist and manager of the User Interface and the Collaborative Systems groups at Xerox PARC (1974–2001) and founding Director of Xerox EuroPARC in Cambridge, England (1986–1990). In 2001 he joined IBM as a Distinguished Engineer, where he leads a multilab research program on Unified Activity Management. He founded the journal *Human–Computer Interaction* in 1984, and continues as its Editor. He is an Association for Computing Machinery (ACM) Fellow and recipient of ACM SIGCHI's Lifetime Achievement Award.

Donald A. Norman is cofounder of the Nielsen Norman Group, Professor at Northwestern University, Evanston, Illinois, and former Vice President at Apple Computer. He serves on many advisory boards, such as those of Chicago's Institute of Design and *Encyclopedia Britannica*. He is a member of the Industrial Design Society of America and fellow of many organizations, including ACM, the American Association of Arts and Sciences, and the Cognitive Science Society. He is an Emeritus Professor of the University of California, San Diego, and has an honorary degree from the University of Padua. He wrote *The Design of Everyday Things* and most recently, *Emotional Design: Why We Love (or Hate) Everyday Things*.

Andrew Ortony is Professor of Education, Psychology, and Computer Science at Northwestern University, Evanston, Illinois, and codirects its Center for the Study of Cognition, Emotion, and Emotional Disorders. He is a cognitive scientist whose primary research interests are in the relation among emotion, cognition, behavior, and personality. He has conducted theoretical and empirical studies on emotion, and is the major author (with Gerald Clore and Allan Collins) of a landmark book, *The Cognitive Structure of Emotions*, the core of which—the OCC (Ortony, Clore, and Collins) model—is widely used in computational models of emotion.

Lucio Picci directs the Internet Economics degree program at Bologna University. Most of his scientific contributions cover the intersection of economics and information and communications technology. As a consultant, he has represented the Italian Prime Minister's Office within the World Wide Web Consortium's Advisory Committee.

Sharon Helmer Poggenpohl is a faculty member at the Institute of Design, Illinois Institute of Technology, Chicago, where she teaches masters and PhD courses in design practice, research, and theory, and, with Keiichi Sato, coordinates the PhD Design program. Her research interests relate to new media opportunities in communication, and she edits and publishes the journal *Visible Language*. She has served as vice president for education on the boards of the American Institute of Graphic Arts and the American Center for Design, has been honored as a Master Teacher by the Graphic Design Education Association, received the Education Award from the American Center for Design, and been recognized as an Outstanding Teacher by Mortar Board.

Oliviero Stock is Senior Fellow of the Institute for Scientific and Technological Research (IRST), Trento, Italy. He has been chairman of the European Coordinating Committee for Artificial Intelligence (ECCAI), president of the Italian Association for Artificial Intelligence (AI*IA), president of the Association for Computational Linguistics (ACL), and a ECCAI Fellow. His research concentrates on intelligent multimodal interfaces and natural language processing artificial intelligence, human–computer interaction, cognitive technologies, and knowledge-based systems. He has authored over 100 articles and authored or edited nine volumes. He is also on the editorial boards of a dozen scientific journals, including *Artificial Intelligence,* and associate editor of *Applied Artificial Intelligence.* He has lectured at several international universities and schools.

Philip Tabor is Visiting Professor at Interaction Design Institute Ivrea and the Bartlett School of Architecture, University College London (UCL). A registered architect, he worked in the studios of James Stirling and of O'Neil Ford (San Antonio), and as a partner in Edward Cullinan Architects (London). He was later the Professor of Architectural Criticism and Theory at UCL, and the Director of the Bartlett. His publications focus on the psychological and cultural effects of information technology.

Federico Vercellone is Ordinary Professor of Aesthetics at Udine University, Italy. Recent publications include *Introduzione al nichilismo* (1992), *Nature del tempo: Novalis e la forma poietica del romanticismo tedesco* (1998),

L'estetica dell'ottocento: morfologie del moderno (2002), and *Storia dell'estetica moderna e contemporanea* (2004). He is on the Scientific Committee of *Estetica* and is consultant for *Iride* and editor for *Filosofia e Teologia*. He is also on the Scientific Committee for the Stuttgart publication series *Wissenschaftskultur um 1900*.

Massimo Zancanaro is a researcher at the Institute for Scientific and Technological Research (IRST), Trento, Italy, where he is responsible for the Intelligent Interactive Information Presentation (i3P) research line in its Cognitive and Communication Technologies division. He recently coedited the volume *Multimodal Intelligent Information Presentation*.

Series Foreword

With the rapid introduction of highly sophisticated computers, (tele)communication, service, and manufacturing systems, a major shift has occurred in the way people use technology and work with it. The objective of this book series on Human Factors and Ergonomics is to provide researchers and practitioners a platform where important issues related to these changes can be discussed, and methods and recommendations can be presented for ensuring that emerging technologies provide increased productivity, quality, satisfaction, safety, and health in the new workplace and the Information Society.

This volume is published at a very opportune time, when the Information Society Technologies are emerging as a dominant force, both in the workplace, and in everyday life. For these new technologies to be truly effective, they must provide communication modes and interaction modalities across different languages and cultures, and should accommodate the diversity of requirements of the user population at large, including disabled and elderly people, thus making the Information Society universally accessible, to the benefit of mankind.

The 22 chapters of *Theories and Practice in Interaction Design* provide pointers to interesting perspectives on human activity from disciplines surrounding interaction design—such as psychology, philosophy, economics, design—which can enrich our understanding of what it means to design interactive systems for everyday life. This book provides valuable insights for professional designers of interactive systems into how to integrate the social, psychological, and emotional variables that contribute to

the effectiveness of system operation. The volume is also intended as a ho-
rizon-expanding textbook for graduate students in the various disciplines
involved in interaction design.

—Gavriel Salvendy
Series Editor
Purdue University, United States
and
Tshingua University, P. R. China

Introduction

Sebastiano Bagnara
Politecnico di Milano

Gillian Crampton Smith
Interaction Design Institute Ivrea

THEORY AND DESIGN

Interaction design is a practice: It intervenes in the world. And the term *practice* conjures up, as its complement, *theory.*

In its narrowest sense a theory is an imaginative conjecture refutable by empirical observation, usually in the form of quantified experiment, or a coherent structure of such conjectures; thus defined, theory is properly the realm of the physical and life sciences, and those aspects of the human sciences susceptible to measurement. But "theory" is also commonly used to mean the constantly evolving configuration of epistemological assumptions, conceptual constructs, methodologies, and critical values that flow around and through individual practices and fields of study, contributing to their wisdom and power. To avoid confusion this introduction refers to this looser meaning of "theory" as *discourse.*

This distinction is needed here because interaction design (the design of systems in which people and artifacts engage each other in usually computer-assisted interactivity), due to its association with "logical devices," might wrongly be assumed particularly amenable to theory in the narrow sense of scientific method.

The last half of the last century, especially the 1960s and 1970s, saw a strong movement to apply scientific method to design. Modeled on the positivist "scientific world-conception" of the Vienna Circle, it envisaged design as a kind of experiment in which a design problem is defined in terms of "performance criteria," rather like truth criteria, open to unambiguous, usually quantitive, evaluation. Empirical knowledge structured by "theory" is then applied to it, a process that supposedly generates a range of hypothetical solutions, which in turn are tested against the criteria. Eventually the designer converges on an optimal, or at least satisfactory, solution. If this process fails to deliver, the assumption is that it needs to be conducted more rigorously, the problem must be defined more tightly, more data is required, or the theoretical principles need reframing. But there is strong resistance against admitting into the equation larger, more complex "metaphysical" phenomena, such as esthetic or moral values.

This model resulted in a disastrous, if temporary, reorganization of the curriculum in some architecture schools; but it had little impact on the practice and study of design generally. The reason lies in the nature of design. Its broadest definition, "changing existing situations into preferred ones" (Simon, 1972, p. 55), catches design's active, interventionist nature and hints at its value-laden intentionality; but it omits the association of design, in normal parlance, with material outcomes and social and symbolic use. A more common definition of the activity of design is therefore something like "to shape and make our environment in ways without precedent in nature, to serve our needs and give meaning to our lives" (Heskett, 2002, p. 7).

It is characteristic of most design tasks, although not all of course, that although they may be initially presented in terms of the utilitarian ends to be achieved, those ends are formulated in terms either too general to define the criteria of success, or so narrow that they ignore the human context in which any solution must operate. Nobody is to blame for this. Most design, and perhaps especially interaction design, concerns human reaction. It is thus about mental phenomena, such as the intermingling of intellectual and affective response or of esthetic and moral value, which can only be described in fragmentary and provisional form, whose dimensions are many and intrinsically incommensurable. The context of a design task of any interest is therefore complex and ambiguous, the criteria of success contradictory and unstable. This is why the positivist model of knowledge, so productive in some endeavors, has proved inadequate when applied to design. It is also why theory in the narrow sense can play only a partial role in the discourse of design generally, and of interaction design.

INTERACTION DESIGN

The complexity of interaction design's domain is evident in its parentage. On one side of its family is human–computer interaction (HCI), whose history (Baecker & Buxton, 1990) is the coming together of hardware and software engineering, and physiological and cognitive ergonomics; on the other, a range of design practices and discourses including those of industrial design, graphic design, architecture, and film—each of which has a medium requiring a particular set of skills and mental attitudes. (The term *HCI*, incidentally, was more widely used in the United States than in Continental Europe, where interaction with computers was studied in the broader context of cognitive ergonomics—indicating a different approach.)

HCI emerged, mostly in the 1960s, to fill the lack of techniques and tools available suitable for designing the digital world made possible by the development of computers. Its mating with design to produce interaction design, which can probably be dated to the 1980s, was because digital products were increasingly aimed at a lay consumer (rather than professional) usership, a diffusion later dramatically accelerated by the development of networks and telecommunications. The brief history of interaction design can thus be seen as a collective effort to translate the wealth of tools already available for designing the physical world to enrich the design of the digital world.

The initial problem was to design the computer, the machine itself and its interface. Later, when computing became distributed and embodied in many everyday appliances, people sought tools and models for designing digital devices that were easy to understand and use. More recently, now that digital technology has transformed how we get food delivered, buy a book, order and pay the bill at a restaurant, the need is for tools to design digital services. Nowadays, when not just single components but a whole environment may become digital, as happens in some immersive virtual entertainment environments, the need is for tools that allow us to design complete digital *experiences*. Indeed, the design of experiences, through the medium of interaction design, may be seen to represent a growing proportion of the design industries' production.

However, interaction design is more than directing traditional design techniques to new applications: Interactive technologies are a new medium that also requires its own techniques. Some it has borrowed, such as ethnographic observation and role-play scenarios. Others it has developed for itself to model the interactive experience, like "paper prototyping," software simulations (using Director or Flash), and video scenario-sketching.

There is another perspective from which one may look at interaction design: the number and character of its users. It is well known that *interactive sys-*

tems (from now on, we use this term to refer to all interactive and digital machines, objects, services, and experiences), unlike previous technologies, have become diffused very rapidly and the type of user has changed equally rapidly. As David Liddle recounts (quoted in Bill Moggridge's chapter in this book), the original, expert users, who shared the expertise of the technology's designers, were soon outnumbered by enthusiast users, who, although not sharing the designers' background, like them enjoyed technological novelty and did not mind, and even enjoyed, the effort necessary to learn the systems. These two groups, experts and enthusiasts, were not large and raised no problems for the designer. The experts used the new tool in a working situation similar to that of the designer, while the enthusiast played with it in a context culturally homogeneous to that of the designer. Between these groups was no cultural, cognitive, or competence gap.

A technology becomes widely diffused when employed in many working contexts by professionals, and reaches mass diffusion when sold in the consumer market and used in home and leisure contexts. For these two groups, professionals and laymen, it is vital that the new tools are designed to be easy to use. Enthusiasts are willing to put much energy into discovering what a new tool is good for and how it operates. But professionals, and consumers even more, are unwilling to spend effort discovering its possibilities. They want it to be "usable"—a term usually denoting cognitive transparency and ergonomic facility: They want the tool to make clear what it can do in various contexts and what they need to do to operate it successfully, and to be easy and quick to learn. A tool without these characteristics will probably fail in the market. So the "usability" of interfaces plays a crucial role in achieving real diffusion of interactive systems.

Except in those computer games and other contexts where mystery and challenge are essential elements of the experience, good interaction design makes interfaces usable in this sense. And, as usability is indeed one goal of interaction designers, it might be assumed that interaction design is simply design for usability. This assumption implies a restricted and conservative view of interaction design. Restricted, because the interface (the component of the interactive system by which it and the user communicate with each other), although essential, is only one of its components; and interface design, although by no means culturally neutral, has a more limited purview than interaction design as a whole, which must consider the wider cultural, social, and organizational contexts of use. This view is conservative, because as usability aims to provide users with effortless interaction, minimizing learning time, a new object must as much as possible be designed to use only the existing skills and knowledge of users.

From the point of view of interaction design, the design-for-usability approach largely underevaluates users' dynamics in learning and experiencing, which are of course rooted in the past but guided by the future because any such experience is undertaken to reach a goal or fulfill an intention. Interaction design considers past experience as a set of possible constraints on the future possibilities they aim to imagine. It is both realistic, in that it anchors the future in the opportunities offered by the past, and proactive and risky, because it anticipates future behaviors. This is why, in interaction design, any scenario of future activities or experience (essential for building successful systems) must be deeply rooted in a detailed ethnographical analysis of a current, dynamic situation: Interaction design never builds a scenario from a single snapshot view of a context.

In a more comprehensive perspective, interaction design takes into consideration that any interactive system also develops and transforms the dynamics of human behavior, changes experiences and the way people make experiences. So interaction design, like most other kinds of design, is often a conscious and explicit endeavor to imagine and design human behaviors, in this case behaviors in the digital world, by creatively exploiting the potential of digital technologies.

It goes without saying that interaction design would be inconceivable without information and communication technologies, which have changed radically over interaction design's short life. When these technologies started to develop, they were typically used to automate an existing task within an existing context. Designers had a relatively straightforward objective: to simplify how people interact with computers. They designed sets of simple actions for commanding, controlling, and monitoring the computer—a new machine, certainly, but still a machine. At this point, in short, interaction design was still in the industrial era. But the context soon changed: The integration of computing and telecommunication technologies altered how time and space were experienced in work and everyday life. People no longer had to carry out actions required by the machine—whether computer or not—at the time it demanded (Zuboff, 1988) and colocation became less critical. Designers no longer aimed only to make operations faster and more reliable: Technology was instead seen as a chance to invent new activities, to design new ways of living, hitherto not only unrealizable but almost unimaginable.

Interaction design now has the opportunity to design technological artifacts for human freedom rather than ameliorating the inconveniences and demands associated with their use. Interaction designers, fortunately, are no longer expected to design only operational, strictly instrumental, activities. Nowadays, they are concerned with designing contexts of living where *desires* as well as needs must be satisfied. This is a unique opportunity and one implying great social responsibility.

SEVEN THEMES AND AN OVERVIEW

A young practice like interaction design is in a hurry. It must do more borrowing than inventing. So this volume aims to map out the surrounding context of discourse—the assumptions, concepts, categories, methods, and indeed theories of other practices and disciplines—from which interaction design might profitably beg, borrow, or steal.

Our map therefore includes contributions from the study and practice of cognitive psychology, computer science, discourse analysis, economics, engineering, linguistics, philosophy, psychology, semiotics, and sociology. But it is not panoptic. Most obviously (the invisible elephant in our lounge) it makes little reference to the vast body of codified discourse on art and design that has developed since the Renaissance and, in the case of architecture, over the past two millennia. For this book we thought it more valuable to welcome areas of endeavor external to design as it is usually defined.

A symposium, held at Interaction Design Institute Ivrea in November 2003, related issues in interaction design to ideas and methods developed in other, largely non-design, fields. Although not its transcript, this volume shares the symposium's structure. It identifies seven themes (activity, emotion, situatedness, community, conversation, memories, and market) to each of which two or three authors contribute an essay. Each essay discusses some aspects of a discourse drawn from another field, and their possible application to interaction design. Together they constitute a rich commonwealth of ideas and methods. Equally interesting, for the interaction designer, they reveal the wide range of (perhaps unconscious) epistemological assumptions and even ethical priorities that drive these fields.

Activity.   The name of this first theme makes reference to activity theory, a cultural-historical conceptual system first developed by the psychologist Alexei Leontiev (1978), active in the 1930s in the Soviet Union. The theory took on new life when further developed in the 1980s by European and American researchers (Michael Cole, for instance) to explain the mutual relation between mind and artifacts. It maintains that any artifact is the outcome of a process by which mind builds up objects having personal sense and cultural meaning. This process is social because, although subjective and intentional, it uses existing artifacts and its outcome is available in the social environment. Furthermore, the same process modifies mind itself because it acquires, in building the artifact, the cognitive and social skills embedded in the artifact it constructs and the artifacts it uses to do so. Mind is social in nature. To be meaningful, any artifact has to be embedded in an activity system: a complex, interactive, organized set of actions, artifacts, and people. Any activity system adapts to and influences its living context. Both activity system and context are dynamic: They have a history. Intervening in an artifact means to intervene in the history of the context and in that of the activity system.

The first contribution to our book, Yrjö Engeström's, describes activity theory and illustrates it through its application to the design of a medical care system for chronic patients with multiple illnesses. Thomas Moran, in his essay, uses the term activity in a more precise and restricted sense: a set of actions, not necessarily contiguous in time, but coherent in that they support the same goal. He describes how activity management tools can be designed following this definition. Massimo Bergamasco adopts an even more restricted definition of activity: "the actions of the hands in doing or performing a task," such as manipulating an object, exploring shapes in the real or virtual world, or in remote operations. He applies this definition in a thorough description of the achievements in the design of haptic interfaces for virtual environments.

Emotion. Until very recently the HCI field did not consider emotions, perhaps because one of its main influences, cognitive psychology, had devoted scant attention to affective phenomena. Emotions were seen as at best unwelcome factors that could disturb the cognitive processes involved in using a designed object. HCI sought effectiveness and efficiency, and, unlike more established design fields, paid little attention to feelings and esthetic response. In the past few years, however, emotion has become a topic of increasing interest to engineers and HCI experts. They have conceded that there can be no experience without emotions.

The philosopher Roberta De Monticelli introduces the theme here with a thorough analysis of the complex "phenomenon of the affective or emotional sphere." By adopting the phenomenological approach, she discriminates between sensibility, a "disposition of reception," from feeling, "essentially perception of the positive or negative values of things." A feeling is an evaluation. The qualities considered for evaluation can be many or few, and the evaluations can be ordered by ranking the weight of their values. Feeling may vary in breadth and depth inasmuch as emotions are manifestations of feeling. Both feelings and emotions are related to personal identity that, through experience and socialization, allows us to discriminate diverse feelings and assign them either positive or negative value. De Monticelli also distinguishes emotions, mainly short-lived and reactive, from passions, long-lasting "habitual channels of volition."

This structured view of affective life is intriguing, stimulating the reader's curiosity and offering the designer many insights, but does not pretend to be a tool for immediate use in designing emotional interactions. The contribution by Donald Norman and Andrew Ortony, by contrast, more closely addresses designers' needs. It simplifies the emotional life to three levels of emotional response: visceral (perceptually based), behavioral (expectation based), and reflective (intellectually based). These distinctive responses are rooted in either biology, experience, or the personal, subjective elaboration of one's culture. With these distinctions in mind, the designer can try to induce or avoid particular emotions in a product's users.

Patrizia Marti concludes the theme by analyzing the emotions that take place in the interactions between humans and robots, especially when the robots stimulate feelings like tenderness, docility, and caring.

Situatedness.   As already noticed, the notion that any object needs to be considered in its context of use is common throughout interaction design practice. Situation is often thought as anchoring and providing sense to human knowledge when engaged with previously distributed pieces of knowledge (cognitive artifacts). Learning, too, is always rooted in a situation (Suchman, 1987). However, Claudio Ciborra—by academic background and professional role a computer scientist, but by mindset and passion a theoretician—argues here against the interpretation, in HCI discourse, but also in that of artificial intelligence (AI) and cognitive science, of "situatedness" as referring to just the network of relations an object has in the world. This, he believes, impoverishes Heidegger's original concept of situatedness, which also includes the inner life, the intentions as well as the emotions and mood, of the actor who understands and lives the situation. Situatedness is the finding of oneself in a situation: This notion captures the invisible ties between the inner and the outer world on which the understanding of meaning is based.

Gillian Crampton Smith and Philip Tabor's chapter examines the notion of inner situatedness by distinguishing emotions, which have particular objects, from moods, which in Heideggerian terms respond to the individual's "throwness" in life. They distinguish both emotions and moods, however, from the frame of mind or mental stance with which a user relates to an instrument or device. They then demonstrate that the complexity and instability of most design tasks has defeated attempts to reduce designing to methodology. So designers—through their training, practice, and discourse—have traditionally inhabited a situatedness especially oriented toward the inner world of values and meanings. And they do so both introspectively, as designers, and through empathetic imagination, as surrogates of prospective users.

Conversation.   Interaction is often commonly understood as a prolonged and dynamic exchange of communication among human and artificial agents to reach a goal. Because most human exchanges are verbal, conversation is generally accepted as the metaphor, although difficult to imitate, for designing interactions. Cristiano Castelfranchi's chapter challenges this view, introducing behavioral implicit communication (BIC) theory, which considers communication to consist of actions where messages are not sent explicitly but are instead traces left in a shared environment. He shows that much social control, collaboration, monitoring, and coordination are based on these kinds of nonlinguistic behavior. He also argues that interaction design in domotics (digital technologies applied to the home) and robotics should rely heavily on BIC and move toward a collaborative paradigm where artificial and human agents coordinate with each other on the basis of their actions rather than their (verbal) interactions.

Mario Mattioda and Federico Vercellone, by contrast, do indeed see dialog as a useful metaphor for human–system interactions, allowing those interactions to be perceived as "natural." They also suggest that the dialog metaphor can be used to identify the user's real needs and leads to a "case-centered" methodology. The contribution by Oliviero Stock, Marco Guerini, and Massimo Zancanaro, finally, describes the characteristics of persuasive intelligent interfaces and discusses the use of persuasion in HCI, and the ethical issues it raises.

Community.   The idea of community, introduced in the late 19th century by German sociologists, has gained new popularity at the end of the 20th. The word originally meant a quasi-autonomous group of people, usually living in one place, that depend on and help each other—whose roles, that is, are complementary. The modern meaning is again a group of people, but now defined not by location but by a shared interest or activity: the "scientific community," for instance. People stay in a community because their shared interests are better served communally than alone.

Anthropology and sociology have studied at length how communities originate, live, and eventually disappear, and how newcomers become competent members of a community. Charles Goodwin's contribution uses a linguistic-anthropology framework to describe the process through which members of very restricted communities demonstrate their understanding of the events they are engaged in by actions contributing to the further progression of the same events. He unfolds the process by which they participate in communal activities, and shows how cognition is linked to interactive organization. The framework and methodology adopted, applied here to a clinical case, offer valuable insight for designing tools for supporting communities in general.

Participation is the topic of Pelle Ehn, a prominent figure in the field of participatory design since the 1970s. His contribution outlines the advance in participation from the work-related design of computer artifacts, centered on industrial democracy and quality of work, to the design of information technologies in context, culminating in a manifesto for a creative and socially useful meeting of art and technology. It provides an historical view of what might be seen as a particularly Scandinavian approach to interaction design: social, participatory, and esthetically concerned.

Memory.   When HCI started, human memory was seen as a problem because of its untrustworthiness and the difficulty of accessing it. The first guidelines for designing interfaces underlined the spatial and temporal limitations within which memory operated, and stressed its unreliability in recalling complex procedures at the precise moment required. Interfaces greatly improved when designers realized that human memory is better accessed through images and icons than by alphanumeric strings.

The study of memory has greatly advanced since the 1980s. Giovanna Leone's essay reviews recent research. She challenges the once prevalent view that memory is an essentially individual faculty, by showing how much social interaction influences memory processes. Theories, intuitions, and even methodologies, developed in the 1930s by eminent figures like Bartlett and Vygostsky, have recently been brought back to productive life to describe the social dimensions of remembering. She also shows the value of shared social memories in keeping communities alive.

Community memory is also discussed by Giorgio De Michelis, a computer scientist who is very much oriented toward the humanities. His chapter offers recommendations, derived from both theoretical considerations and the practice of a large project, for designing interactive systems supporting the memories of large communities. Maria Brandimonte discusses prospective memory, how we remember to act in the future. She recalls that Winograd (1988) noticed that if retrospective memory fails, the person's *memory* is seen as unreliable, but if prospective memory fails, the *person* is seen as unreliable. So people have for a long time designed tools for remembering actions to be done in the future: One might call this the origin of distributed cognition, which plays a crucial role in everyday life and, inevitably, in interaction design.

Market. Interaction design is also an activity that produces economic value, because its outcomes are products and services that are eventually sold in the market. The two economists contributing to this volume both start from the notion of interaction cost, a fundamental dimension in economics for establishing the price of goods. Lucio Picci reminds "those designers whose mission is to do their very best in the service of the final user of a product" why the market often prefers intentionally badly designed goods; this line of reasoning is intriguing in that it may help explain why some high-quality interaction design products and services have not met with commercial success. Within the same framework, Maurizio Franzini notices that interaction costs may not be visible and clear-cut, preventing consumers from balancing them against the benefits obtained by buying interactive systems. He speculates that the benefits of interaction design may be of particular value to a segment of consumers who need but cannot afford them. If so, he suggests provocatively, access to interaction design's products might be offered not just through the market but as a social service.

The contribution by Bill Moggridge, finally, a pioneer of interaction design, discusses relations with the market, but also the evolution of the field, through six stories of creative innovation. It is both an autobiography and a reasoned review of the problems and solutions interaction design has met along its short but rich history of failures and successes.

In Search of a Framework.   Our book ends with three essays that do not fall neatly within any of our seven themes but seek, each in its own way, a more panoramic view of the field. Sharon Helmer Poggenpohl suggests the elements for a comprehensive framework for interaction design, reminding us of the relevance of games to interaction design, indeed of the spectacular interactivity of computer games. Thomas Erickson advances an alternative view of the field's organization, suggesting five "lenses" through which to look at interaction design: mind, proxemics, artifacts, the social, and the ecological. He concludes, wisely in our view, by recommending that interaction design should accumulate its theoretical apparatus eclectically but cautiously, taking care that it is sufficiently articulated to accommodate nuanced distinctions for use within the field, but not so elaborate that it cannot be understood outside. Borrowed constructs, similarly, should be not be so cumbersome that they import the irrelevant debris of another field's debates, yet not so slight that they cannot express a useful complexity.

The last contribution, by Isao Hosoe, a successful designer, Japanese by birth, who for more than 30 years has worked and lived in Milan, offers a taste of a different way of feeling and thinking. The idea of having eight senses, the notion of behavioral energy, and the mixed references to Western and oriental philosophies may motivate readers, as it did us, to confront a different way of theorizing and practicing not only interaction design but living.

TERRITORIES AND LANGUAGES

Most of the contributors to this volume are associated with teaching, research, or both, and can thus be said to inhabit a particular territory of discourse or practice. When such a territory defines itself, or is defined, as a distinct discipline or a guild of common interest—and especially when it becomes institutionalized as a university department or professional association—its members feel a strong pressure to cluster around a limited set of axiomatic assumptions and a uniform apparatus of taxonomies, terminologies, and techniques. This pressure derives from the common work to be done but is also, as Michel Foucault (1970) pointed out, social in origin and effect. To that extent, disciplines and professions tend to resemble nations, as Lucio Picci mischievously points out in his chapter. Each seeks to maintain internal solidarity and hierarchy, guard its territorial integrity, and project its presence externally.

Much of what we have called discourse, as the contributions to this book show, is taxonomic. To pin down a phenomenon or critical nuance in which it has a special interest, but for which no existing name quite fits, a field will assign it a label, sometimes a new word or phrase but usually a familiar word used in a special way, and assemble it with others into an explanatory, or at least descriptive, tool. It is tempting to call such constructions a language, or at least a dialect (a language, it has been said, is a dialect with an army and a navy). It is certainly true that the proprietorial claims associated with a real language come into play: a feeling that a discipline owns the moral copyright of the terms, particularly if they are neologisms that it invented— like *affordance* or *situatedness,* two words whose application to interaction design is queried in this volume. The fear is that such usages, once they escape their home discipline, might lose their original meaning and build an interdisciplinary Babel.

We acknowledge the danger of misunderstanding and irritation when a term is used in different ways in different fields of practice, and the need therefore to define one's terms carefully. But we would argue that a linguistic free market, in which words' meanings can adapt themselves to different circumstances and purposes, is more likely than linguistic protectionism to provide an activity with the precise conceptual tools it needs for the job in hand. It is, moreover, an exaggeration to characterize a discipline's discourse as its "language" when one is really referring—even in the highly technical sciences—to a small, if specialized, fraction of the total (and commonplace) lexicon it actually uses to communicate with itself and the outside world.

Where, then, is interaction design located in this metaphorical world of "territories" and "languages?" Although interaction design is relatively young, the education and experience of most of its practitioners make them deeply versed in at least one strong and older discourse like architecture and other forms of design, anthropology, computer science, or film-making. As well as inheriting some specialized terms and ideas from these original activities, they also borrow and bend some more for their own purpose, and invent a few of their own. They have their own journals and conferences, and are beginning to find a place in the academy. Interaction design, in short, is forging its own discourse—necessarily a robust one, since forged in the unforgiving fire of practice, but also one that overlaps with other discourses.

But this does not make it a "discipline" in the established and semi-autonomous sense that architecture, anthropology, or psychology are disciplines (although established disciplines, seen from the inside, are of course never as coherent and homogeneous as they seem from the outside). And, as we peer into the future, it is not entirely clear whether interaction design, as a "discursive community of practice," would gain more than it would lose from a significantly greater conceptual uniformity and guild solidarity.

It would be misleading—to return to the territorial metaphor—to characterize any contributor to this book as some kind of ambassador of his or her disciplinary nation, bringing the civilizing influence of its older language and culture to the younger (and less orderly?) tribe of interaction design, and nervously hoping that the indigenes will not steal and misuse its carefully honed tools. For interaction designers are not islanders. They inhabit the same continent of culture and knowledge as everyone else and—again like everyone else, if they are wise—seek, use, and if necessary refashion whatever conceptual tools best suit the current task.

To encourage such beneficent eclecticism this book offers a glimpse of part of that conceptual continent that adjoins the usual hunting grounds of interaction designers. But its contributors are not ambassadors, still less missionaries. They approach each other, instead, as generous traders in a free market of ideas.

The chapter by Claudio Ciborra was his last work. He died just a few days after formally approving its translation and editing. We miss a friend, and, as his chapter shows, the world has lost a brilliant mind.

Part

I

ACTIVITY

Chapter 1

Activity Theory
and Expansive Design

Yrjö Engeström
University of Helsinki, Finland

FIVE ARGUMENTS

In this chapter, I approach interaction design from the viewpoint of cultural-historical activity theory (Engeström, Miettinen & Punamäki, 1999; Leontiev, 1978). My central arguments may be condensed in the following five:

1. Interaction design needs to be embedded, integrated, and made visible in the activity systems within which the targeted products and services are produced and used.
2. Interaction design needs to be directed not only at products but also at relations, processes, services, organizations, and, most importantly, at "germ-cell" concepts or visions of the future activity.
3. Expansive interaction design creates integrated instrumentalities, not only isolated products. It operates by anchoring its ideas and outcomes upward, downward, and sideways.
4. Expansive interaction design is best performed jointly by producer-practitioners and their key customers, supported by interventionists. This requires special reflective intervention methods, "microcosms" which combine joint negotiated decision making, joint future-oriented envisioning, and simulation of future modes of interaction across boundaries.

5. Expanded in these ways, interaction design tends to merge with implementation and learning; expansive design, expansive implementation, and expansive learning are (so to speak) three sides of the same coin.

To open up and substantiate these arguments, I first introduce some central concepts of cultural-historical activity theory. After that, I sketch steps in the evolution of design, and, in particular, the emergent mode of coconfiguration work as a landscape where interaction design will increasingly take place. Next, I present the case of interaction design within the medical care of chronic patients with multiple illnesses. I then move on to discuss certain key features of expansive design in the context of coconfiguration work as exemplified in the medical care case. This leads me to discuss the interventionist methodology needed for enhancing and studying expansive design. I conclude by discussing briefly the relations between design, implementation, and learning.

CENTRAL CONCEPTS OF ACTIVITY THEORY

Cultural-historical activity theory looks at artifacts and people as embedded in dynamic activity systems (see Fig. 1.1). If we think of a designer as the *subject* of his or her design work, the initial *object* would be an idea, order, or assignment that triggers the design process. The initial object is necessarily ambiguous, requiring interpretation and conceptualization. Thus, the object is invested step-by-step with personal sense and cultural meaning. The object goes through multiple transformations until it stabilizes as a finished *outcome*, for example a prototype or even a commercial product. This process is only possible by means of *mediating artifacts*, both material tools and signs. The designer may use pencil and paper, clay models, or three-dimensional computer graphics, along with internalized images and concepts that seem relevant for the forging of the object. The process alters, sometimes even generates, entirely new mediating artifacts.

The preceding paragraph describes the uppermost subtriangle of the activity system depicted in Fig. 1.1. The bottom part of the figure calls attention to the work *community* in which the designer is a member, for example a product development unit or an in-house design unit of a corporation, or perhaps an independent design firm. Within the community, the members continuously negotiate their *division of labor*, including the distribution of rewards. The temporal rhythms of work, the uses of resources, and the codes of conduct are also continuously constructed and contested in the form of explicit and implicit *rules*.

For designers, as for any practitioners involved in complex organized activity, making sense of their own work as a collective activity system repre-

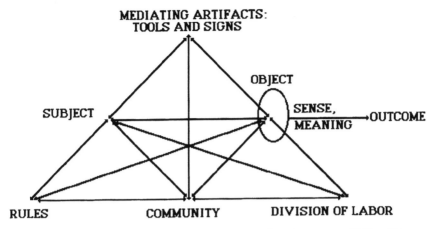

FIG. 1.1.   The structure of a human activity system (from Engeström, 1987, p. 78).

sents an expansive challenge of "visibilization." But this is only the first step. Opening up and making visible the activity systems of key customers or users is the logical second step of expansion. This requires that the unit of analysis be extended to include, minimally, two interconnected activity systems. In Fig. 1.2, the triangle on the left represents the activity system of the designer, the one on the right represents the activity system of the customer or user. Of course, the network of relevant activity systems is often more complex, including for example multiple design teams, subcontractors, internal client units within the corporation, and external end-user customers.

The formation of a partially shared object between the designer and the customer or user is a crucial challenge. In Fig. 1.2, Object 1 represents the initial problem, assignment, or "raw material" of the design process. Object 2 represents an elaborated image, vision, or prototype of the object. Object 3 stands for the potential common ground or synergy between the two perspectives.

Design is an activity that easily becomes self-absorbed. The emerging object of design tends to become an object of affection, an end in itself (Engeström & Escalante, 1996). This may lead to the assumption that the object will have the same centrality and appeal for the end user as it has for the designer. From the point of view of the user, the designer's product is commonly expected to be simply a tool, an instrument among many others. If the product turns into an object that requires constant attention, it often becomes a source of frustration rather than affection for the user, especially when the user is given no or minimal tools to handle, understand, and modify the supposedly self-explanatory object. Expansive design is demanding

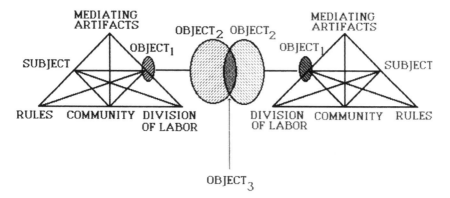

FIG. 1.2.   Two interacting activity systems as a minimal unit of analysis for expansive design (from Engeström, 2001, p. 133).

as it requires successive radical shifts of perspective in critical transitions between design and use.

THE EVOLUTION OF DESIGN

Without a substantive understanding of the historically changing character of the work done in a given organization, theories of design are likely to remain too general and abstract to capture the past vestiges and the emerging possibilities of design. Victor and Boynton (1998) provided a useful historical framework for such a reintegration of organization, work, and design. They identified five ideal types of work in the history of industrial production: craft, mass production, process enhancement, mass customization, and coconfiguration.

Each type of work generates and requires a certain dominant type of knowledge and design. In craft, the worker and the designer were essentially merged in one and the same person, the master craftsman. In mass production, design was concentrated in the hands of engineers and radically separated from the execution of work. Mass production also creates the demand for professional designers whose task is typically focused on giving a "final touch" of visual and tactile consumer appeal to products.

In process enhancement, frontline workers are given responsibilities for the continuous improvement of processes and products, whereas the development of new products and processes is still firmly kept in separate design units. As quality becomes of crucial importance, professional designers are increasingly used in assisting development projects with their particular insights.

In mass customization, the customer is brought into the design process by being offered the chance to put together a unique "personalized" combination of available standard components. Even in mass customization, the development of truly new products and processes remains separate from the actual production, but feedback from customer choices has an increasingly speedy and direct impact on product and process development efforts. Professional designers often become true members of product development teams.

ON THE EMERGENCE OF CO-CONFIGURATION WORK

In his ingenious film of 1967, *Playtime*, Jacques Tati followed in great detail the final steps of construction and the opening night of a restaurant. As the opening approaches, workmen are still finalizing various details of the physical structure. Already before the first guests arrive, some details begin to fall apart, and frantic efforts are made to fix them. As the evening progresses, there are increasingly dramatic breakdowns. Workmen, restaurant personnel, the architect, various customers, and accidental passers-by get involved in intricate schemes of repair and improvisation, dispersed in unexpected ways in time and space. It becomes evident that this restaurant will never be "finished"; the activities of serving and dining become saturated with simultaneous actions of coping with the falling apart and reconfiguration of the restaurant itself.

Tati envisioned something not unlike coconfiguration work, as defined by Victor and Boynton (1998). An observer characterizes the coconfiguration efforts she witnessed in the planning, design, and implementation of an information infrastructure for a city district as follows: "The actors are like blind players who come eagerly to the field in the middle of the game, attracted by shouting voices, not knowing who else are there and what the game is all about. There is no referee, so rules are made up in different parts of the field among those who happen to bump into one another. Some get tired and go home" (oral communication).

A critical prerequisite of coconfiguration is the creation of customer-intelligent products or services which adapt to the changing needs of the user:

> The work of co-configuration involves building and sustaining a fully integrated system that can sense, respond, and adapt to the individual experience of the customer. When a firm does co-configuration work, it creates a product that can learn and adapt, but it also builds an ongoing relationship between each customer-product pair and the company. Doing mass customization requires designing a product at least once for each customer. This design process requires the company to sense and respond to the individual customer's needs. But co-configuration work takes this relationship up one level—it brings the value of an intelligent and "adapting" product. The company then

continues to work with this customer-product pair to make the product more responsive to each user. In this way, the customization work becomes continuous.... Unlike previous work, co-configuration work never results in a "finished" product. Instead, a living, growing network develops between customer, product, and company. (Victor & Boynton, 1998, p. 195)

We may provisionally define coconfiguration as an emerging, historically new type of work that has the following characteristics: (a) adaptive "customer-intelligent" products or services, or more typically integrated product-service combinations; (b) continuous relations of mutual exchange between customers, producers, and the product-service combinations; (c) ongoing configuration and customization of the product-service combination over lengthy periods of time; (d) active customer involvement and input into the configuration; (e) multiple collaborating producers that need to operate in networks within or between organizations; and (f) mutual learning from interactions between the parties involved in the configuration actions.

In other words, coconfiguration is more than just smart, adaptive products. "With the organization of work under coconfiguration, the customer becomes, in a sense, a real partner with the producer" (Victor & Boynton, 1998, p. 199). Coconfiguration typically also includes interdependency between multiple producers or providers forming a strategic alliance, supplier network, or other such pattern of partnership which collaboratively puts together and maintains a complex package which integrates material products and services and has a very long life cycle. Coconfiguration requires flexible "knotworking" in which no single actor has the sole, fixed authority—the center does not hold (Engeström, Engeström, & Vähäaho, 1999).

Coconfiguration is typically needed in divided multiactivity terrains, or multiorganizational fields, in which the different activity systems have critically important shared objects or customers but little evidence of productive collaboration across organizational boundaries. In such terrains, design needs to take shape as self-reflective renegotiation of collaborative relations and practices.

A precondition of successful coconfiguration work is dialog in which the parties rely on real-time feedback information on their activity. The interpretation, negotiation, and synthesizing of such information between the parties require dialogical and reflective knowledge tools as well as collaboratively constructed functional rules and infrastructures.

Although partially similar, the concept of coconfiguration must not be confused with the notion of coproduction, put forward by Normann (2001), who pointed out (p. 97) three aspects of co-production: customer participation (or "prosumption"), customer cooperation (or customer communities), and value constellations (or cooperation between providers). These characteristics correspond to the idea of coconfiguration, and Normann's emphasis on customer communities actually enriches the concept of coconfiguration. The difference between the two concepts becomes manifest in his argument about time:

> From being primarily sequential in time, they [co-productive relationships] tend—as a result of connectivity and interactivity—to become *simultaneous, synchronous,* and *reciprocal.* By this process we can compress time—we can, in fact, *create time* since we can package activities more densely into given time slots, thus liberating other time slots for other activities. And we can also proceed by occupying time slots that used to be "unproductive." (p. 96)

Normann's emphasis on compression of time is in line with the general postmodern argument about compression of time and space. In contrast, the idea of coconfiguration is based on analysis of specific historically new objects, namely customer-intelligent products and services which have very long half-lives and require constant collaborative reconfiguration, never resulting in a "finished product." This means that the time perspective must be radically expanded, not just compressed.

The expansion of objects of work in coconfiguration happens along four dimensions. Social-spatial expansion means that a radically wider circle of activity systems is directly involved in the construction of the object. Temporal expansion means that the constant reconfiguration of the object requires a mastery of its history and a long-term plan for its future evolution—along with very quick improvisation of collaborative action when needed. Moral-ideological expansion means that responsibility and power are constantly redistributed and renegotiated among the participants. And systemic-developmental expansion means that seemingly singular or routine everyday actions are increasingly problematized and connected to their systemic consequences and developmental potentials.

In coconfiguration, "products" are to be understood as complex configurations of organizational arrangements, services, and technologies. Thus, product design, process design, and organization design become increasingly integrated, and management itself is penetrated by design language (Boland, 2004). Professional designers may in these conditions gain a strategic role as scouts, negotiators, and boundary-spanners who bring together previously separate activity systems and domains of expertise, facilitating the formation of expanded objects and novel partnerships.

EXPANSIVE DESIGN OF MEDICAL CARE FOR CHRONIC PATIENTS WITH MULTIPLE ILLNESSES

In Helsinki, the capital of Finland, 3.3% of the patients used 49.3% of all health care expenses in 1999. A total of 15.5% of patients used 78.2% of all resources. This is an example of the well-known "20/80" rule, implying that roughly 20% of patients in industrialized countries use roughly 80% of the resources.

Health care in a large city is typically a divided field of multiple, poorly coordinated activity systems, with historically formed hierarchical relations and turf tensions between them. Many of the patients who belong to the "20%" become so expensive because they have multiple serious chronic illnesses which cannot be dealt with by any single specialty alone. These patients often drift and bounce from one caregiver to another without anyone having an overview and overall responsibility for their care. Coconfiguration work is a strategic priority because the different caregivers and the patients need to learn to produce together well coordinated and highly adaptable long-term care trajectories.

The design challenge in this field is to construct a new, negotiated way of working in which patients and practitioners from different caregiver organizations and specialties will collaboratively plan and monitor the patient's trajectory of care, taking joint responsibility for its overall progress. This is easier said than done.

The design of an entire new way of working across and between multiple activity systems is typically a task that may be best approached by generating a simple germ-cell concept of the foundational relations on which the new practice will be built. This in turn requires that the existing contradictions within and between the key activity systems are identified. These steps were accomplished in the winter of 1998 in an intervention project, Boundary Crossing Laboratory, which we conducted with approximately 60 representatives of the Children's Hospital on the one hand and the local primary care health centers on the other hand. The resulting image of the contradictions may be summarized with the help of activity system models (see Fig. 1.3).

To make analytical sense of the situation, we need to look at the recent history of the activity systems involved. Since the late 1980s, in municipal primary care health centers, the personal doctor principle and multiprofessional teams have effectively increased the continuity of care, replacing the isolated visit with the long-term care relationship as the object of the practitioners' work activity. The notion of care relationship has gradually become the key conceptual tool for planning and recording work in health centers.

A parallel development has taken place in Finnish hospitals. Hospitals grew bigger and more complicated in the postwar decades. Fragmentation by specialties led to complaints and was seen to be partially responsible for the rapidly rising costs of hospital care. In the late 1980s, hospitals began to design and implement critical paths or *pathways* for designated diseases or diagnostic groups.

HEALTH CENTER   **CHILDREN'S HOSPITAL**

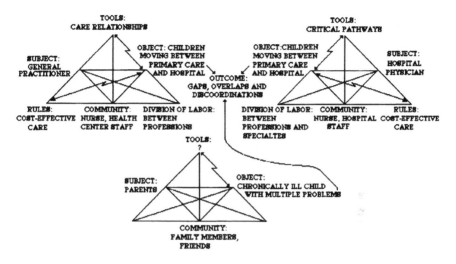

PATIENT'S FAMILY

FIG. 1.3. Contradictions in children's health care in the Helsinki area (Engeström, 2001, p. 145).

Care relationships and critical paths respond to contradictions internal to the respective institutions. Care relationships are seen as a way to conceptualize, document, and plan long-term interactions with a patient inside primary health care. Their virtue is that the patient can be seen as having multiple interacting problems and diagnoses that evolve over time; their limitation is that responsibility for the patient is practically suspended when the patient enters a hospital. Correspondingly, critical paths are constructed to give a normative sequence of procedures for dealing with a given disease or diagnosis. They do not help in dealing with patients with unclear and multiple diagnoses, and they tend to impose their disease-centered worldview even on primary care practitioners. Fundamentally, both care relationships and critical paths are linear and temporal constructions of the object. They have great difficulties in representing and guiding horizontal and sociospatial relations and interactions between care providers located in different institutions, including the patient and his or her family as the most important actors in care.

In both the hospital and the health center, a contradiction emerges between the increasingly important object of patients moving between primary care and hospital care, and the rule of cost-efficiency implemented in both activity systems. In Helsinki, the per capita expenditure on health care is clearly above national averages, largely due to the excessive use and high cost of services provided by the central university hospital of which the Children's Hospital is a part. Thus, there is an aggravated tension between the primary care health center and the university hospital. Health centers in the Helsinki area are blaming the university hospital for high costs, whereas the university hospital criticizes health centers for excessive referrals and for not being able to take care of patients who do not necessarily need hospital care.

A contradiction also emerges between the new object (patients moving between primary care and hospital care) and the recently established tools, namely care relationships in primary care and critical paths in hospital work. Being linear-temporal and mainly focused on care inside the institution, these tools are inadequate for dealing with patients who have multiple simultaneous problems and parallel contacts to different institutions of care. In the activity system of the patient's family, the contradiction is also between the complex object of multiple illnesses and the largely unavailable or unknown tools for mastering the object.

As concrete patient cases were discussed and different aspects of these contradictions were articulated in the Boundary Crossing Laboratory, we observed a shift among the participants from initial defensive postures toward a growing determination to do something about the situation. The determination was initially fuzzy: a need state (Bratus & Lishin, 1983), looking for an identifiable object and corresponding concept at which the energy could be directed:

> *Hospital physician:* I kind of woke up when I was writing the minutes [of the preceding session].… What dawned on me concerning B [name of the patient in the case discussed] is, I mean, a central thing … for the mastery of the entire care. How will it be realized and what systems does it require? I think it was pretty good, when I went back through our discussion, I think one finds clear attempts at solving this. It is sort of a foundation, which we must erect for every patient.

> *Researcher:* That seems to be a proposal for formulating the problem. What is … or how do we want to solve it in B's case? I mean, is it your idea that what we want to solve is the mastery of the entire care?

> *Hospital physician:* I think it's just that. I mean that we should have … or specifically concerning these responsibilities and sharing of responsibility and of practical plans, and tying knots, well, we should have some kind of arrangement in place. Something that makes everyone aware of his or her place around this sick child and the family. (Excerpt 1: Boundary Crossing Laboratory, Session 5)

Step by step, the idea of care agreement took shape as a "germ-cell" concept with the potential to resolve the contradictions. The practitioners formulated the idea with the help of a diagram (see Fig. 1.4). Subsequently, the model was enriched, tested in practice, and concretized in our next intervention project, where we used a method we called Implementation Laboratory (Engeström, Engeström, & Kerosuo, 2003).

The crucial point of the care agreement model is that at least the three key players of care—the patient, the health center general practitioner responsible for the patient, and the hospital specialist in charge of the patient's care—negotiate an overall framework for the patient's care for the next year. They sign a mutual agreement that obliges them to inform each other of any significant care events and changes in the plan.

In October 2002, the chief executive officers of the Helsinki-Uusimaa Hospital District and the Helsinki City Board of Health declared the care agreement model as an official framework for the coordination of the care of chronically ill patients in the Helsinki area. The administrative decision contains an algorithm, designed in the Implementation Laboratory, which describes the basic steps of this way of working (see Fig. 1.5).

CHARACTERISTICS OF EXPANSIVE INTERACTION DESIGN: INSTRUMENTALITIES AND ANCHORING

Expansive interaction design is oriented at complex configurations of people, organizational arrangements, and mediating technologies, including language, concepts, and patterns of discourse. This implies a shift from designing well-bounded singular products to designing tool constellations or instrumentalities.

A tool constellation or instrumentality is literally the toolkit needed in an activity. The tools of a skilled carpenter may fill multiple boxes. They offer the practitioner multiple alternative access points to a task. Thinking is performed with the tools. Thus, the tools open a window into the mentality of the trade. In their study of a blacksmith's use of tools, Keller and Keller (1996) pointed out the variability and flexibility of tool constellations: "It is important to note that the ideas constituting the mental components of a

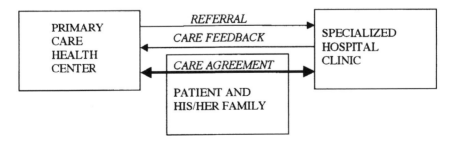

FIG. 1.4. A germ-cell model of care agreement practice.

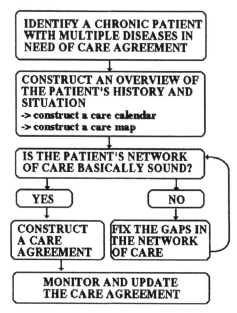

FIG. 1.5. The basic algorithm for negotiated care agreement practice.

constellation often include procedures for correcting or repairing deviations from the image of the desired outcome of a particular step in production. Therefore, tools may well be used in multiple ways even within a given
constellation" (p. 103).

The design of instrumentalities is obviously a stepwise process which includes fitting together new and old tools and procedures. In the design of
an instrumentality for the negotiated collaborative care of chronically ill patients in Helsinki, three central tools were developed to be used by the practitioners and the patients: the care calendar, the care map, and the care
agreement.

The care calendar is a simple template for listing the most important
events of the illness and care of the patient for the past few years. The idea is
to condense the often prohibitively voluminous historical information
stored in the medical records, and in the patient's own recollections and interpretations, into one or two pages that may be easily reviewed in any encounter or planning situation. The construction of the care calendar
requires conarrating between the medical professional and the patient.
The care calendar serves the temporal expansion of the object.

The care map is a one-page template for representing graphically the
different caregivers and institutions involved in the care of the patient. Ideally the doctor and the patient together construct the first version of the

map, marking down also problematic or missing connections between the various parties. Thus, the care map becomes not only a memory tool but also a device for identifying and diagnosing gaps and ruptures in the network of care. The care map serves the sociospatial expansion of the object.

Finally, the care agreement is a one-page document template that asks the practitioners and the patient to write down the diagnoses and the patient's main concerns, the division of labor in the care (what problems are treated where and by whom) during the next year, the procedures for informing one another, the date by which the care agreement is to be reviewed, and finally the signatures of the involved parties. The drafting of a care agreement requires exchange and negotiation between the caregivers and the patient. Requiring a renegotiation of responsibility and power, the care agreement serves the moral-ideological and systemic-developmental expansion of the object.

In the practice of negotiated collaborative care, the instrumentality becomes alive and new ad-hoc tools are created and used in situ. An example from an Implementation Laboratory session demonstrates this nicely.

At the beginning of the laboratory session, the researchers showed a 4.5-min video clip from the preceding consultation between the general practitioner and the patient. The actual discussion of the patient case began after that with the case presentation given by the general practitioner. She immediately presented an overhead diagram she had prepared to summarize the patient's situation (see Fig. 1.6).

The general practitioner explained her diagram as follows:

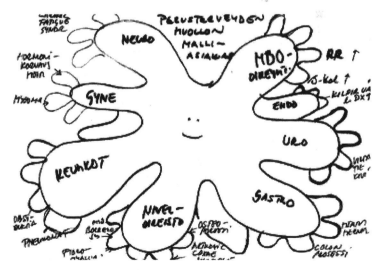

FIG. 1.6.   Diagram depicting the patient's overall situation, prepared and presented by a general practitioner.

General practitioner: … And we met for the first time only in April, after Easter. But already then in our first encounter it became clear that there is a lot of ground to cover. For this patient, perhaps the most central consequence of our contact was that we found the metabolic syndrome, then we got to treat sugar hypertension parameters, the lady had herself already done a dietary intervention. And here [on the overhead] is the whole range of different kinds of diagnosis and their connections the lady has. (Excerpt 2: Implementation Laboratory, Session 7)

The general practitioner subsequently called the image in Fig. 1.6 "an amoeba." The different legs of the amoeba represent the patient's various diagnoses and their particular symptoms. Above the amoeba, the general practitioner had somewhat sarcastically written "Model client of primary health care." The amoeba figure graphically captures the gist of lateralization: the search for an overview of and interconnections between multiple parallel threads of illness and care.

The physician went on to present the care calendar:

General practitioner: As requested, I then prepared this care calendar, and I found it to be extremely good and helpful in representing the overall situation. May I present it next?

Researcher: Please do.

General practitioner: So if we think about Rauni as a person, she has made a tremendous career abroad. And this is manifested in the list of diagnoses … (Excerpt 3: Implementation Laboratory, Session 7)

Subsequently, the general practitioner also introduced on another overhead a care map she had constructed, depicting the different caregiver organizations involved as boxes grouped around the patient. She also introduced a draft for a care agreement she had prepared:

General practitioner: Then the last overhead, it is this care agreement. And I didn't squeeze the whole amoeba into it, into this field for listing the diagnoses, since we agreed with the patient that we'll concentrate this year on these issues … (Excerpt 4: Implementation Laboratory, Session 7)

The different tools may be categorized with regard to the questions they are meant to ask and answer. The hierarchy of Fig. 1.7 indicates that on the top, one germ-cell model opens a very wide landscape of applications, whereas at the bottom, images and stories are typically quite specific and bound to a particular situation or case.

In the health care case discussed earlier, the general practitioner's amoeba model represents a situated image or prototype (what is wrong with the patient?). The care calendar represents partly a narrative (what happened when to whom?), partly an algorithm (how has the illness devel-

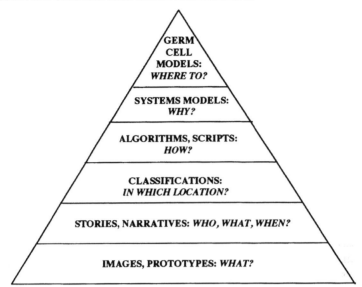

FIG. 1.7. A hierarchy of tools.

oped?). The care map represents a classification on the one hand (in which locations are the caregivers related to one another?), and a systems model on the other hand (why is coordination between them not working?). The care agreement represents an instantiation of the germ-cell model (where are we moving to in the care of this patient?). Interplay between the different tools makes the instrumentality robust.

This entire health care instrumentality evolves and revolves around the germ-cell model depicted in Fig. 1.5. Without such a vision for the future of the activity as a whole, the intermediate instruments may easily become mere techniques, or, in the worst case, empty forms or rules imposed from above. The germ-cell model is an example of anchoring a newly designed instrumentality *upward.*

On the other hand, the new instrumentality must be anchored in daily actions and decisions with immediate consequences. The general algorithm of the steps of negotiated care, presented in Fig. 1.6, anchors the germ-cell model *down* to practical procedures and gives a general format for the actions taken by practitioners. The emergence of the amoeba model in the example given earlier is a situated indication of such anchoring downward: the general practitioner took on her own the action of crystallizing her image of the patient's overall situation in a durable and transferable graphic form (see also Engeström et al., 2003).

Perhaps the most crucial form of anchoring in coconfiguration work and expansive design happens *sideways.* This implies that the emerging new instruments are negotiated and shared in use with partner activity systems, above all customers or users. In the medical care case, this anchoring sideways, or cross-appropriation (Spinosa, Flores & Dreyfus, 1997), happened both between the different medical practitioners, and—most important— between the doctor and the patient. Joint storytelling, or conarrating, is a typical action used to achieve anchoring sideways. To facilitate such sideways anchoring in design, specific "trading zones" may be constructed (Galison, 1997). These are physical and discursive spaces that offer neutral ground for exchange between members of different activity systems.

INTERVENTION, AGENCY, AND RESISTANCE

In a well-known article, Brown (1992) put forward the concept of design experiments. She wanted to create a methodology that would simultaneously generate elements toward a general theory of learning and facilitate practical formation of intentional learning environments, radically different from the passive receptive tradition of school classrooms:

> [C]onsider the design experiment that my research team is currently trying to engineer in the classroom. This includes effecting basic change in the role of students and teachers, modifying assessment, introducing a novel curriculum, establishing a technologically rich environment, setting up cooperative learning situations, establishing a classroom ethos where individual responsibility and group collaboration are the norm, and so forth. In short, we intervene in all aspects of the environment. Our interventions are deliberately designed to be multiply confounded. Although I was taught to avoid such messy things like the plague, I do not see an alternative. (pp. 166–167)

Brown's acceptance of complex confounded constellations as the object of design experiments is commendable. At the same time, I find her treatment of agency in design experiments very problematic.

She maintained that in her interventions, students are "designers of their own learning" and "partially responsible for creating their own curriculum" (p. 150), even "coinvestigators of their own learning" (p. 165). Agency is shared in some unbelievably benign and harmonious way between very different actors, namely university researchers, school teachers, and school students. The foundational differences and tensions between the objects, motives, and activity systems of researchers, teachers, and students are blurred or disappear in Brown's account. However, as Long (2001) convincingly showed, interventions never take place without struggle and resistance:

> Intervention is an on-going transformational process that is constantly re-shaped by its own internal organizational and political dynamic and by the

specific conditions it encounters or itself creates, including the responses and strategies of local and regional groups who may struggle to define and defend their own social spaces, cultural boundaries and positions within the wider power field. (p. 27)

Long pointed out that it is crucial to identify and come to grips with "the strategies that local actors devise for dealing with their new intervenors so that they might appropriate, manipulate, subvert or dismember particular interventions" (p. 233). From an activity-theoretical perspective, I would add that it is necessary to dig into the historical contradictions taking shape and generating disturbances within the activity systems at which the interventions are aimed (recall Fig. 1.4, shown earlier).

In other words, the conduct and study of design experiments is necessarily a tension-laden process of negotiations itself. If the researchers and designers seriously want to engage in such negotiations, they need to create spaces where their authority can be contested and their ideas can be overridden. This requires that such spaces are not alien to but fit the work practices of the participants.

The laboratory methods mentioned earlier were attempts at creating such spaces. The laboratory sessions focused on concrete patient cases, so they resemble usual patient-centered shift-change meetings. Second, the laboratory sessions included practitioners from multiple clinical settings and specialties, they dealt with clinical work practices beyond the particular patient case, and they followed a script prepared well in advance—features that resemble clinical practice-centered meetings common in medical settings. Third, in laboratory sessions, the participants envisioned and drafted strongly future-oriented organizational changes, resembling management meetings in hospitals. Putting together these three aspects led to a hybrid form.

However, to make the hybrid work, we added four important new elements: (1) the presence and participation of the patient in the session, backed up with medical records and videotaped excerpts from the patient's recent care experiences; (2) the systematic development and use of new models and conceptual tools to envision and represent the expanded object (in this case, the care calendar, the care map, and the care agreement); (3) the repeated articulation of the historical challenge and mission of the sessions (in this case, the challenge of coordinating the care of chronic patients with multiple illnesses and multiple caregivers); and (4) the presence of several researcher-interventionists who engage in the laboratory debates, also disagreeing among themselves. Thus, the laboratory sessions represent a blend of elements familiar from existing practices and new elements brought in by the researchers. They were designed to serve as "microcosms" where potentials of collaborative care and "knotworking" negotiations could be experimented with and experienced: "A microcosm is a social test-bench and a spearhead of the coming culturally more advanced form of the activity system ... the microcosm is supposed to reach within itself and propagate outwards reflective communication while at the same time expanding and therefore eventually dissolving into the whole community of the activity" (Engeström, 1987, pp. 277–278).

Obviously, our laboratory sessions were marginal microcosms in the sense that only a limited number of practitioners were involved in them and they were not meant to become a permanent feature in the routine functioning of the organizations. However, there are two kinds of marginality: centrifugal and centripetal. In one, the marginal practice is pushed out and tends to disappear. In the other, the marginal practice finds inroads and tends to spread into the central structures and interactional routines of the organization. The decision of the chief executive officers to adopt the negotiated way of working and care agreement instrumentality as system-wide practices in the care of chronic patients with multiple illnesses in Helsinki is an indication of the centripetal potential of the laboratory sessions.

DESIGN, IMPLEMENTATION, AND LEARNING

In their classic study, Pressman and Wildavsky (1984) pointed out that the implementation of complex new programs is a creative process of design and learning: "As programs are altered by their environments and organizations are affected by their programs, mutual adaptation changes both the context and content of what is implemented. The study of implementation is shaken from its safe cognitive anchorage in prior objectives and future consequences that do or do not measure up to original expectations" (p. xvii).

Correspondingly, expansive design should be seen as a longitudinal process which includes implementation and learning. It is in these lengthy processes that we typically see the importance and productive potential of resistance and turning points. These are nicely demonstrated in Table 1.1, which depicts the implementation and appropriation of the key tools of negotiated care over the span of 10 patient cases discussed in successive Implementation Laboratory sessions.

Table 1.1 shows that not all the new tools were adopted in patient Cases 1 to 4. The care calendar and care map were used, but the care agreement was not used until Case 5. In other words, the members of the physician pilot group resisted the use of the care agreement.

Kindred (1999) pointed out that in addition to open objections, resistance may also be silent. Rejection of the care agreement during discussions of Cases 1 and 2 may be interpreted as an expression of silent resistance. Adoption of the care agreement was not openly objected to, but neither were agreements completed. More open ways of objecting, as well as dilemmas in the use of tools, emerged in Case 3:

TABLE 1.1

Implementation of tools of negotiated care in ten successive patient cases

Patient Case, Main Ailment	Care Calendar	Care Map	Care Agreement	"Own Tools"
Case 1 Rheumatoid Arthritis	X	X	—	Patient's care map:— Problems in the flow of information
Case 2 Heart Ailment	X	—	—	—
Case 3 Heart Ailment	X	X	—	Care calendar a list of epicrisis
Case 4 Diabetes	X	—	—	—
Case 5 Diabetes	X	X	X	Combined care calendar and care map. Care agreement proposal on a hospital referral.
Case 6 Nephropathy	X	X	X	—
Case 7 Diabetes	X	X	X	Depiction of a model client at the health center as an amoeba.
Case 8 Heart and Pulmonary Ailment	X	X	X	Care map as a flowchart.
Case 9 Nephropathy	X	X	X	—
Case 10 Pulmonary Ailment	X	X	—	—

Note. Taken from Kerosuo & Engeström, 2003, p. 347.

Researcher: Do you really have a feeling that one does not need a kind of written anything here? That this goes well enough [without the documented agreement].

Physician: Yes, in a way now, it is that at the moment the medication as a whole, the treatment of the coronary disease is undertaken at the health center, Marevan medication is at the health care center, it is at the moment. So, if we want to document it, yes, but there is nothing to negotiate about. The patient herself agrees that it is like this and we all agree. But right now the examinations of her stomach troubles are under way over here, and that—but that is something we cannot make an agreement, because it is not finished. (Excerpt 5: Implementation Laboratory, Session 3)

Another doctor thought that the agreements are "dead documents" that are signed, sent ahead, put into archives, and have no value for practical use. At this point, another member of the pilot group took up the missing treatment of the patient's leg as well as a follow-up visit not completed with lung specialists that the patient had continued to mention during the laboratory session. She proposed a care agreement that would include information on the treatments, as well as the visits, where they were provided, and when. Despite this lengthy, multivoiced discussion, the care agreement was not completed.

Resistance is often interpreted as an obstacle to development and learning. However, resistance is not only an obstacle but also a dynamic force that may be triggered to generate learning. The "foreign" or "unknown" must be made one's own. This requires attacking, testing, and questioning the new.

As shown in Table 1.1, the care agreement was adopted from Patient Case 6 onward. However, the turning point occurred already in Case 5. A general practitioner sent a "home-grown" care agreement proposal to the hospital, inquiring about the patient's diabetes follow-ups. She did not use the template of the care agreement suggested by the researchers; she simply used a copy from the patient's medical record. The hospital endocrinologist gave a formal reply confirming the prevailing rules about the division of care responsibility between primary and secondary care. However, the general practitioner was not after the formal rule—she knew it well. She wanted a specific reply and negotiation about conducting the follow-ups.

In the laboratory session, the problem of diabetes follow-ups was discussed. At first, the endocrinologist again offered the rules regarding the division of labor between primary and secondary care. However, the general practitioner, a visiting nurse, and the patient himself insisted that there had been problems in information exchange that could not be solved formally. Finally, the endocrinologist admitted the necessity of negotiation and even suggested that it might be worthwhile to sign an agreement which the patient could bring along when entering the different care locations:

Endocrinologist: This [the information exchange] is, as I said, a never-ending question. And it has been recognized, and also admitted, the same thing, that we should inform, the information should flow, but it becomes continuously disrupted on and on. So, I think that until we all have computers, a kind of, what could it be, an agreement, a paper that the patient could carry with him, where …

Researcher [speaking over]: One page.

Endocrinologist: … one has documented of what is being treated, and where, I consider it to be quite a good thing. (Excerpt 6: Implementation Laboratory, Session 5)

After the comment by the endocrinologist, the atmosphere in the laboratory changed. Members of the pilot group began to generate new, practical ideas about the contents and uses of the care agreement. A "home-grown" version of the designed instrumentality served as a springboard for a turn from resistance to further design.

Chapter 2

Activity:
Analysis, Design, and Management

Thomas P. Moran
IBM Almaden Research Center, San Jose, California

ACTIVITY IN INTERACTION DESIGN

The notion of "activity" is a foundational concept in interaction design. Interaction design is about creating artifacts that are computationally empowered to provide interactive capability. The artifacts are designed to be used (or experienced) by people. We call this the *use activity* (or *experience activity*) that surrounds the artifact, in which the designed artifact is a resource for carrying out the activity. This is the narrowest perspective on the notion of activity, where the focus is on the interaction between the person and the artifact. The concern in this perspective is for such issues as the usability of the artifact, with minimal regard for the context of the use activity.

A broader perspective on activity includes the larger context of the many different activities in the organizational and social setting. We call these the *context activities*. The focus here is the relation between the different activities and the people involved. In this perspective, the concern is with such issues as the usefulness and the role of the use activity in the activity context.

Various user-centered design methodologies in human–computer interaction provide techniques for studying use activity (e.g., Preece et al., 1994). Various perspectives provided by the behavioral and social sciences, such as distributed cognition (e.g., Hutchins, 1995), activity theory (e.g., Engeström, Miettinen & Punamaki, 1999), or ethnography (e.g.,

Suchman, 1987), provide frameworks and methods to study context activities and methods for designing for context (e.g., Beyer & Holtzblatt, 1998). Although there are a broad array of theories and perspectives (e.g., Carroll, 2003), in all of these activity is treated as an object of analysis for the purpose of designing an artifact, as well as redesigning some of the activity context of the artifact. This is how we usually regard the concept of activity in design.

ACTIVITY MANAGEMENT AS META-ACTIVITY

Let us step back and consider activities more generically, not from the perspective of the analyst or designer who is concerned with designing artifacts, but from the perspective of the person carrying them out. Imagine a person reflecting (looking down from above, as in Fig. 2.1) on the multitude of activities in which he or she is involved. To carry out the various activities, the person has to manage them, to plan and prioritize them, for example. Activity management consists of generic meta-activities that act on different specific activities. Such meta-activity is an inherent part of any activity.

People carry out activities to accomplish specific objectives, from the serious (running a business meeting) to the playful (doing a crossword puzzle to relax). The actions that accomplish the specific objective are the "real work" of the activity. We can call these "real work" actions the execution of the activity. The management meta-activities support the execution:

- Before one can execute an activity, it must be *set up* (assembling the resources to be ready to use, for example).
- Before one can execute an activity, one must *remember*, or be reminded or alerted, or *seize an opportunity* to do it.

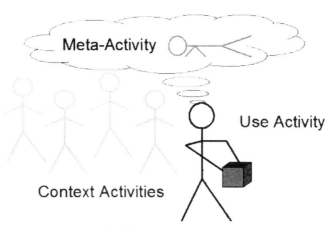

FIG. 2.1.   Various senses of activity.

- When someone is doing an activity, he or she must often peripherally *monitor* his or her other activities to be aware if they are in need of attention.
- One must *respond to the unanticipated* activities that require attention.
- When another activity needs attention, one must *switch contexts* from the current activity to the other activity.
- When one has many activities, one must *plan, organize,* and *prioritize* them and *arrange to be reminded* of them at the appropriate time or in the appropriate context.
- When one has many activities, it is often beneficial to have an overview of the relations among, and the states of, the various activities.
- Finally, one must often account for activities by reporting or documenting them.

We are exploring the idea of providing generic support for these kinds of activity management meta-activities. Given this broad definition, many different computer tools address activity management, but they do so in a localized and ad hoc way. Personal information management (PIM) tools provide to-do list facilities to help the individual keep track of tasks. Groupware tools provide shared repositories that include shared task-list tools for coordinating group efforts. Project management applications provide tools for scheduling and synchronizing the activities of project teams. Workflow systems provide facilities for developing and running formal business processes, thus "choreographing" the activities of people across an organization. Although these tools are all in use, people have problems with them. The personal and group tools tend to be tedious to use. The project and workflow tools tend to be overly formal, overly rigid, or coercive (Abbott & Sarin, 1994). Perhaps the biggest problem is that all these tools take a limited view of activity, each focusing on particular meta-activities. From a user's perspective, his or her activities cut across all these tools, and having to deal with many tools makes activity management more difficult.

WHAT IS AN ACTIVITY?

Let me start by offering a partial definition of activity considered generically, for the purpose of capturing what people mean by activity and providing a basis for systems to represent it. An activity is a set of (mental or physical) actions carried out by people:

- The actions do not have to be contiguous in time, but they do have to have coherence to be considered an activity. It is useful to distinguish two kinds of coherence. An activity is conceptually coherent if its actions are directed to the same goal, such as a project. The actions may be diverse, but they are coherent because they support the same goal. An activity is contextually coherent if the actions share a context, such as a meeting where many conceptually-distinct issues may be discussed. The coherence is of a different kind; the convenience of the shared context gives coherence, at least temporarily.

• Activities are related to other activities. The most important relation is *composition*. An activity can have subactivities, which can have sub-subactivities, and so on. We call the spawning of subactivities "articulation." It is a practical matter how far we want to articulate the decomposition. An activity provides a context for its subactivities. Also, an activity can have multiple superactivities, because of the different kinds of coherence (for example, a discussion activity can be a subactivity of a meeting activity as well as of a project activity).

• Thus it follows that activities can be at different time scales—months, days, minutes. Again, it is a practical matter of activity management as to how finely to articulate.

• An activity utilizes a set of resources—people, tools, objects, information—to carry out the activity. Accessing and organizing the resources is a large part of managing an activity. (See Kaptelinin, 2003, who proposes a scheme for automatically linking resources to activities by monitoring user actions.)

• Another relation among activities is that they share resources. In particular, different activities must be carried out by the same person. This, for the individual, is the activity management problem—how to distribute his or her limited time and attention among the various activities.

Note that this definition of activity is person-centered, from the viewpoint of a single individual. This does not imply that activities are not social. A social or collaborative activity is characterized from the perspective of each of the individuals involved. A social activity involves the intertwining of activities of individuals. Often the interrelations can be characterized by the notion of different roles that the individuals play.

EXAMPLE OF AN ACTIVITY

To better understand the concept of an activity, consider a concrete example of a fairly complex activity—chairing an awards committee for the ACM SIGCHI (Special Interest Group for Computer–Human Interaction). This was a real activity that I carried out during 2002 and 2003. The particular activity described later occurred over a period of four months in 2003. The analysis presented here was derived from the various materials that I created *during* the activity. Most of the work was done via e-mail, plus two phone conferences. The activity involved several people—committee members, SIGCHI officers and administrators, CHI conference officers, and an assistant.

Figure 2.2 shows a data display of the activity. The shaded items on the left in outline format represent my e-mail folder structure, which roughly corresponds to the activity-subactivity structure. One subactivity, "decision process," is expanded in the middle of the figure. To the right of these subactivity items is a table showing the resources for each subactivity. The first column shows the number of e-mails involved in each subactivity. The second column gives icons for the people I had to contact or work with in each subactivity. The third column lists the key documents created and used in each subactivity.

Reflecting on my experience in chairing the awards committee, it is clear that the basic structure of this activity consists of four subactivities:

Chair Awards committee

    1 Set up the committee

    2 Decide on the award winners

    3 Announce, coordinate, present, ...

    4 Hand off chairmanship to the next chair

These subactivities were logically sequential; for instance, I could not announce the winners until the committee's decision process was complete.

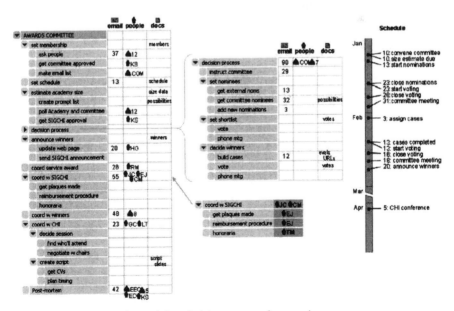

FIG. 2.2.   A complex activity: chairing an awards committee.

But there were also overlaps; for example, parts of the third subactivity (the coordination with CHI) was done during the first two subactivities. Each of the four subactivities had a distinctly different character:

1. Setup—First I had to set up the committee by recruiting people for the committee, and, in parallel, working out a fair and efficient decision process for the committee. This process was a refinement of the process I had used the previous year. The materials from the previous year were used as a sort of checklist for defining the current process.

2. Decision process—I had to manage the sequential decision process itself, which was fairly formal and had to be run on a tight schedule (shown on the right of Fig. 2.2). Committee members were assigned specific research and voting activities during the process. Their work was communicated by e-mail and collated into a few key documents. This was used to structure two phone meetings for discussing and finalizing the decisions on the award winners.

3. Announcements, and so forth—There were several postdecision subactivities, which were run mostly in parallel with each other. These involved making announcements, working with SIGCHI on award plaques and reimbursements, negotiating with CHI conference officers for an award presentation slot, creating the presentation, and finally presenting the awards at the CHI conference.

4. Handoff—A couple months after the presentation I handed off the job to the newly-appointed committee chair. This was important for the continuity of the awards process. The handoff involved giving him the key documents and contacts, and thoroughly discussing the various subactivities, key issues, and problems, and possible revisions in the process.

Running the awards committee required considerable effort, of two kinds. First, the parts of the work that I did as an individual came in bursts of subactivities. But my main effort was activity management, which required continual attention to keep the overall activity progressing. From a personal point of view, the committee activity was only part of my activity management. I had to interleave the committee activity with my job and home activities during this period. This is a typical activity management situation for most knowledge workers.

STUDIES OF ACTIVITY MANAGEMENT

When we started this project at IBM in 2003, we were focused on personal time management and calendaring. We conducted three ethnographic studies, consisting of over 30 interviews of people inside and outside IBM. From these, we identified a dozen time management issues, and decided to focus on the issue of planning—how people cope with mostly unscheduled tasks.

The interviews revealed that people put a lot of effort into planning. Planning ranges between two extremes. At one end, people engage in long-term planning, laying out goals and subgoals, milestones, budget, people, resources, risks, and so on. Long-range planning is a considerable task in itself. People only do this occasionally—yearly planning, creating proposals, project startups, and so forth. At the other extreme, every day people juggle what they had planned to do with unanticipated daily demands. This hardly seems to be planning, but more like continuous adaptation. Some people step back every week or so to look at the intermediate term. There is great variability in people's planning practices.

People use a multiplicity of tools, both electronic and physical, for planning. Electronic tools—from standard calendar and to-do tools to quirky ad hoc customizations—are generally problematic. The electronic tools are not coordinated with each other, and are available only when people are at their computers and with particular applications open. Physical tools—from paper to-do lists to Post-Its® stuck on computer displays, to plans written on whiteboards on walls, to paper piles on desks—seem to be more satisfactory, or at least more comfortable. To-do items function mainly as reminders and are thus distributed in the natural flow of work, in both the physical and electronic world, which is a prospective memory heuristic (Brandimonte, Einstein, & McDaniel, 1996). A task list on a whiteboard is seen every day, often by several people, which helps with awareness and coordination.

Many people use their e-mail inboxes to manage their activities. It is interesting to consider why:

- E-mail is the place where new activities are initiated, such as requests in incoming e-mail.
- Many activities are carried out *through* e-mail. That is, e-mail serves as the main resource for many activities.
- E-mail headers serve as a reminder of the current state of various activities conducted through e-mail.
- E-mail is where many people spend much of their online time; it is their "habitat" (Ducheneaut & Bellotti, 2001). Thus the representation of activity in e-mail is usually "at hand."

That is to say, e-mail provides baseline support for many of the meta-activities involved in activity management. See Bellotti, Decheneaut, Howard, and Smith (2003) for a proposed redesign of e-mail to support task management more directly.

We can illustrate some of the important temporal properties of activities that emerged from our studies by laying out a time line, as shown in Fig. 2.3.

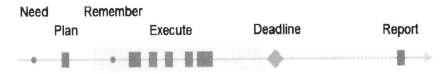

FIG. 2.3.   Life cycle of an activity.

An activity comes into being when a need to do something arises. If a person cannot do it immediately, then he or she may plan to do it at some later time. Often this takes the form of a "planning interval" during which the person intends to do it. The planning interval is not a scheduled time to do it, but a fuzzy time interval during which it can be done (for instance, "do it next week sometime"). But the person must remember to do the activity. This can be purely a memory recall, or the person can be reminded by some sort of to-do artifact, as discussed earlier. Now the person can execute the activity. Often an activity cannot be completed in one "sitting." The execution must be accomplished by intermittent bursts of work on it. Often the activity is constrained by a deadline, an externally-set time for the activity to be completed. (Note that the planning interval should be distinguished from the deadline.) Finally, an activity often has to be accounted for, such as by a report on it, which can be done long after the execution.

Most people are engaged in multiple activities at any point in time, illustrated in Fig. 2.4 as parallel, overlapping activities. Multiple activities are complex to manage. This is why the execution of a given activity is intermittent—one is switching between activities to keep them all going at once. This presents the issue of dealing with multiple activity contexts. Also, when one is focused on a particular task, there is the need to peripherally monitor the other activities to know when they need attention.

REPRESENTING ACTIVITIES

Our goal is to provide generic support activity management. To do this we claim that activities have to be explicitly represented electronically. Let us

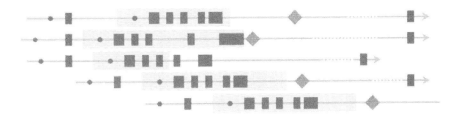

FIG. 2.4.   Multiple overlapping activities.

call the representation of an activity an *activity structure*, which at minimum consists of the following:

- An informal description of the activity.
- A hierarchy of subactivities (also allowing for multiple superactivities).
- A collection of resources (people, tools, documents, etc.).
- Properties for process semantics (planning times, deadline, status, dependencies, etc.).

(There are many other important generic properties of activities, but we just consider these in this chapter.)

Activity structures vary from informal to formal and from simple to complex. An example of the simplest activity structure is "Buy wine." Jotting down these two words (with an appropriate tool) creates a full-fledged activity structure. This simple activity structure functions as a reminder to stop at a store. This activity structure could be elaborated slightly by a pointer to a recent article recommending wines, which serves as a resource for the buying activity. An example of a complex activity structure is the one representing the awards committee activity described in Fig. 2.2. Activity structures typically evolve and become more elaborate as the activities progress. For example, I had the simple item "awards committee" on my to-do list for months to remind me that I had to get the committee work going. Once started, the activity structure (or rather, the set of artifacts I used in place of an explicit activity structure) expanded quickly.

Activity structures inside the computational system are resources for managing the activities in the outside world (compare Suchman, 1987). It is work for people to create explicit activity structures. People will not do any more work than they have to, and they will describe the activity only to the extent that descriptive accuracy helps them manage the activity. Ease-of-use is one issue. A fluid and flexible tool for jotting and organizing activity structures will lower the barrier to creating them. Some situations naturally call for more elaborate activity structures. For example, if an activity structure is shared among several people to coordinate their efforts, then more descriptive accuracy is useful to make clear their respective roles.

The basic hypothesis of our work is that providing a platform for representing activities as personal and social activity structures will provide many benefits, from personal productivity to interpersonal coordination to organizational learning and adaptation. This is a tall order. We are beginning by exploring a personal activity management tool.

ACTIVITY MANAGEMENT TOOLS

As mentioned earlier, we began by considering time management. An early mockup is what we called the *Planning Tableau,* shown in Fig. 2.5. In this envisionment, the calendar was the principal tool, and it was located in the middle section of the tableau. A space for goals and to-do items was in the left section, and a space for other calendars was in the right section. The theme of the tableau was to lay out a person's intentions (goals and to-do items), commitments (scheduled events), and possibilities (events from other calendars that might be of interest). The idea was to facilitate the easy movement of items across the sections of the Planning Tableau, such as events being dragged from other calendars to the central calendar. But we became most intrigued by the relation between scheduled events and unscheduled, or vaguely scheduled, goals and to-do items, which we have come to call activity structures. Let us call the visual representations of activity structures (in this case, just textual labels in the tableau) activity icons. The intentions section of the tableau has a rough two-dimensional spatial semantics. Activity icons on the left represent higher level goals that are not very time bound (for instance, an activity for this year). Activity icons on the right of the section (near the calendar) are more closely related to time. For example, the activity structures bunched up near the Today section of the calendar are intended to be dealt with today. As an activity icon is moved leftward, it is less bound to a specific time. Activity icons can also be linked, as subactivities or superactivities. An activity can also be linked to a calendar

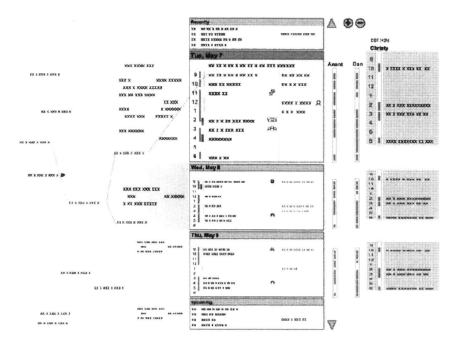

FIG. 2.5. The Planning Tableau interface for time management.

event. For example, the activity icon representing the preparation for a meeting can be linked to the meeting event in the calendar as its deadline. Last, if a meeting event becomes unscheduled, it can be dragged from the calendar to the intentions section, where it becomes a to-do (an activity icon) to reschedule the meeting.

Our work after this mockup focused on the idea of an activity space for jotting down, monitoring, and organizing activity icons. An early running prototype of the Activity Tableau is shown in Fig. 2.6.

The Activity Tableau is a freeform space containing activity icons that represent activity structures, which are stored in an activity database. Icons can be created, arranged, and edited in the freeform space. (Actually, there are many spaces, because the tableau has multiple pages.) The user can just click and type into an empty spot on the tableau, and a new activity icon is created, along with a new activity structure in the database. The user can edit the properties of the activity structure through the activity icon: for instance, add or change a deadline, planning time, or priority status ("completed," say). An activity icon can be dragged to any location in the tableau. When it is dragged near another activity icon, they lock together in an outline structure that represents an activity–subactivity relation.

We noted previously that an activity can have multiple superactivities. This is represented by "cloning" an activity icon. Activity icons that are

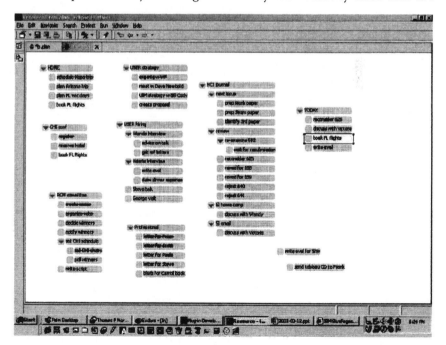

FIG. 2.6.   The freeform Activity Tableau.

clones of each other point to the same underlying activity structure. When a user changes the label on an activity icon, this changes the label of the underlying activity structure, which propagates the change to all of its activity icons. Multiple superactivities are represented by placing activity icon clones in different outline structures, where each clone shows a different parent. An example can be seen in Fig. 2.6, which shows three yellow icons. The icon on the right is selected, and it and its clones are highlighted. They all represent the activity "book flights." This is a subactivity of three different activities—"CHI conference," "Home," and "Today"—because the trip being booked is both the conference and for a vacation (a home activity) and the booking is intended to be done today. We are working on other features for the Activity Tableau to better support the temporal aspects of activities, such as a graphical time line to set and show deadlines and planning times and an automatic "Today" activity that dynamically groups activities planned for the current today.

Another version of the Activity Tableau brings time back into the spatial representation. Figure 2.7 shows a time line on the left, which provides a temporal context to the spatially arranged activity icons. When an activity icon is selected, the temporal properties of the activity structure are shown in the time line—green for planning interval and red for deadlines. Con-

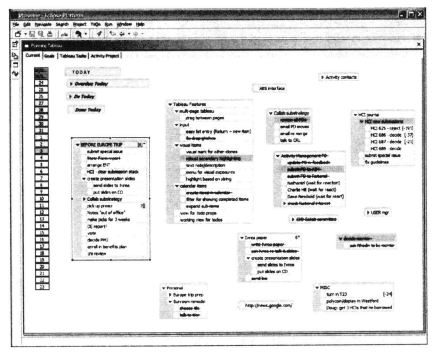

FIG. 2.7.   The Activity Tableau with a timeline.

versely, when a range of days are selected in the time line, the activity icons for the activities planned on those days are highlighted.

ACTIVITY REPRESENTATION PRINCIPLES

These flexible kinds of interactive representation tools support some fundamental features of human activity: emergence and incremental growth. Sometimes activities can be planned from the top down, first laying down a broad goal (an activity) and then deciding steps to accomplish the goal (subactivities). But often activities emerge from the bottom up. One engages in different activities that seem only vaguely related. These can be represented in the tableau by spatial clustering. Over time, as the relation between various activities becomes apparent, they can be organized into nested activity structures to represent the emerging coherence.

It is important to be reiterate the relation among activities, which are carried out in the external world, and their activity structures, which are representations internal to the system. Activity structures are not meant to be analytic products, such as ethnographic accounts of activities. They are more like checklists or task lists, which are not full descriptions of the activities they represent, but are still useful artifacts. People will represent activities only to the extent that it is useful for managing them. From our studies we see that people only need terse descriptions to remind them. The articulation of activities into subactivities is often only partial, and thus there are unrepresented subactivities. Sequential dependencies are not articulated. People seem to understand the logical dependencies in a list of activities, and they are opportunistic in the order in which they actually carry them out. Thus, any tool for managing activities must tolerate informal and partial descriptions and must give the users flexibility in carrying the activities. This is an issue in workflow systems (see, for instance, Dourish, Holmes, MacLean, Marqvardsen, & Zbyslaw, 1996).

An interesting related phenomenon we observed in our studies is that people sometimes do elaborate their activity descriptions *after* they are completed, to be able to better report on them later. But the principle is the same: they do the work because they understand the benefit.

THE BIGGER PICTURE

A facility to capture a rich representation of activities gives us a basis on which to provide many more services.

In addition to planning functions, we want to support the carrying out of activities. Essential to this is handling resources for carrying out activities. Thus, each activity structure can collect various resources. This provides an activity-based "foldering" capability, only more flexible, because activities can have multiple superactivities and can collect heterogeneous resources. (See Geyer, Vogel, Cheng, & Muller, 2003, for an excellent example of this kind of capability in their Activity Explorer system.) In the Activity Tableau prototype, we have implemented a version that works with e-mail. When the user drags an e-mail message from the e-mail inbox to the Activity Tableau, it is converted to an activity icon representing an activity with the message as a resource. From this icon, the user can activate the e-mail system to redisplay the message.

We also want to provide lightweight coordination by sharing activity structures among people. When an activity structure is shared, all the sharers have equal access to viewing and changing the activity structure; and these changes are synchronized between people. An example is shown in Fig. 2.8, which shows an activity structure for me that is shared with two others, indicated by the person icons on the right. All three of us can see and edit the activity structure. Each subactivity can have one of its sharers assigned to it (either by claiming it or by another sharer assigning it), or the subactivity can be left unassigned. Any of the sharers can mark activities done or add or delete activities. There is great flexibility in this scheme. For example, I can share different subactivities with different people to delegate them; I can see all the subactivities, whereas each of the sharers only sees the one subactivity they share with me. Or I can decide to decide share the overall activity with all the people, so that they each have more context for their individually assigned subactivities. Erickson, Huang, Danis, and Kellogg (2004) discussed the social issues of sharing activity information through activity icons, which they called "task proxies."

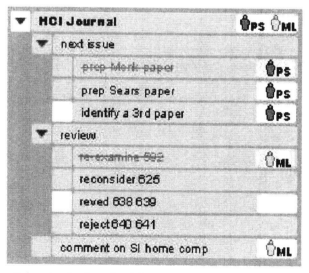

FIG. 2.8.   Example of a shared activity structure.

We are also exploring connecting these activity structures to workflow systems, where a person who plays a role in a process can "share" an activity (representing that role) with the workflow process. Finally, we are exploring the notion of reusing and refining activity structures and creating activity patterns. For example, standardized activities can be catalogued as activity patterns. When a person wants to do one of the standard activities, he or she creates an instance of the activity pattern from the catalog and uses the activity structure to guide him or her through the activity (similar to the way I used the previous year's committee work to plan the current year's committee work).

Perhaps the most powerful aspect of supporting generic activity management is the potential to integrate the many systems dealing with activities around a unified concept of activity, and to bridge the perspectives of business process workflows, team collaboration tools, interpersonal coordination tools, and personal productivity tools. The Unified Activity Management project at IBM is exploring the development of such a unified activity representation.

BACK TO INTERACTION DESIGN

This chapter has presented a view of generic human activity from the point of view of the person managing activities and initial explorations into tools to support the meta-activities of activity management. In the beginning of this chapter, I argued that the activity management perspective was in contrast to the usual view of activity in interaction design, where the activity surrounding an artifact is analyzed to understand context of an artifact being designed. But of course it turns out that designing tools for activity management is really a case of interaction design. The artifacts we are designing are generic activity representations, and activities surrounding these artifacts are the generic meta-activities of activity management. So, in the end, this chapter describes an exercise in interaction design.

ACKNOWLEDGMENTS

The author of this chapter thanks the following colleagues on this project at the time of this writing (November 2003) for participating in the development and exploration of the ideas in this chapter: Beverly Harrison, Andreas Dieberger, Leysia Palen, Alex Cozzi, Joann Ruvolo, and Dan Gruen.

Chapter 3

Virtual Environments and Haptic Interfaces

Massimo Bergamasco
Scuola Superiore Sant'Anna, Pisa

MANUAL TASKS IN VIRTUAL ENVIRONMENTS

The role of new digital technologies in design practice is becoming nowadays ever more important (Cruz-Neira, Sandin, & DeFanti, 1993; Sastry & Boyd, 1998). Activity in interaction design can also be seen in the light of these technologies, particularly those of virtual environments (VEs) that allow the designer to exploit directly features such as immersive visualization and haptic (that is, touch-related) interaction with virtual objects (Gomes de Sa & Zachmann, 1999). However, although VE technologies offer powerful tools for design practice, the most exciting aspect of their use in design concerns the direct implementation of new paradigms of interaction between the user and the product or service, and between the designer and the object to be designed.

If the term *activity* means here the action of the hands in performing a task, its transposition from a real to a virtual design space must take into account how the manipulation of objects and exploration of shapes are usually carried out in real space (Sastry & Boyd, 1998). The importance of the hand in everyday life is undoubtedly recognized in interaction studies (Moran & Carroll, 1996) because the sensation of contact with objects, and with the external world generally, is of great importance in evaluating the ultimate design characteristics of the tools, objects, and products with which we interact daily.

41

In recent years, the term *interaction* has acquired a large significance in the research framework of virtual environments. Simulator systems, single- or multiple-user immersive installations, collaborative or cooperative virtual workspaces—all fundamentally require the user to interact in real time with virtual entities. In all these contexts, the problem of interaction usually refers to the functional ability of the VE system to manage the bilateral exchange of sensory-motor data between the user(s) and the simulated environment. The research is then focused on designing interface systems and applications for rendering to the user a perfect sense of virtual presence while interacting with the VE.

The sense of presence inside a specific environment—either real (the actual physical place where the human operator is performing a task, or a remote physical environment where the human operator wishes to extend his or her action by means of a robot) or synthetic (an artificial virtual environment generated inside a computer)—is becoming an interesting research issue in VE (Slater, 1999; Slater, Usoh, & Steed, 1994; Moran & Carrol, 1996) and robotic teleoperation (Sheridan, 1992a, 1992b).

Three different conditions of presence are possible:

1. Simultaneous presence—The human operator works in a specific real workspace by collaborating with other humans, machines, or, in particular, autonomous robotic systems. In the last case, it is assumed that the workspace of the robotic system is colocated with that of the human (see Fig. 3.1).

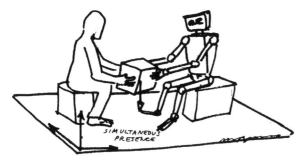

FIG. 3.1.   The concept of simultaneous presence.

2. Telepresence—The human operator, although acting in a real control space, feels himself or herself as present in a remote real environment where a teleoperated robot performs actions under the operator's direct control. The functional condition of telepresence implies that the interface system between the human and the remote robot has been designed for the bilateral exchange of motor actions and sensory data between the two physical spaces (see Fig. 3.2).

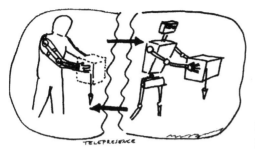

FIG. 3.2. The concept of telepresence.

3. Virtual presence—The human operator interacts with a virtual environment where actions are performed by means of an avatar (see Fig. 3.3). In this context the role of the interface system is essential for generating appropriate sensory stimuli to the human operator.

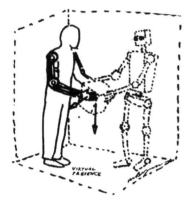

FIG. 3.3. The concept of visual presence.

From an engineering point of view, to generate a perfect sense of presence to the user, the design of the interface system must take into account three determinants (Sheridan, 1992a):

1. The extent of sensory information the user is able to acquire from the environment.
2. The mobility of the user's sensory systems inside the environment to explore it, according to the task and application in which the user is participating.
3. Control of the environment—the user's ability to grasp and manipulate objects belonging to the environment, and to modify their relative positions.

At present, the level of development in VE technology has mainly focused on the first two determinants: Visual and auditory aspects are well developed and VE systems are now available with acceptable levels of visual and auditory immersion and involvement. However, there still exists a high technological threshold for developing the third component: that part of technology devoted to implementing haptic interaction with virtual objects.

To indicate how the control component of presence for interaction with a VE might be designed, the following section presents a general analysis of manipulative and exploratory procedures performed by the human hand.

SENSORY FEEDBACK

Interface system design is a critical issue in teleoperation and virtual environments. The concept of telesymbiotics (Vertut & Coiffet, 1985), recently transposed to telepresence (Aviles, 1990), has emphasized research on those aspects of human–machine interface systems related to replicating sensory information to the human operator.

Because telepresence and virtual presence rely on the operator receiving a perfect sensation of being present in the remote or virtual operational environment while executing a task, a fundamental requirement of the interface system is the ability to replicate adequate sensorial stimuli at the control site (where the human operator is located). Although direct or indirect visual feedback from the remote or virtual task has been considered a primary source of information on how the performed operation is progressing, there are other kinds of sensorial information inputs for the operator. In teleoperation, as was already outlined during the development of the first mechanical teleoperators (Goertz, 1964), sensations related to the forces exerted by the robot "end-effector" (which performs the actual manipulation) on other objects in the operational space is an essential requirement for controlling the whole operation (Vertut & Coiffet, 1985). Other types of information related to the task being performed in the remote or virtual environment can also be conceived: tactile (Srinivasan, 1991) and auditory (Lehnert & Blauert, 1991) feedback have recently acquired interest both for teleoperation and manipulation of virtual objects. The fact that a complete sensation of telepresence and of presence is still far from being achieved, however, is mainly due to two factors:

1. The difficulty of achieving effective functional behavior by the artificial systems devoted to replicating adequate sensorial sensation to the operator. Although good performances have been achieved for the visual modality by immersive visualization systems such as the CAVE (Cruz-Neira et al., 1993), other sensory modalities, such as the haptic, still show limited results.

2. The objective difficulty for the human operator of directly integrating and interpreting different sensorial information coming from "unnatural" sources.

The rest of this chapter discusses the problem of rendering to the human operator as much information as possible about the course of manipulative operation performed in a remote or virtual environment. It focuses on the problem of replicating, on the human operator's hand, the contact forces generated during manipulative or exploratory procedures performed by the robotic or "virtual" hands on remote or virtual objects. In particular, by considering various situations encountered in VE applications, it investigates the possibility of integrating hand force feedback (HFF) systems on advanced glove-like interfaces. Later, we present design considerations regarding the HFF system conceived in the PERCRO laboratory of the Scuola Superiore Sant'Anna in Pisa.

GLOVE-LIKE ADVANCED INTERFACES

The morphology of human–machine interfaces used for teleoperation purposes covers a wide spectrum of systems, ranging from faithful kinematic replicas of the remote robotic manipulator (Srinivasan, 1991), to arm exoskeletons (Bergamasco et al., 1994), and different types of hand controllers (Bergamasco, De Micheli, Parrini, Salsedo, & Scattareggia Marchese, 1991; Smith, Armogida, & Mummery, n.d.).

In these cases, robots at the "slave" site of the operation usually have end-effectors with only one degree of freedom (DOF), so teleoperation control is limited to pick-and-place tasks or simple grasping operations. The intrinsic mechanics of the slave robot end-effector, in other words, do not allow control of effective manipulative actions.

So a basic requirement for operations requiring direct manipulation of the object is more kinematically complex end-effectors, allowing motion of the grasped object relative to the rigid "palm" of the end-effector. For this purpose it is increasingly attractive to use, as the slave robot's end-effectors, so-called "multifinger, multifreedom grippers" (Fisher, Daniel, & Siva, 1990) such as the Stanford/JPL (Salisbury, 1984) or Utah/MIT hands (Jacobsen, Iversen, Knutti, Johnson, & Biggers, 1986).

Clearly, when a multifingered, multifreedom gripper is used at the slave end, the manipulative procedure can no longer be controlled by a simple input device like a handle system or a joystick with different buttons for the grasp command. Because the end-effector's DOFs are greater, it is essential to use similar kinematics for the master controller too. Hand controllers (interface systems capable of recording the motions of the human operator's hand) can do this.

The same applies to controlling manipulative tasks inside a VE. The human operator should have an interface system capable of recording the fingers' movements, and should use this data to control the corresponding movements of the virtual hand. A practical design solution for hand controllers is sensor-equipped glove-like devices or compact hand exoskeletons that allow the operator to perform manipulative movements in a natural way. By means of a glove structure, the human hand is free to move and perform the command action for manipulation without being attached to external devices; here the hand "is" the interface (Bergamasco et al., 1991).

Exploiting a glove as a human–machine interface has been considered since 1960, especially for teleoperation purposes (Johnsen & Corliss, 1995). Other types of glove-like interface have been designed to assess and measure hand movements (Grimes, 1983), and more sophisticated sensor technologies have been used to develop the present-generation glove-like advanced interfaces which are largely accepted as interactive devices in the new field of VEs (Fisher, 1986; Marcus, Lucas, & Churchill, 1989; Kramer & Leifer, 1990; Sutter, Iatridris, & Thakor, 1989). However, their use for telemanipulation and virtual manipulation purposes has been unsatisfactory for two reasons (Hong & Tau, 1989): their limited precision and accuracy in recording finger motions, and their lack of means for controlling contact forces at the remote or virtual site.

Appropriate recording of finger motions requires not only good performance in terms of sensor technology but also a well-studied arrangement of the sensors on the glove's surface (Bergamasco et al., 1991). The role played by the kinesthetic sensors in the control procedures for force replication to the fingers is fundamental, and so therefore is the contribution of the glove structure to the kinesthetic sensor system's accuracy in recording the hand configuration. The presence of a glove structure, in fact, allows a stable relative positioning of the kinesthetic sensors when putting the glove on and off, and during different working sessions (Bergamasco et al., 1991).

Replicating to the operator the forces sensed by the remote robot "hand," or modeled in the virtual environment when the virtual hand reaches the contact with virtual objects, could improve the operator's control over the manipulative action; thus the concept of proprioceptive (force) feedback for the hand. But in the current stage of VE technology little attention has been paid to developing glove-based force feedback systems. The main reason is the difficulty of the technological problems associated with the need to locate actuating systems and complex mechanisms on the glove structures (Loomis & Lederman, 1986).

HAPTIC INTERFACES

Telecontrolled and exploratory operations imply interaction between the robot or virtual hand and remote or virtual objects. This usually means that generalized forces are generated at the contact between the environment (real or virtual) and the finger surfaces. The kinds of force generated during exploratory procedures differ from those generated during manipulation because several factors—such as hand configuration, number of contacts between fingers and object, amount and type of motion at each contact—depend strictly on the purpose of the action. Furthermore, during a general task in which robot or virtual hands are involved, other types of force such as random collisions and inertia forces are present, and, once replicated to the human operator, contribute to the general feeling of physical interaction with the remote or virtual environment.

Let us try to analyze all possible situations by referring, for clarity, to interaction with virtual environments, for from the point of view of the interface system there is little difference between considering a virtual environment and considering a remote one. The only difference resides in the fact that for controlling virtual manipulation it is necessary to model the behavior of the virtual entities, whereas in telemanipulation the procedure must be performed in a real physical environment.

Exploratory procedures (EPs) are usually considered when the human operator wants to derive information about the external environment (objects) haptically (Loomis & Lederman, 1986)—that is, by means of cutaneous (skin-detected) and kinesthetic inputs. Usually exploration procedures involving active motion by the hand seek knowledge about the external object—its hardness, temperature, texture information, and so on. And for each kind of sought information it is possible to derive a particular hand movement pattern (Lederman & Kltazky, 1990).

If exploratory procedures must be executed in a VE to acquire information about unknown virtual objects, it is necessary to replicate to the human operator's hand the sensation of the contact forces occurring between the virtual hand and the virtual objects.

Figure 3.4 gives examples. It shows the areas of the human hand in contact with the explored object, taking as a reference the hand movement patterns associated with specific exploratory procedures proposed by Lederman and Klatzky (1990). Note that for certain types of exploratory procedure the contact areas over the finger and palm surfaces remain almost unchanged even if the hand moves relative to the object. Figure 3.4 (a), for instance, depicting the exploratory procedure of "lateral motion," demonstrates that although the left hand is used to sustain the object in a fixed position so as to present the object's face of interest, the index and middle fingers of the right hand are used to touch the object's surface gently and actively to acquire texture sensation. In this case, although the fingers are moving with respect to the object, the contact area remains localized at the fingertips (where it is possible to elicit maximum tactile sensitivity). The same applies to the exploratory procedure "pressure" in Fig. 3.4 (b), by which it is possible to acquire information about the object's hardness.

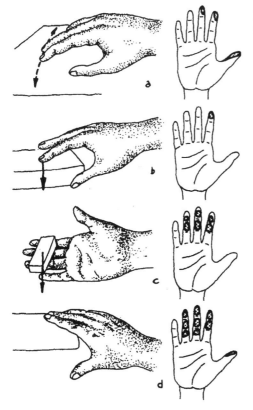

FIG. 3.4. Contact areas during exploratory procedures.

If we were considering a remote manipulation, knowledge of the contact forces could have been obtained directly by reading tactile and force sensors located on the robot multifreedom, multifinger gripper. But because we are dealing with a virtual hand contacting a virtual object, appropriate modeling can calculate the forces acting on the virtual hand from knowledge of the geometrical parameters of each contact area.

The forces that should be replicated on the human operator's hand during exploratory procedures are as follows:

1. Low-intensity forces acting on the fingertips, or also on the palmar surface of the other phalanges (the articulated elements of each finger), in the direction of extending the finger. These forces can be considered the resultant forces at the contact regions. The intensity is very small if we remember that the humans are here using their hands for apprehension, not prehension.

2. Forces replicating the weight of the object when held or supported to extract weight information or to present a specific surface to the other hand.

3. Forces of collision between parts of the object (especially in the case of exploratory procedures such as "function test" or "part motion test") and regions of the hand.

Manipulative procedures: Observation of common manipulative actions performed in everyday life demonstrates that the spectrum of forces to be replicated is very large and quite unlike the one identified for exploratory procedures, especially if we refer to the forces acting on the fingertips and the palmar side of the other phalanges. Manipulation implies that, once the object is held in a stable grasp, it is possible to move and reorient it with respect to the palm by means of finger movements. Usually, precision grip movements of the hand are primarily considered (Napier, 1956), where the hand is required to exert small forces and impart delicate movements to the grasped object (Mason & Salisbury, 1985). However, it is also possible to take into account power grip actions (Napier) by which the object is held by the finger in a stable position within the hollow of the palm. In this case, although the contact regions between the object and the hand are augmented in terms of both the number of contacts and the total area, for manipulative procedures the magnitude of the resultant forces for each contact generally spans a larger range than in exploratory procedures. Moreover, as the manipulative operation evolves in time, the contact areas change their trace on the surface of the hand, justifying the dynamic character of the action (Fig. 3.5).

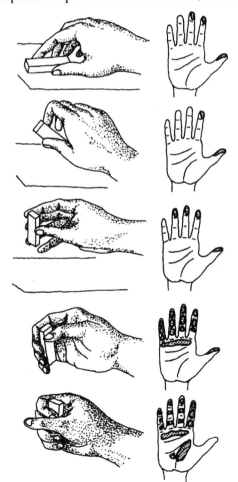

Taking these considerations into account, in addition to the forces related to the object's weight and to the collisions with external objects already specified for exploratory procedures, the replication of the forces occurring during manipulative procedures must consider forces that primarily act on the fingertips during a precision grip, but can also extend their contact area to the whole palmar and lateral surface of the fingers during a palmar grip.

In fact, distribution of the local contact forces is not usually localized on the same area of the hand but, because the object can move relative to the fingers, the contact

FIG. 3.5. Contact areas during manipulative procedures.

area can vary geometrically over time. The fact that the range of intensities and directions of the local contact forces is extremely diverse represents a tight requirement for the interface system devoted to replicating the forces.

A GENERAL SYSTEM FOR HAPTIC INTERFACES

The objective of the previous qualitative analysis has been to determine what kind of forces are involved in common tasks of manipulation and exploration. We can now introduce the functional specifications for the HFF system by considering which set of the previously mentioned general contact forces can be effectively replicated.

The HFF system can be considered as a part of a general haptic interface system. As its name implies, the HFF system is designed to replicate forces on the fingers of the hand. These forces can be internal forces, generated for example during grasping operations, and external forces on the fingers such as those arising during "pressure" exploratory procedures.

This implies that only a subset of all the forces identified in the previous analysis will be taken into account for replication by the HFF system. This is because of design constraints in terms of the power/volume and weight/volume ratios required for the actuation components of the HFF system; problems of allowable volumes and weights for the mechanical and electronic hardware must be solved, taking into account the need to allow the operator to maneuver the fingers naturally. We have assumed that high-intensity forces which act on the object (weight and external collisions, for instance) or directly on the hand (collisions, for example) can be replicated to the operator at the level of the hand by means of a complementary component of the haptic interface system, called the *external force feedback* (EFF) system.

One possible design solution for the EFF system is a 7-DOF arm exoskeleton connecting the hand to the operator's shoulder or a 6-DOF external robotic arm connecting the hand with an absolute ground base. The external forces are reflected to the human hand by applying motor torques onto the arm's exoskeleton joints. These torques have been calculated by considering the projection of the momentum generated on the EFF motion axes by the desired force components (static balance equilibrium). Splitting the replication of forces on the hand between the two systems (HFF and EFF) takes into account not only the requirements in terms of the power of their respective actuation systems but also the functional capabilities of the human body parts (the hand and forearm-arm) to which they are applied. In fact, during real manipulative actions the fingers must impart small movements to the object to control its stability. This exploits the action of the intrinsic muscles, whose sensors are responsible for the sensation of force during fine manipulation, and which mainly work when precision movements are required. However, when the load of the manipulated object increases, the prehensile movements of the hand are controlled by extrinsic muscles located in the forearm, which are in fact less precise but more powerful than intrinsic muscles.

The same considerations apply in the opposite direction in terms of force intensity: When very delicate exploratory procedures must be implemented, for example to detect the texture of the grasped object, the HFF system cannot be used to replicate this kind of sensation. A third subsystem of the haptic interface, the *tactile feedback* (TF) system, will be involved to replicate adequate patterns of spatio-temporal stimuli. Sutter et al. (1989) have assessed evidence of the functional difference between tactile and proprioceptive force sensors in perceiving force sensation during manipulation tasks.

The role of the HFF system is to replicate midintensity forces on the fingers, like those recorded during common manipulations of small-to-middle-size objects. In other situations—when the object is heavy or collision forces are high-intensity or conversely, when very fine micro-indentation cues must be replicated—the HFF system can work in conjunction with complementary components of the whole haptic interface system, such as the EFF and TF systems. Figure 3.6 shows a possible situation.

Although the HFF system has been defined from the functional point of view, a more detailed analysis of the forces to be replicated on the fingers is needed. The previous section only considered contact areas: An ideal HFF system must be able to replicate exactly the local forces acting on a specific contact area. We have seen that the contact area can, over time, vary its relative position on the palmar surface. If for design purposes we ought to take this fact also into account, it is necessary to consider a movable actuation system capable of replicating the exact pressure distribution in each point of the phalanx. At present, however, such actuating technologies are not available and the problem of HFF system design must be addressed by introducing some simplifications.

A first assumption is to keep fixed the contact areas between the glove and the operator hand. This implies that, by considering the HFF system capable of replicating forces on each phalanx of the hand, there are a minimum of 14 contact areas for the phalanges and other areas to be located on the palm (at least two, on the thenar and hypothenar regions at the base of the thumb and of the little finger). Over each of these areas, the HFF should exert a force system equal to the homologous actions modeled for the same contact area in the VE (or detected by force sensors of the robot end-effector). However, if we consider that the contact between the virtual hand and the VE could be considered as a soft finger (or *soft phalanx*) type of contact, the HFF must be able to replicate at least a wrench 4-system (three forces and one torque along the contact-surface normal) for each fingertip contact (Mason & Salisbury, 1985) and a wrench 5-system for each phalanx

FIG. 3.6. Possible force feedback actions distributed among the three components of the haptic interface: external force feedback, hand force feedback, and tactile feedback.

and palm region. Yet, even reduced to this form, the problem presents severe technical difficulties because the number of actuators is still very large.

A practical solution could be to reduce the wrench 4- or 5-system to wrench 2-systems: the qualitative analysis in the previous section showed, in fact, that for most heavyweight external collisions specified in prehensile and nonprehensile movements, the forces generated at the interaction between the hand and the object possess the following: (a) an application point near the center of the contact area, and (b) a resultant force of contact almost normal with respect to the phalanx's longitudinal axis, in a direction producing an extension movement of the finger. So a possible solution for the HFF system is an actuation system capable of generating on the phalanges action forces with the following features:

1. A line of action normal to the longitudinal axis and belonging to the sagittal plane of the fingers (that is, which travels across their short axis).

2. A direction such as to produce an extension movement of the finger.

3. An intensity limited by the mechanical performances of the actuators.

Figure 3.7 represents the characteristics of the action performed by the HFF system we are designing. A first prototype of the HFF (see Fig. 3.8a and Fig. 3.8b) has been designed and realized.

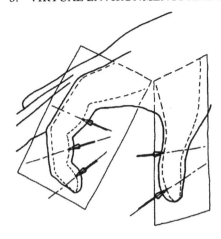

FIG. 3.7.   Representation of the action forces on the sagittal plane reproduced by hand force feedback.

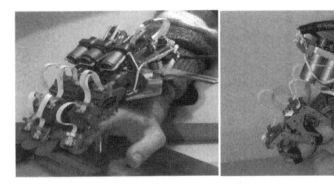

FIG. 3.8a (*left*) and FIG. 3.8b (*right*).   The PERCRO hand force feedback system.

CONCLUSION

This chapter has introduced the functionalities and components of haptic interfaces. Replication on the human hand of the forces acting on the virtual hand during manipulation procedures can improve the effect of virtual presence. The possibility of integrating all three different components on the same interface structure results in a condition very favorable for the effective test of haptic perception procedures in both teleoperation and virtual environment conditions.

The problem of haptics interface development is not only tied to HFF, kinesthetic, and tactile system hardware implementation on a glove structure. It requires defining adequate control procedures, especially in the case of virtual active exploratory procedures, where the modeling of contact and force replication plays an essential role.

The effective power of virtual environments as a research tool for haptic perception can be completely exploited only when these interface technologies have achieved an acceptable degree of implementation; efforts in this direction are in progress. The haptics component will acquire ever more importance in the future development of VEs. Immersive spaces for collaborative and cooperative work will witness such developments.

ACKNOWLEDGMENTS

The work described was conceived and carried out at the PERCRO Laboratory of the Scuola Superiore Sant'Anna, Pisa, Italy. The author of this chapter thanks Professors Carlo Alberto Avizzano and Antonio Frisoli for their continuous support and help in defining the vision, mission, and objectives of the laboratory.

Part

II

EMOTION

Chapter 4

The Feeling of Values: For a Phenomenological Theory of Affectivity

Roberta De Monticelli
Università Vita-Salute San Raffaele, Milan

OF THE IMPRECISION OF PHILOSOPHERS, AND OF TWO ERRONEOUS THEORIES

Philosophers have always wreaked a bit of havoc on the common language. But in the case of the vast population of the *emotions* (we tentatively employ this term as a class name, synonymous with "phenomenon of the affective or emotional sphere"), this violent simplification is endemic and surprising. In the age of Descartes and for centuries after, the term *passions* played the major role. All sorts of things, along with others that we today have come to designate in the same way, were grouped under this single term: feelings, moral dispositions, moods, states of mind, and emotions. "Emotions," in the lingua franca of contemporary philosophy, is a word which has come to be employed with the same all-inclusive generality. Even the recent, beautiful book by Martha Nussbaum, *Upheaveals of Thought: The Intelligence of Emotions* (2001), is no exception to this general trend. Despite the wealth of empirical and literary examples provided, Nussbaum does not seem to feel the need for a terminological and conceptual distinction between, say, a fit of anger and a long-term affection, such as love for one's parent—qualified, however, as a "background" emotion. (Compare the following, for instance:

Calabi, 1996; Damasio, 1994; De Sousa, 1987, 1995; Goldie, 2000; Magri, 1999; Oksenberg Rorty, 1980; Tappolet, 2000.)

On the other hand, the choice of the term matters little. Something decidedly worse than the conventional adoption of a general term could occur: the adoption of an erroneous theory which reduces all the diverse affective phenomena to one, and furthermore provides it with a general characterization which is mistaken.

With no pretensions to historical or polemical precision, I believe that at least two erroneous or unsatisfactory general characterizations of emotional experience can be identified. Both of these, generally speaking, tend to contrast reason with passions in the same way that objective power is contrasted with subjective influence, clarity with confusion, control with *sauvagerie,* and mind with body. Both, however, accord a certain orientational value to emotions; this value is not cognitive, but biologically functional. No truth here, but great utility.

The first theory, common to a great part of both the rationalist and empiricist traditions relegates affective experience to the sphere of "subjectivity" pure and simple, depriving it of any access to the true, whether moral or generally axiological. In certain versions of this way of regarding things, for example in Cartesian or empiricist versions, a practical value is accorded to emotional experience, "emotions" being an indicator of sorts as to what is useful or harmful in the life of the organism. How in the world, then, do profound and noble sentiments, like veneration or benevolence, or fine esthetical emotions, like gratitude for the beauty of a sonnet by Petrarch, or even vicious passions such as those of Dostoyevsky's Gambler, fit into this scheme? In the brilliant but theoretically doubtful Cartesian treatise, this question is left open. Here, too, common sense and common language are more precise than philosophy in dealing with "moral sentiments," "esthetic emotions," and so on. There is no place in an "economic" philosophy of emotion for refined, delicate feelings, not to mention utterly useless, gratuitous, or even harmful ones. Dostoyevsky may as well head for the psychiatric ward along with all of his characters—with most of the rest of us not far behind.

The second theory, characteristic of the Freudian model, or perhaps of his vulgate, seems to share this thesis of cognitive inanity or of an at most economic-practical value of the emotions. But it also reduces all affective phenomena to one of the two dimensions of emotional life: that of *feeling* and *inclination* (it remains to be clarified whether this inclination is to be distinguished essentially, or merely differentiated modally into appetitive and aggressive-defensive, inclination toward and inclination away from, concupiscent and irascible). A reduction of emotional states to drives is a phenomenologically groundless and incomprehensible reduction of modes of feeling to modes of inclination. Indeed, *sensibility* is a disposition of *reception* par excellence and as such cannot be reduced to a function of the sphere of drives. If sensibility and excitability correspond, respectively, to feeling and inclination (desires, drives, reactions), an analysis of the structural difference between them—rather than a reduction of one disposition to the other—would be necessary. And this psychoanalytic vulgate, as it were, is also philosophically less precise than common sense and common language, which are significantly more attuned to the distinctions in these phenomena. If we say that a person is "sensitive," for example, we do not by any means intend to say that he or she is excitable, nor that he or she is lacking in objectivity. On the contrary, what we mean is that he or she is more capable than others in terms of discrimination, and consequently of *truth* in the exercise of *feeling*.

THE PHENOMENOLOGICAL APPROACH. FEELING: EXTENSION AND DEPTH

For the phenomenologist, nothing *appears* without a reason, although not all that is real appears. We receive an extremely rich series of the apparent qualities of things by means of *feeling*. I feel the unpleasantness of a sting on my skin, the discomfort associated with a state of illness or weariness, the agreeableness of an arrangement of colors. But I also sense the nobility of a gesture, the vulgarity of an attitude, the wickedness of an act, the beauty of a masterpiece. The harmonious fitting of a tool or a piece of furniture to our body, the placid physiognomy of a teapot, are among the "affordances" of an object—to use J. J. Gibson's (1966) terminology: "I have coined this word as a substitute of *values*, a term which carries an old burden of philosophical meaning. I mean simply what things furnish, for good or ill. What they *afford* the observer, after all, depends on their properties" (quoted in Jones, 2003, p. 111).

These qualities too are somehow "perceived": we suggest that feeling is the appropriate mode of this perception. The functional and esthetic qualities of artifacts are not only "seen": they are felt. This feeling is always accompanied by the exercise of other functions, both sensory and otherwise (phenomenologically, it is "founded" on other acts: sensory perceptions, acts of psychological or empathic perception, acts of linguistic comprehension, and so on).

Feeling is essentially perception of the positive or negative values of things. We may call this the *first phenomenological thesis on emotional life*. Feeling is the mode of presence of the most diverse axiological qualities, or values. This opening to axiological qualities is *intentionality* (that is, the sort of relation to reality), characteristic of all the experiences of the emotional sphere. Thus, for the phenomenologist this is a dimension of our experience of the real, by no means a realm of subjective arbitrariness—and it is not a mere system of functional alarms or incentives necessary for the survival of the organism. In fact, if it is true that this face is beautiful, that this action is horrible, then the evidence by means of which we sense this is evidence in the epistemological sense, that is, proof of truth (until proven otherwise) and access to reality (until denied, if such be the case). These parentheses are not superfluous. Opening to reality, in phenomenology (and, we believe, in truth) is synonymous with *fallibility*. Errors of affective feeling are equally if not more frequent than those of sensory feeling—in fact it would be more precise to speak, in both cases, of possible *illusions*—and therefore equally open to correction and "proof to the contrary." Here again, as in the case of the sensory domain, reflection may help to correct an erroneous perception; but it is, in fact, a new perception that will tentatively confirm the correction. It is only that here (where the opening to truth is an opening to values), errors and illusion, inappropriate perception, myopia, idiocy, or obtuseness, are at the base of the inappropriate axiological responses—and among these, of the *ethically* inappropriate responses. The first phenomenological thesis implies a profound attention to the problem of the *education of feeling*, or emotional culture. This education of feeling *with precision* is at the heart of culture from the phenomenological point of view.

With this thesis, the basis is proposed for a criticism of the two theories (erroneous from a phenomenological point of view) described earlier. Far from a reduction to excitability, *emotional experience is based on sensitivity*, and excitability, when the former is in effect, tends to be conditioned by it. Far from being *opaque* and *irrational*, emotional experience is the mode of presence of the qualities of value of things and is also the more or less appropriate response, and therefore the more or less correct, proportionate, or "rational" response to the *requirements* that these present.

This thesis adds to the first, which characterizes feeling *essentially*, the role of feeling in the complexity of emotional life. In one way or another, feeling represents its basis, but emotional life is certainly not reducible to feeling, this primary, irreducible mode of *reception*.

Feeling, on the other hand, even when erroneous or approximate, indeed obtuse or atrophic, even when conspicuous in its absence, "serves as a foundation" for the rest of emotional life, in the precise sense that it *motivates everything in this life which is not mere reception, but response and spontaneity*. In this sense, feeling, even when it is "missing," seems always to be determinant in every aspect of emotional life, as opposed to inclination, and to the drive-based and appetitive-self preservation side. We call this conjecture the *hypothesis of the omnipresence of feeling*. A paradoxical example is the inverse proportion often observed between sensitivity and excitability, or between the appropriateness of feeling and *emotional agitation*. There are countless examples from everyday life but there is a little literary gem, a sort of theorem of this inverse proportion, in Chekhov's early play, *Ivanov* (1887/1997).

We could comment further on the second thesis and the hypothesis of omnipresence in this way: the experience of values is at the basis of the responses to the demands of the real (the responses of which a life essentially consists)—and thus at the basis, for example, of choices, of close personal acquaintances and of behaviors, both habitual and exceptional. *In the final analysis, feeling motivates volition.* But feeling *manifests itself to us and to others first in phenomena of emotional life.* Emotions, states of mind, sentiments, and passions are the principal modes of manifestations—and thus of the concrete existence—of feeling: extremely diverse modes, which stand in diverse relation to feeling.

To deal with the analysis necessary to clarify this point, however, the other dimension of feeling is missing. Thus far we have considered one, that of its objects: the *breadth* of its domain. The qualities of value—positive and negative to varying degrees—are *many*. The extraordinary *richness* of the negative or positive qualities of value to which we are sensitive is one of the elementary data for any phenomenological reflection on feeling. There is virtually no situation in life in which at least some of these qualities are not currently or habitually present. I walk along the road and the atmospheric conditions present me with an environment that is pleasurable to a greater or lesser extent; the urban architecture affords me esthetic joy or suffering with every step; the beggar on the corner draws my attention to the horrors of poverty; in words and images the newsstands scream violence and beauty, ferocity and injustice; every event along my way, whether grave or trivial, manifests qualities of value from some point of view: esthetic, ethical, legal, economic, ecological, ergonomic, hygienic, gastronomic. Viewed *a parte subiecti*, this axiological richness of the world is the quantity of values that *afficiunt* us, touch us and move us, physically or emotionally. But equally immediately given is the *quality* of these, *the way in which they touch us* to greater or lesser levels of depth or intimacy.

The other dimension that reveals itself immediately to phenomenological observation is therefore that of the greater or lesser *depth* at which we are touched when we feel a quality of value, positive or negative—the ugliness of a pair of shoes on a friend's feet, or of one of his gestures toward us. We also express this difference—responsible for the (not merely metaphorical) character of the notion of greater or lesser *depth of the phenomena* of the emotional sphere (it is a notion that could, in fact, be associated with a criterion of measurement)—in speaking of the difference in *importance or weight* that things have. We are usually willing to distinguish, in general, the importance or weight that things have in themselves—the ugliness of a pair of shoes, for example is generally less serious than that of an action, the esthetic value of an instrumental object less important than the ethical value of an action—from the importance or weight that things hold for us under such and such circumstances. For example, if it is important to me that my friend should make a good impression on someone in a position to help him, his ugly shoes will disturb me more than a nasty remark he directs toward me.

The difference of "objective" importance which we recognize both generally and in many specific cases (killing is more serious than insulting, defacing a Fra Angelico more serious than scrawling graffiti on a supermarket wall, giving one's time or one's life for a friend is greater than giving her money for medical treatment) may be called *the rank of a value*. As difficult as it is to sketch out a hierarchy of spheres of value in the abstract, all of us, in our everyday affective responses, and more so in those far rarer cases of personal or historic events, that are decisive in our lives, have the evidence (often painful, astonishing, unexpected) of what is in effect our *value preference system*, our *ordo amoris*, to use the Augustinian expression that Max Scheler (1957) reproposed with his notion of *ethos*. A person's ethos is his or her moral identity, but this moral identity, manifested secondarily in a person's choices and behaviors, manifests itself primarily in the emotional life that motivates these choices and behaviors, and in which, ultimately, his or her uniquely and unmistakably *own* mode of feeling is expressed.

FEELING AND PERSONAL IDENTITY

The emotional life can, therefore, be defined as a "manifestation" of that feeling that is experience of values in their variety and in their personal importance or incidence.

From what we have said, it seems that it would be impossible to bring order to the apparent chaos of affective phenomena if not for one fact, ordinary insofar as it is massively attested to in our everyday lives, extraordinary for its philosophical importance, for what we could call its *key* nature in any serious personology. This fact is the role of feeling in personal identity. A person who is wounded in her feeling is a person wounded in her deepest self. Without hesitation we say "in the soul" (without implying in the least by this expression any dualistic or spiritualistic idea of the "soul," or the *personality* of a person). It is not surprising that poor Phineas Gage, the construction foreman from Vermont who suffered extensive cerebral lesions in an accident at work, and whose case was narrated by Antonio Damasio (1994) in his highly successful book *Descartes' Error*, "was no longer Gage." For the psychosomatic functions disrupted by the accident were exactly those that are indispensable for the modulated exercise of feeling.

We have redefined emotional life as the "modulated exercise of feeling," taking another step forward relative to its previous characterization as a manifestation of feeling. As elsewhere in personal life, phenomena based on feeling do not "just happen," in spite of their being involuntary. We *exercise* the functions of affective feeling by acts of *response* to the environment, as we do in the case of sensory feeling. Seeing or not seeing does not depend on us, but the act of withdrawing our gaze, at least to a certain extent, does. And feeling, likewise, is in some measure always a case of suffering, or of joy, and as such does not depend on us—this polarity is characteristic of feeling precisely insofar as it represents a mode of presence of negative or positive qualities of value. We may say that this is the essentially receptive side of feeling, in particular of its passive—"pathic"—character. But by *which qualities* among the host of situations and things *are we struck*, and, once struck, do we *continue to allow ourselves to be touched*? *To what depths* are we touched— and what hold *do we allow* such qualities to have over us? Throughout all of this there is a living set of yesses and nos by means of which we discover for ourselves the order of our value preferences, which activates and modifies itself throughout life. Not directly in the feeling, but in the *granting or withholding of consent relative to this feeling* and to the needs that this feeling presents—and to the degree and manner in which the spontaneity *of a self* and an *order of one's own axiological preferences*, otherwise unknown to us, is manifested and affirmed.

In this sense, not only acts of volition, decisions, choices, and actions, but also those same affective phenomena at their base, motivating them—emotions, moods, feelings, sentiments, and passions—are to be understood not as mere occurrences of psychic life, "mental events," but as *personal responses to the experience of values*, responses which taken together, are manifestations, at times even discoveries, of self—and stages of human maturation. If affective phenomena are taken in this way, then the most ephemeral of the emotions and the most deep-seated of sentiments, the passion of a lifetime and a morning fit of crying can no longer be put on the same level. We can no longer throw veneration and a panic attack together into the same pot, but will have to accord a founding character *to certain responses*, that of a matrix of other responses. Sentiments, in the sense of relational feelings, establish a new condition and a new profile of affectivity as a whole; they become *matrices of responses*.

Sensitivity is not activated all at once, from the very beginning of an emotional life. How does a properly "personal" layer of affectivity, constituting the core of personal identity and the substance of an *ethos*, come to exist? Answering this question is important even from the point of view of practically oriented research. Emotional responses to artifacts or other objects charged with value qualities will be quite different within the frames of different *ethoi*.

Emotional life has a tendency to progress in stages, or rather in peaks, with each peak representing a characteristic discovery or rediscovery of *what truly, and to a greater degree than other things, we attach value to*. One of the factors of positive peaks in the growth and axiological structuring of a personal sensitivity is love.

PHENOMENOLOGY OF LOVE

Love is, in principle, *the spontaneous activation of a layer of sensitivity to positive values that had hitherto lain "dormant" or inactive:* hence the characteristic *wonder* with which its resurgence announces itself, along with a sense of expansion of the vital breath and mental horizons. Hence also the characteristic sensation of "new life" that accompanies it, at times perceived as having "begun to live only now," not before. And it is, in its essence, *the acknowledgment of an essential individuality*—curiously, *the only full and complete acknowledgment of the existence and uniqueness of an "other"* that is accessible to us—that is, not only as represented to the mind, but as actually given to the evidence of feeling. (On the notion of essential individuality, see De Monticelli, 2000a, 2000b, 2003). Persons, naturally, are the paradigm of individuals in the essential sense—that is, endowed with an individual essence. But "other" in this phrase is intentionally neutral. We can love the opera, or the work of an author to the point of making it our reason for living.)

Strangely enough, the more this acknowledgment represents *unconditional assent* to full existence of the other, the more it seems capable of insight or vision. In terms of *unconditional,* you do not love someone because he or she is a person of value; you love that person because he or she is *that person* and because you, mysteriously, perceive his or her true worth more clearly than others. In terms of *assent,* the "yes" that rises from the depths is a spontaneous stance taken in relation to the position of feeling, a sort of second order position assenting to the first "yes" of pleasure aroused by some appeal of the being to be loved. To whatever sphere this appeal may belong—beauty, sweetness, elegance, nobility (or more simply, in the non-elective forms of love, the call of blood, whether filial, fraternal, and so on)—it can act as a "call" only insofar as it actually calls us personally, "touches" us at such a depth that we feel that we ourselves, and not anyone else, are being called. But the "assent" in which love is acknowledged, and which is infinitely rarer than the call, is a true and totally new or renewed *identification* with this "heart" to which the other and his or her world truly matter, and is, in this sense, a discovery or a rediscovery of self, which has a strange, irreducible affective positivity. Assent is *happy.* And the more authentic the acknowledgment of love, the happier the assent (that is, the more *contented* with the existence and full realization of the other), in conformity with the other's essence.

This happiness of assent—in which the acknowledgment of the other and gratitude toward another is also an immediate yes to oneself, a joyous yes to a new or renewed part of oneself—is undoubtedly an enigma, aside from its being so well known to us all. It is something so striking as to be always noted, but too often in a distorted way. Happy is love, but why happy, when it is nearly always a source of suffering due to the constant threat to which life and the happy realization of the other are subject? And it is here, in an attempt to explain this enigma, that love has been mistakenly identified with desire.

This remark might be of some interest to producers of all sorts of gadgets, or even to artists and craftsmen. The more developed the personal layer of emotion (whose root is not desire, but gratitude), the more personal (i.e., coherent to a personal style of consumption) the "user perspective" on products of design. Standardized objects of desire, such as those incessantly presented to us by commercials, are addressed to the sensory and vital, yet impersonal, layer of emotions (discussed later). Personality constitutes itself as a *revelation of a higher level of being:* the manifestation—in the awakening of not yet or no longer aroused layers of sensibility—of new and deeper value horizons, and hence of one's previously ignored potential for life, knowledge, and action. Desire is the opposite of loving fullness of personal life, being a kind of *absence of being*, as it manifests itself in the modes of tendency and appetite: *need, demand, hunger, libido, tension to satisfaction, drive.* Unsatisfied desire, along with everything it brings with it—restlessness, worry, anxiety, apprehension, quarrels, and conflict—seems to be the exact opposite of happy assent.

According to this phenomenology of mature, personal emotional responses, we should set about dethroning this false god, desire, if we wish to shed more light on the phenomena of emotional life and on the reasons for our ordinary unhappiness. Desire is an eternal objection to *real* assent to the other's and to one's own existence. It is the true root of the endless negotiations and never-ending wars of acquisition that have to be gone through before assenting, usually unhappily, to our own being. It is the sole root of envy, jealousy, or, at best, emulation and competition. *It is a pole in emotional life, opposed to that of feeling, always ready to contend for vital energy.* This vector of tendency and action can refine itself, and the more it does so, perhaps, the unhappier it renders us, when, having exhausted the attraction of "having," it turns to being. Desire—even the desire to be what we are not, or not yet, or not any longer, or not in time—is perennial dissent within ourselves and perennial objection to the yes of gratitude. In gratitude lies the essence of beatitude.

THE STRUCTURE OF FEELING AND ITS CREATION: DRAFTING A THEORY

Let us resume the two phenomenological theses on affectivity and emotion:

1. Feeling is essentially the perception of values, positive or negative, of things.

2. Affective experience is founded on sensibility.

To these we can now add the result of the considerations just proposed regarding the new or renewed activation of sensibility that is love, in which the entire profile of a personal identity is defined or redefined. What else, on the other hand, are the first stages of life if not the discovery or assumption of self through these progressive *reawakenings of new layers of being-feeling, in contact with new spheres of reality and of values*. Now, generalizing, we can formulate the third and fourth phenomenological theses on emotional life:

3. Sensibility has a structure of layers of depth that corresponds to the order of a person's axiological preference as activated and revealed by the *structuring responses or matrices of responses*.
4. Structuring responses that *extend* and *reduce* the depth of the activated sensibility are, respectively, love and hate, because they are potential and fundamental dispositions of the sphere of relational sentiments (respect, benevolence, gratitude, admiration, veneration, and so on, and their opposites).

We cannot develop here a phenomenology of hate and its effect of reducing the activated sensibility—although Desmond's observations (1995, pp. 130–150) are very relevant to this point. The acquaintance we all have with certain benumbed souls is perhaps insufficient to illustrate this notion with adequate clarity. For the time being, however, it is important to bear in mind the *dynamic character* of the structuring responses, which become matrices of other responses precisely as they temporarily or permanently modify (extend or reduce) the structure of the underlying sensibility and the overall willingness to exercise it—apart from accordingly modifying the volitional disposition and thus the modes of behavior.

The phenomenon of "maturity" is connected, among other things, to the dynamic character of the structuring responses. Philosophers have so far neglected this incredibly important aspect of the personal emotional life, which has no *comparable* counterpart in animal life. An advantage arising from the distinction between *structuring responses*, or matrices of responses, and *consequent* responses is that emotional and affective "maturity" can now be defined in terms of at least three parameters: (1) greater or lesser contact with reality and with values, (2) greater or lesser extension of the layers of sensibility that have been activated (that is, the "depth" activated), and (3) the at least implicit existence, subsequent to certain structuring responses, of an order of axiological preferences.

The examples from the previous paragraphs are perhaps sufficient to illustrate points 1 and 2. So let us then turn our attention to 3.

An axiological order of feeling is actually given only for partial segments. But for *some order* to be established, it is necessary to have *felt*, at least once in life, that there *are* differences in value: that, for instance, the suffering or even death of a person is more serious than my present malaise, due to, let us say, my desire for financial autonomy. It is necessary for someone to have felt such obviousness and *assented with his whole being* to this obviousness. Some recent stories of young killers, for instance, raise the chilling doubt that in certain lives no axiological structuring of sensibility has ever been realized, that no structuring response has activated this permanent sense of the distinction.

It is now easier to see that the sentiments structuring the personal layer of affectivity require something more than the unintentional stance, in which a positive or negative quality of value is appreciated (such as suffering or joy that to a certain extent always accompany feeling, and represent its positive or negative valence). Sentiments—or emotional attitudes of "reflective level"—require, as suggested earlier, *second level stances:* they represent varying degrees and intensities of assenting and dissenting, relative to feeling of the first degree. An individual's entire system of value preferences manifests itself, and at the same time is activated, through this assenting and dissenting. For this reason, all emotional responses are also modes of experience of self, and in particular, to a greater or a lesser depth, of one's own moral identity. They are discoveries of what we attach value to—a bit, a good deal, a great deal.

Our emotional life is not at all chaotic, and is not an unstructured flow of "contents" like those in a well-known science fiction film of 1995, Kathryn Bigelow's *Strange Days,* which can presumably be saved on diskettes so as to be retransmitted to the mind of another by means of headphones that directly stimulate the cerebral cortex. We should be grateful to such films, true thought experiments which demonstrate the incongruence of the philosophy, perhaps current—but ultimately substantially Humean:

There is no personal reality *other* than a stream of experience—on which they are based. This incongruence stems from the fact that there is no place in this type of philosophy for both the notion of *real* personal identity and for the correlated one of emotional *life* as activation-constitution-experience of this reality (and of the correspondent value qualities).

Thus, first, any real emotional experience of a mature individual is a personal *response* to a given set of real circumstances. For instance, *I* cannot really experience through direct cerebral stimulation what the act of the sadistic torture and murder of a girl is like, unless this direct cerebral stimulation magically extracts me from myself, to be replaced with that sadistic murderer whose action "I" would be re-experiencing: except that in that case it would no longer be me, me the way I am, who experienced it. If I remain myself, the way I am, with my own *structure of perception,* I cannot experience that action differently from the way I experience it at the cinema: by way of the imagination or of the possible, perhaps with "compassion and terror." But without that stance, which is characteristic of *real* emotional experience, a second level stance occurs in which *I reveal to myself, at least partially, what I am really like.*

Second, any real emotional experience usually enters a vertical dimension of importance as well as a horizontal dimension of temporality. Various of these may thus occur simultaneously, but not chaotically.

For instance, I perceive this noise that I hear coming from the street as acutely unpleasant, even painful (it gives me a headache)—that is, as negative. I respond to it with some level of annoyance that may affect—or not—my present peaceful state of mind. Suppose this state endures, together with the irritation. But neither annoyance nor mood can modify a more profound and simultaneously present disposition, an ongoing feeling of gratitude toward this or that author, for example. Gratitude awakens in me as I read her well-written pages or I transcribe her clear, well-founded ideas; it is certainly positive in that it responds to the presence of two values of knowledge, but furthermore, it certainly has more motivational weight than the annoyance or the present mood. I will not, for instance, go down into the street to shout insults at the owners of motor scooters; I prefer to go on studying my author. And I will not be more than temporarily distracted from this reflection by the fleeting sensation of disgust I experience at the sight of a jeans commercial, alerting me to the vulgarity of the image in question.

The notion of axiological profundity and structure of feeling must account for all of this evidence. In effect, without an axiological stratification and structure of sensibility it would not be possible for different—possibly simultaneous—emotional experiences to find their "correct" position, arousing more or less appropriate responses. In every moment of our life we assume the entire axiological range of information "in force," distributing and redistributing *relative motivational weight* to individual experiences.

Thus—regarding the reality that surrounds us, fully loaded with *demands,* a reality that clips of recorded experience cannot carry with them— there are responses of *circumstance:* For instance, the negative value of physical pain will have greater or lesser weight according to the relative weight that the physical well-being of the moment has for me. This well-being will surely count less than will rescuing my child from a fire.

But there are also habitual responses that are relatively independent from the specific situation, and these highlight in an even better way the axiological layout of a personality, or its moral identity. For instance, my characteristic response to physical pain, the way I endure it (or avoid it), will be different depending on who I am: a fakir, an ascetic, a dandy, a masochist, and so forth.

LAYERS OF AFFECTIVITY AND LAYERS OF SELF

In everyday emotional life we are ordinarily oriented toward things, specifically toward the qualities of value and the demands of the realities to which we respond. We have already said that it is by means of these responses—along with what we, to a greater or lesser degree, attach value to—that "our heart," and who we are, is revealed. Let us further clarify this point.

Not *all* of experience is *also automatically* experience of self. In looking, in listening, in thinking and reasoning, of course I live, and yet I do not "live *myself*"—I do not "feel" myself living—*unless this exercise of sight, hearing, and intelligence involves emotions, feelings, passions, and choices.* On the contrary, from the slightest headache to the most profound grief, it is impossible to feel something without *experiencing oneself* affected in some way or other. From getting up in the morning to making the decision of a lifetime, it is impossible to act, or to decide to act, without *experiencing oneself* as the actual subject of these acts—the one on whom the action and decision inevitably *depend.*

Affectivity and actuality or effectiveness, "to suffer"—broadly speaking—and to act, these are the two spheres of experience that phenomenologists refer to as "egological." In short, acts of the cognitive sphere (from perception to thought) are nonegological. Acts of the affective-volitional sphere are egological. Needless to say, it is extremely rarely that the exercise of cognitive functions does not involve affectivity and action.

This distinction, however, furnishes the phenomenological concept of the constitution of self with a precise content. It shows that some experiences, but not others, are constitutive of what each of us refers to as himself or herself, in the sense that they *are in fact the modes in which this self is presented.*

But undoubtedly there is a difference between the emotional sphere and the conative-effective sphere. It is in the most interesting phenomenon of the conative sphere, the act of volition par excellence, the *decision,* that *I affirm myself.* But what I refer to as "I" is present in the punctual act of endorsing a project of action, which at a single blow transforms a possible motive for action into an actual one: "Yes, I'm tired, I'm going now." What I call "I" seems to affirm itself especially in this punctual way. Certainly a decision may be experienced as having greater or lesser importance, but it will in fact be *felt* as such. As with any other event, it touches us and involves us to a greater or lesser degree of depth: it puts a greater or lesser part of ourselves at stake. Although *consent* can "rise" from greater or lesser depths ("I'm going now," said at the bar as you are paying for your coffee; "I'm going now," said to your wife as you abandon her), the decisive *assent,* the move that renders the consent effective or ineffective, remains a punctual act.

The emotional sphere is precisely the sphere of experience that we may call the depth or interiority of the person himself or herself.

Let us take up the thread of common language: One is involved at levels of greater or lesser "depth," of greater or lesser "intimacy," in an emotional experience. Or, as it is sometimes expressed, one is involved with a greater or lesser part of one's self.

To a certain extent, all of feeling is suffering or enjoying. Let us consider the emotional value of feelings, which are to a greater or lesser degree, pleasant, unpleasant, or even painful.

Unfortunately, analytical precision here requires some terminological artifice. Let us refer to all shades of pleasure and unpleasantness of the five senses as affective sensations.

What is "experienced" in the sphere of affective sensations is the "surface" or "outer" part of one's being: the *parts of the body* where pleasure or discomfort or pain are localized, but more generally the parts of the body involved in the fulfillment of vital and sensorial functions. However, it is only in the passivity of pleasure and of physical pain that I am aware of the *physical* borders of my being, from my aching head to my foot as it is trodden on.

BODILY STATES AND STATES OF MIND (MOODS)
OR OF THE VITAL SPHERE

The layer of self experienced in the "vital feelings" in which one's global vital state announces itself is more "profound": The "physical" (modes of "bodily feeling," tiredness, freshness, well-being, malaise) and the "psychic" (moods or "states of mind," in their continuous range from the pole of depression to that of euphoria, through the infinite degrees of anxiety and trust, insecurity and ease, restlessness and serenity) tell us "how we feel," and constitute the sphere of what Heidegger, elaborating from common language (*Wie befindest Du dich?*: "How are you?"), called *Befindlichkeit*. Vital feelings, in effect, aside from "bleeding into reality" and coloring the atmosphere around the subject pink or black, indicate my actual vital state in a more or less reliable way. They do not indicate what I consist of, but only how I feel, what my current state is—the level, we might say, of my vital energy. In this way they also signal an essential characteristic that the person shares with every living being that depends on external elements to live, whose life must be continuously fueled. If I don't eat, I'm hungry; if I don't sleep, I'm sleepy; if I'm forced to remain alone, I'm soon depressed.

We draw a distinction between "life" as a sequence of functional states of an organism, and life as personal history—we would not find any mention of the former in a biography, except to the extent that it interferes with or conditions the latter. The distinction that phenomenologists suggest between what "fuels" the former and what "fuels" the latter is to be understood in the same way, regardless of doctrinal prejudice. And, in effect, this is what we mean when we ask ourselves: "What do I live on?" In fact we say, "not by bread alone." Let us take a further small step and reflect on the fact that some experiences do in fact "consume" us. This is commonly said of love, and also of pain, but the plainest example is to be found in every state of "stress." Other experiences "fill us," "give us life," "recreate us": It could be love itself, from a different viewpoint, but also reading a poem or meeting someone, even a simple conversation. We would then naturally be led to discover the connection between those vital feelings that are states of mind, or moods, and the states of vital energy on which our personal existence feeds. We shall thus distinguish two levels of the dynamics of living, and two corresponding and relatively independent types of energy that fuel them: one at the basis of physio-psychic development and of the normal operation of all functions (cycles of variation in energy, from the vital-biological cycle to those of daily biorhythms), the other that is needed for the unfolding or "realization" of the existential potentiality of each person that fuels, so to say, a person's maturation and inner history.

A phenomenology of tiredness, even of exhaustion—of "emptiness," of aridity, of inner inertia—remains to be undertaken, as does a phenomenology of recreation. And these are the phenomenological nuclei of the most fundamental—and also spiritual—experiences of man. How is it that a conversation is sufficient at times to restore us; that a simple esthetic emotion can provide us with the strength to act or create? How is it that positive feelings may seem to "fill us up" and to allow something like a current of new life to flow into us, whereas negative feelings have the apparent effect of emptying what was inside us, of consuming it? What is the relation between depression and weariness, exhaustion? The dependence of a part of our emotional life on *causal circumstances* is manifested through these phenomena and more. These phenomena fill the notion of psychic causality with intuitive, and specific, content. I feel tired because I haven't slept. "Because" has a different sense here from that in the sentence "I am indignant because you lied to me."

Yet another small step and we shall discover that the sources of these energies are different: that one is fueled and consumed through exchanges with the physical environment, the other through exchanges with the surrounding world and especially through interpersonal exchanges, direct or indirect. *Moods are, so to say, indicators of what we live on;* therefore they orient us precisely as to our *ubi consistam.*

Shaping the environment of a person in such a way as to promote positive moods can be seen as a part of designers' aims: The preceding analysis might shed some light on the ways of correctly identifying the causal links between environmental elements (landscape features, colors, atmosphere, style of furniture, and so on) and affective responses in the user's perspective. Yet a further study in the personal layer of feeling might prove necessary to understand the possible variety of responses in the field of moods—what a person *needs* depends on what he or she *is*.

The dynamic of vital feelings and of moods, in its more or less cyclical course, does not differ so greatly from one individual to another—at least there have always been identified "types" of temperaments within which no great variations are exhibited. Wherever causality reigns, a certain degree of uniformity reigns as well. This is why it should not be surprising that chemistry, at this level, seems to work wonders, for instance in the treatment of depression—which can kill, preventing any chance of personal renewal: It can block any response, but is not, in itself, a *personal* response to values and requirements of the real.

FEELINGS AND PASSIONS

Personality is, in its very essence, an axiologically structured system of sentiments and related attitudes, an individual "order of one's heart." Hence it is not "revealed," but at most hinted at, by the sphere of vital affections and moods. Thus we find, as a deeper layer than the one just mentioned, the layer of self that seems to be awakened by contact with other people, bearers of values and demands, the standing of which is felt to be superior to that of values of the vital sphere (all forms of physical and psychic well-being). The modes of this awakening of self, which, if unimpeded, is accomplished through the creation of new dispositions and matrices of affective and active responses, are positive sentiments, all of which have a characteristic we might call "consonance": the tendency that they stimulate to know and promote the being and well-being of the other; a tendency that is itself perceived as liberating, as flourishing, as "an increase of one's being." Positive feelings are the locus of encounter simultaneously with others and with oneself. Negative feelings have the opposite characteristic. Precisely because it is the most real, the truest self which is in question here—unlike the affectivity of the sensorial and vital layers—these negative feelings are subject to a characteristic mode of the illusion that thrives wherever there is social life: inauthenticity.

There are finally those *habitual channels of volition* that are the *passions*—if, using this term in its proper sense, we mean an affective disposition constituted by a sentiment associated with an inclination of some sort. A passion, unlike a sentiment, is already a mode of volition: (wanting) to have, to do, to be. Greatness is never accomplished without real passion—as both ancient and current wisdom have it. There is greatness for both good and evil, but what is important here is the role this aspect of emotional life that we have thus far left in the background plays in the constitution of a passion: *inclination* in all of its forms (drives, tendencies, desires, aspirations). A sentiment is not a *vector* of action, although undoubtedly it motivates choices and actions, but a passion *is*. Unlike a pure feeling, it constantly stimulates in us that decision-making center that we refer to as "I." In this sense a passion is simply the concrete form taken by the will of a structured personality, the profile—of greater or lesser duration—of its decisions.

EMOTIONS

Among the many elements still lacking in this outline is a characterization of emotions in a proper sense, which we can arrive at only now, because however frequent and varied this phenomenon may be, it presupposes *at least* the presence of sensorial and vital sensibility: the succession of causally conditioned affective states which are the sensorial affections and the cycles of vital feelings, and of moods in particular. In its richness, variety, articulation, and peculiarly human finesse, the life of the emotions also presupposes a sensibility which is structured in depth, with its relatively long-lasting heritage of feelings and passions.

Let us consider the first point. We can always answer the question "How are you?"—that is, we are always in some state. And an emotion is always, as the word itself suggests, a sudden *alteration* of the affective state, constituting an unintentional and *reactive* response to events or situations provided with positive or negative quality of value. Emotions are certainly the most common and frequent manifestation of feeling, but their peculiarity is that of being events that occupy the present, or the consciousness of the subject who experiences them, and as such of consuming themselves entirely in the present: They do not have, like sentiments and passions, or even like moods and other states, a dispositional or habitual dimension allowing them to exist without being currently "in force." They most certainly can trigger a structuring response, thus paving the way for the creation of a sentiment; shame for instance—as an emotion that presents a new sphere of precious values inherent in sexual and emotional intimacy, threatened and to be protected—may give rise to the feeling of modesty, which is in short an extension of the activated depth of perception, an extension that children do not experience before a certain age. Or they can be consequential responses, possible only by way of a certain structure: for instance, the joy that the presence of a loved one arouses in me, which I would not experience were the sentiment in question not there. Or, finally, they may be simple activations of a sensibility completely lacking in personal structure, existing at merely the sensorial and vital levels. This explains why we clearly share certain types of rather primary emotions with the higher animals, whereas some other types of familiar emotions are inconceivable for even the most intelligent of dolphins. Regret, remorse, and compassion, for instance, presuppose not only superior cognitive abilities, but also and especially the existence of *sentiments involving attitudes toward others and for ourselves*; esthetic emotions presuppose that esthetic values are felt under certain conditions as more important than vital values, more capable of motivating one's long-term choices, and so on. Emotions of a sufficiently profound layer of feeling (let us say emotions of a *personal* level) are in fact possible only on the condition that the states pertaining to sensorial and vital levels, even continuing to activate themselves, informing us of the positive and negative values of that layer, have received a *relative* collocation in an order of personal value preferences. This explains why persons differ much more one from another, even in their elementary affective reactions, than animals.

But there is another evident characteristic of emotions. As they are *reactive* responses to events or situations of some importance or value, they are also *immediately active or motor* responses, outlines or schemes of action: escape, aggression, regression to more primitive behaviors (crying, screaming), attempts at grasping, and other acts. In other words, they are *simultaneously activations of feeling and of inclination* (impulses, desires, aspirations). Unlike sentiments, they are vectors of (impulsive) action and of *immediate* action (unlike passions). And regarding this we can note an inverse proportionality between motor agitation and depth of emotion—as if it were characteristic of emotions to position themselves along an axis with excitement at one pole and *being moved* at the other, according to the level of depth at which they touch us. Consider certain portraits of saints, caught in a moment of ecstatic immobility. Or the strange and powerful effect that every profound esthetic emotion has of *arresting* us, of virtually taking our breath away. Esthetic contemplation, as well as philosophical reflection, begins with a crisis of immobility.

A TENTATIVE CONCLUSION

As Damasio (1994) has proved, it is certainly tragic that someone's cerebral bases of benevolence and of veneration should be damaged—that a physical lesion should also be a lesion to his *personal identity.* Yet it happens.

Perhaps, however, a theory of the structure of feeling such as the one that we have outlined here, once developed, might account for a fact that seems to have escaped a philosophy inclined to reducing emotional life to the sensorial and vital dimension. How does it happen that affectivity, with its functional bases, should not be lost altogether, that a primary level of emotional life may survive—but at a level that, as reports seem to indicate, would correspond to a regression of personality—to an impoverishment, or extreme restriction of the range of possible personal responses?

Not even a Humean type of theory, some form of "flow of experience" mentalism with no underlying reality or personal identity, a theory of emotion "without depth," seems to be able to account for it. Much less a theory that reduces affectivity, or at least its true foundation, to the pole of pulsional life: because in this case benevolence and veneration are nothing more than internalized social conventions that should remain "in force," given that the same conditions of life remain "in force," in the same environment—contrary to the evidence that Damasio provided.

But if a personal order of feeling is formed, or rather activated and revealed by means of structuring responses that widen, so to speak, and deepen the axiological horizon of feeling, then it is readily conceivable that a "reduction" of the layers of feeling open to activation might lead to a progressive *alteration of personal structure,* to a benumbing of the sentiments, a withering of life at the sensory and vital levels, at the level of mood and state of mind—and to a consequently more accentuated role of the bodily-appetite pole. Sensibility is reduced at the gain of excitability.

More generally, no theory of emotions that would neglect the analysis of the structure and dynamics of personal being can provide the rationale for a discrimination of different affective phenomena. Yet such discrimination seems a necessary premise to any research on the emotional impact that art, craftsmanship, and design can have on persons and their modes of life.

ACKNOWLEDGMENT

For the English translation of this chapter the author thanks Frank Bragg of San Raffaele University, Milan.

Chapter 5

The Contagion of Emotions

Patrizia Marti
Università degli Studi di Siena

A FIRST CONSIDERATION OF INTERACTION WITH MACHINES

In the study of emotional bonds and social interaction, certain notions play a fundamental role, changing according to disciplinary perspectives that are not always convergent.

This is the case with concepts like *empathy*, seen as the emotional response that moves people to interaction with others due to the sharing of their emotions, or *emotional contagion*, a notion that modern ethology has taken from Darwinian positions and molded to mean the set of immediate and involuntary emotions that lead one to respond to another's emotional expressions by imitating them.

Perhaps it is the lack of cognitive mediation that characterizes humans' innate tendency to recognize others' emotional experiences and make them their own that makes such a special social relationship more or less attractive to those who study and design interaction.

Consider the case of consumer goods. The chance that the user or buyer may experience "emotional tuning" with a product may play an important role in positioning it on the market. Obviously, what in an oversimplified way we call here "emotional tuning" is actually a much more complex concept, which Donald Norman and Andrew Ortony (in Chapter 6 of this volume) define on three levels: visceral, behavioral, and reflective emotional response to a product.

For years, the study of human–machine interaction has concentrated on the behavioral aspect, that of usability and "being in control" of the object using it. But the visceral and reflective levels have been almost entirely neglected, or at least not methodically analyzed.

Although more recently "experience design" (Shedroff, 2002) has aimed at the harmonious and elaborate design of the object and the context in which it used, this aim does not seem backed by a real overturning or enrichment of the more familiar approach to design whose focus is the typical user of the human–machine interaction. We believe it necessary to capitalize on the experience of this discipline, rather than confining it to a secondary role in modern design.

The distinctive element of the design of experiences is apparently a deeper focus on the design of "environments" (cultural, relationship- and value-related, aesthetic, or physical) capable of hosting intellectually stimulating and emotionally involving activities and experiences. It involves not only interaction with objects, but also object-mediated interaction—mediated, that is, through their history and the cultural network in which the actors live and operate. This approach to the design of experiences is attractive, but the theoretical principles on which it is based, like the methods and tools needed to carry it out, are still deeply rooted in tradition.

For those who study interaction and its emotional, cultural, and aesthetic elements, the idea of human–machine interaction seems limited, as does the nature of the "objects of design." If we shift the target of consumer object design (the perspective considered by Norman and Ortony) to other objects that are not merely static (a teapot) or reactive (a VCR) but autonomous and physical with decision-making abilities—proactive, dynamic objects capable of showing and catalyzing emotions—we need a multidisciplinary approach, from the point of view of both interaction theory and interaction design, to understand the social dynamics that are established with machines.

We are referring to interaction with robots, and to a special category of them: robots capable of mediating social interaction.

THE NATURE OF MACHINES AND THE INTERACTION WITH THEM

Human–machine interaction has only recently welcomed human–robot interaction within its disciplinary scope. Until 2004, key phrases like *robot* and *human–robot interaction* were not included in the areas of interest listed by Computer–Human Interaction (CHI), one of the most influential conferences of the sector: Robotics was not one of its conference topics. The reason is simple: Robotics had been mainly the object of industrial applications that produced programmable machines capable of carrying out physical tasks, an area of interest far from that of interaction designers.

However, robotics has recently produced technologies that have made interdisciplinary studies possible and opened up more interesting prospects. We mean the designing of machines capable of carrying out tasks, but also of engaging in social relationships with other robots and with human beings. We mean robots with some autonomous decision-making abilities, capable of taking initiatives, negotiating their presence with their operating environment, and mediating communication in social contexts.

This type of machine, however, confronts interaction designers with new and problematic issues that need further consideration. One aspect is that humans tend to perceive robots as different from other machines. In interaction with a computer, the input–output activity—limited to the keyboard and screen as well as by the user's position, sitting most of the time—drastically reduces the chances of interaction. The physical and tangible nature of autonomous robots, by contrast, catalyzes interaction and encourages humans to anthropomorphize them (Friedman, Kahn, & Hagman, 2003), even when their physical aspect is obviously different from that of a human.

A second important aspect is the mobility of robots. These machines can perceive the physical characteristics of their environment and negotiate with other robots or humans the use of the resources in that space. Therefore we see robots moving in space, avoiding obstacles, reacting to environmental input, and moving independently on their own initiative. It is interesting because their dynamic behaviors change according to the environmental conditions with which they are presented.

A third important aspect is the ability of autonomous robots to take decisions, an ability shared by other systems such as software agents. However, unlike software agents, robots combine their ability to take decisions with their ability to create a relationship with the environment, to negotiate dynamically the use of available resources—and have physical components closely integrated with their functional ones. A robot can decide to change its direction, but to do this it needs to know how to manage the "turn right" function according to the temporary setting of its physical components—weight, position, balance, and so on. All this adds up a level of complexity that software agents are not expected to manage.

Moreover, the fact that robots are deeply rooted in the spatio-temporal context in which they operate makes them creatures (or creations?) with their own historical dimension: memory of the past (they can adapt their functionalities to their past experience of the world), ability to recognize the characteristics of the current situation (they are situated agents in a specific spatio-temporal context), and inclination toward producing future configurations (they may show learning abilities).

SOCIALITY AND INTERACTION IN ROBOTS

Attempts to create a robot capable of showing social behavior and interacting with humans have been very popular in the recent history of robotics. Research in this sector has rapidly expanded from the design of robots inspired by the biological and behavioral characteristics of animal organisms—the reproduction of stigmergic communication in robot-ant communities in the studies carried out by Beckers, Holland, and Deneubourg (1996), for example—to the design of social robots inspired by the way human relationships and communication are carried out.

The concept of sociality in robots has taken on a wide variety of nuances and meanings that basically depend on two elements: the ability of machines to support the social model to which they refer, and the complexity of the interaction scenarios they can face (Breazeal, 2003). In line with these two elements there are several kinds of robots, from those which *evoke* sociality *(socially evocative robots)* by placing the accent on anthropomorphic characteristics, to those known as *social interface robots,* which adopt social and behavioral rules to provide their human interlocutors with a "natural interface," and from *socially receptive robots,* which learn through imitation, to *sociable robots* able to interact proactively with humans to satisfy an internal need (desires and emotions).

The variety of ways in which the concept of sociality may be interpreted poses fundamental questions for the interaction designer: How should robots be designed for them to be acceptable to the human interlocutor? What are the advantages or the risks of designing robots with physical characteristics similar to those of humans or animals? And is it possible to apply the "less is more" principle, much referred to in human–machine interaction, to this special case of interaction design?

We now try to answer these questions and offer our reflections on them. Doing so we will perhaps raise further questions, and later sections of this chapter provide some experimental data supporting our statements.

SPACES AND DESIGN NICHES

Let us start with a question. Why are robots with social behavior designed to resemble humans and animals *(lifelike robots)* in their physical, behavioral, cognitive, and emotional characteristics? There may be various reasons:

- Robots capable of showing social behavior similar to those of some animal species may be a new tool of investigation for scientific research into the behavior and mind of animals.
- Robots are often expected to work with or act on behalf of humans. Being similar to them promotes interaction and makes it more "natural."

- Resemblance between robots and animals may promote their integration into everyday life and the application domains in which they operate.
- Imitation is due to an extension mechanism of the properties of the source, which is unique, allowing the production of social models—that is, living beings. There is no such thing as social behavior outside the communities of living beings.

Although you may certainly agree with the first reason, the others are not at all obvious. First of all, the resemblance between robots and humans or animals inevitably creates great expectations in the human interlocutor that are often let down during interaction. For example, a robot with speaking abilities may easily become a boring and annoying interlocutor if it does not keep up with the required complexity of communication. In this respect, Norman (1994) suggested that robots should be designed in such a way as to demonstrate their "being imperfect" and clearly show their limitations. In the case of natural language interaction, the robot should not produce sentences more complex than those it can understand, so that the user knows it has only a partial knowledge of the language; moreover, it should repeat only what it has understood and turn to the interlocutor if it cannot solve a problem. To sum up, it should not be so similar to the human being it is trying to imitate!

But imitation of human and animal characteristics does not only concern physical (humanoid robots, for example) or cognitive aspects (robots with language abilities) but extends to the expression, acknowledgment, and control of emotions. This element of research on *lifelike robots* is not that surprising if we remember that emotions play a fundamental role in human behavior and that there is no such thing as "intelligent" behavior without emotions.

Emotions are fundamental to the development of our intellectual abilities (Damasio, 1994) and an excellent vehicle for both individual and interpersonal communication. For example, they allow the body's independent and subconscious properties to communicate with the conscious ones and the mind (hunger communicates the need for food, and when it becomes urgent, turns into irritation and an exasperated search for food). These are physiological emotions and express themselves through bodily reactions. Other emotions, however, are expressed through cognitive tasks such as planning and control: For example, boredom indicates lack of stimuli whereas attention, a limited cognitive resource in human beings, is often activated and managed by emotions.

The theoretical debate on the criteria used to identify emotions and their functional value reflects the more general debate on the theories concerning emotions, which we only touch on here but are examined more closely by Roberta De Monticelli in Chapter 4 of this book. The phenomenology of emotions is so varied and complex that it is difficult to contain it in a homogenous theoretical framework. Therefore we span from *functionalist and evolutionist* theories (Galati, 2002), or *reactivist* theories concerning discrete emotions, according to which there is a preplanned repertoire of basic emotions which have a high adaptive value and are functional to the individual's survival and independent of any cognitive activity, to *cognitive* theories, whose main focus are cognitive processes that evaluate the situations generating emotional response (Lazarus, 1991). There are also so-called *dimensional* approaches (Bradley, 2000), transversal to both the reactivist and cognitivist positions, which describe emotional responses according to the value in two dimensions: intensity, and positive or negative valence (pleasant or unpleasant).

In the design of lifelike robots the reactivistic, cognitivist, and dimensional perspectives coexist and are articulated by the robots through emotional, physico-perceptive and behavioral expressions. In general, artificial emotions in robots are used to increase the "credibility" of interaction, provide the user with feedback on the robot's internal state and intentions, act as a control mechanism when activating a specific type of behavior, or understand how certain environmental factors influence the robot's behavior.

Even the way emotions are expressed can vary a lot. There are robots that can simply turn on or off LEDs to express excitement, and others, like Kismet, a robotic head that can control the movement of its eyebrows, ears, eyes, eyelashes, lips, and neck to produce a wide variety of emotional expressions. Kismet's design (Breazeal, 2002), obviously inspired by the dimensional theories, produces emotional expressions by interpolating the three values that make up the robot's "emotional space": intensity, valence, and position. There are many other examples of social robots aimed at expressing and controlling emotions: from those that use facial expressions and spoken language, to those that use body language and gestures (Breazeal & Fitzpatrick, 2000; Mizoguchi, Sato, & Tagaki, 1997). However, we would like not so much to provide a phenomenology of artificial emotions and their expression by robotic creatures, but rather determine when the emotional expression of a robot is the goal of interaction (although in this case the design becomes a replica and an imitation) and when it is intended to mediate human activity.

The boost to the development of lifelike robots leads to a paradox. If we accept that social interaction is only possible within a community of living organisms, there are two possibilities. The first is to believe that robots are not living organisms, in which case the development of such machines capable of expressing emotions coincides with the actual aim of designing them; there would be no social interaction but merely a reproduction, a mimicking, of social behavior. In this case the intention is noble, but only if the aim is not to design social interaction but to study simulated behaviors—that is, artificial life. The alternative possibility is to ascribe to robots the dignity of living beings—although we are only considering this to show the nature of the paradox. Since robots are *not* living beings, we cannot but design them as mediators of social communication and therefore as artifacts capable of supporting people's ability to give significance to their experience of the world.

AGENTS AND NONAGENTS: EMOTIONS IN MACHINES AS THE TARGET OF INTERACTION

We now present two different perspectives on social robot design. This section concerns that in which emotional expression is the target of design, whereas the following section describes that in which it is a mediator of experiences.

The possibility of sharing another's emotivity is granted by a psychological process that allows one to recognize others' emotions through a mechanism based on the discrimination of, first, facial expressions, but also body movement, gestures, and language. The ability to recognize emotions becomes a prerequisite for adopting the other's outlook and separating it from one's own. This allows one to share the other's emotions in a secondary way and to represent them. However, for this discrimination to take place, it is necessary to acknowledge the other as an agent, an animated entity with intentions and goals, and capable of expressing its internal state. For social robots, "agentivity" is a fundamental characteristic: Their credibility as independent agents is given by their ability to show intentions and internal states, and to express and pursue them.

But what discriminates between an agent and a nonagent? What are the minimum requirements that need to be fulfilled to design independent agents capable of interacting with human beings? First, the *morphology* of some physical and perceptive characteristics is strictly connected to the notion of agent. A simple stimulus such as a set of points similar to a pair of eyes, a nose, and a mouth, and positioned within a circle representing a face, attracts the attention of a child, whereas the same points positioned in another configuration does not (Johnson & Morton, 1991). The *tactile experience* is another fundamental element that allows one to differentiate between an animate object and one that is not. The studies by Smith and Heise (1992) demonstrate the ability of children to differentiate between animals and inanimate objects through mere tactile impression.

Moreover, there is a set of behavioral characteristics that differentiates an agent from an inanimate object: The *direction of the eyes and look* and the *movement of the head* are just two examples. Children learn their first behavioral patterns by observing and imitating the direction of the mother's look and the way she moves her head—and also the *independent movement* opposite the one generated by external causes (Rakison & Poulin-Dubois, 2001; Scholl & Tremoulet, 2000).

However, the main characteristic of an agent is perhaps its ability to establish *reciprocal and contingent interactions* with other agents and to know how to manage them even where physical proximity is not immediate. Following the other's look, feeling scared after the other's manifestation of fear (Baron-Cohen, 1995), but also children's vocalizations or animal sounds in response to other vocalizations or sounds when there is no copresence, are just a few examples. But if we analyze some well-known social robots on the basis of the agentivity characteristics previously described, we could be surprised.

Take, for example, the Sony Aibo® robot-dog developed by Sony and launched on the market in 1999. Sony Aibo® can respond to voice commands and interact with humans. It moves independently in space and can express some emotional states. Moreover, its learning mechanisms and its evolutionary characteristics strengthen certain traits of its personality. Like many robots in its category, Sony Aibo® totally lacks any physical or perceptive characteristic connected to the tactile impression conveyed by the surface of its body, which is cold, made of plastic, and not really pleasant to pet, despite its dog-like behavior.

Even Feelix (FEEL, Interact, eXpress), the humanoid robot developed by LEGO® in the laboratories of the University of Aarhus, shows clear limitations as "social agent." Feelix can show emotional expressions in response to physical contact, with the aim of providing credible emotional interaction with humans. But its credibility as a social agent is not really convincing. Like Sony Aibo®, it does not convey an adequate tactile impression (what you touch are LEGO® bricks), it cannot move independently, and its range of expression is limited to a predefined set of "basic emotions" in response to external stimulus.

However, one of the main weaknesses of modern social robots is perhaps the morphological characteristics of their bodies, so similar to those of humans and animals but completely betrayed by the sensorial qualities of the material that covers them (plastic, metal, or fur, for example). These weaknesses are often combined with a lack of coordination in the neck-eyes-look movement; for example, Sony Aibo® can move its head but not its eyes. Researchers that have tested Sony Aibo® with a group of elderly patients (Yonemitsu, Higashi, Fujimoto, & Tamura, 2002) had to "dress" the robot to ease interaction that, from the beginning, proved very difficult.

AGENTS AND NONAGENTS: ROBOTS AS EXPERIENCE MEDIATORS—THE "UNCANNY VALLEY" DILEMMA

The unfortunate case of Sony Aibo®, which had to look like a toy to be accepted by elderly patients, is not a unique anecdote for social robots.

Takanori Shibata and Kasua Tanie (2001) have developed a set of "companionship" robots, Tama and NeCoRo, inspired by cat and dog models. Although very similar in their physical aspect and very sophisticated in their behavior, these first examples have received lukewarm acceptance. The reason may be found in what Masahiro Mori (1970) has accurately described as the "uncanny valley," an effect commonly referred to in the literature of film, particularly of animated film. According to Mori, the psychological effect that a *lifelike robot* may have on human beings is predictable: The more the robot resembles a human or animal in its physical appearance and behavior, the more it becomes familiar (and credible) to its observer. However, when the robot is so similar that it may be momentarily mistaken for real, the transition has a local minimum characterized by a sudden decrease of familiarity, the "uncanny valley"—a dip of frustration due to unmet expectations. This is also known as "zombie effect," a colorful but very effective metaphor: A robotic arm may be mistaken for real as long as it is not cold to the touch, made of plastic and metal, and therefore artificial.

Shibata and Tani (2001) experimented with the "uncanny valley" effect using his robot cats and dogs. They later empirically changed their design strategy, choosing rather unfamiliar animal models with which it is difficult to have direct experience. Furthermore, in their search for a model to replicate, he looked for a "catalyzer" of emotional communication, a robot animal that by nature (not by design choice) would stimulate feelings like "taking care," affection, tenderness, and docility. That is how Paro was developed, a high-tech robot in the shape of a newly born seal, covered in soft fur and capable of the movements and sounds typical of very young pups. Its soft white fur hides a complex network of sensors that allows it to react to environmental stimuli. Its back and its front and back paws are covered with pressure sensors, its nose hides light sensors that activate the eyelid motors so that they close in the dark, and its whiskers have sensors that activate the emission of sounds (Fig. 5.1).

Paro responds to pats and other external stimuli by moving its body and

head in a coordinated way, fluttering its eyelids, making sounds, and purring if cuddled (sensors on its belly are activated when the seal is kept in a vertical position or touches the human body). But if it is hit or is stroked in a way it does

FIG. 5.1. Paro, a robot which looks like a seal pup.

not "like," it reacts with "irritation" and stops interacting. Paro is therefore capable of showing a fairly complex type of behavior. Besides reacting to external stimuli, it may take initiatives, such as lifting its head, fluttering its eyes, and making noises, to encourage humans to interact with it. It also has software that allows it to learn a few words and therefore turn around when called. Furthermore, Paro has its own "physiological life": When its batteries are high it acts in a more lively manner, but if it "works" for a long time it looks tired and its movements slow down.

This aspect of the robot is interesting. Although the idea of designing a circadian rhythm for Paro has been entirely fortuitous, the emotional impression this makes on human interlocutors is very strong. They normally sense the robot's "tiredness" immediately and tend to pet it and keep it quiet to help it regain energy. However, the design of the circadian rhythm is not considered a constituent part of its artificial emotions, which are mainly shown through movements and facial and linguistic expressions, but rather an internal state connected to the resources the robot uses during interaction.

Paro seems to have all the agentivity characteristics described in the previous section: The morphology of its body is efficiently harmonized with the tactile experience that one can have through direct contact, the movements of its head and eyes are coordinated, and it can behave in a reactive or proactive way to stimuli whose proximity is either immediate or not.

Paro's qualities as a social robot have been tested in nonpharmacological therapies on children with cognitive, sensorial, and motor deficiencies (Marti, Palma, Pollini, Rullo, & Shibata, 2005) and other studies have been conducted on elderly patients (Saito, Shibata, Wada, & Tanie, 2002). An interesting element of these tests is that Paro seems not to produce the "uncanny valley" effect: Its morphological characteristics, together with the tactile impression it yields, favor exploration of its body and behavioral characteristics. This effect seems even more surprising when observed in child patients with serious deficiencies in interpersonal relationships. In these cases Paro's presence has favored dyadic (patient–robot) and triadic (patient–therapist–robot) exchanges that with normal therapies would only be episodic and extremely limited in duration. However, if we consider the experimental conditions in which they were carried out, these studies give us more to think about.

The experimental sessions (weekly over three months) were carried out in two conditions: (1) with the robot turned off, like a regular peluche (soft toy); and (2) with the robot completely functioning. The aim of this experimental design was to detect any difference in interaction dynamics between a situation in which just the morphological characteristics and the tactile impressions were present, and one in which the robot's behavioral characteristics were also displayed (movement of paws and fins, rotation of head and eyes, emission of sounds). During the dyadic and triadic exchanges, condition 2 resulted in a considerable increase in the patients' production of vocalizations (the children did not have articulated verbal communication abilities) as well as in attention span. But a similar effect, although slightly less intense and durable, was also observed in condition 1, when Paro was used as a simple peluche. This effect had not been previously noticed in any case that involved the use of dolls, for example.

The high occurrence of triadic relations in the sessions with Paro (up to 250 interactive events within the same session, lasting an average of one hour and analyzed in video sequences of 10 seconds each) suggests the robot's role as interaction mediator and catalyzer of social activity. Its physical, perceptive, and behavioral characteristics, together with its specific circadian rhythm, offer the human interlocutor the possibility of filling the interaction experience with private and personal significance. As an entity to be explored and discovered, the robot mediates the relation between what is inside and outside the individual, both in the direct relationship between the human and the machine, and in the human–human exchange mediated by the machine.

MORE ON EMOTIONS AND DESIGN NICHES

Our discussion so far has brought us to consider a few issues connected to interaction design and to agree with the importance and value of designing robots as experience mediators rather than mere replicas of the expressive and emotional abilities of humans and animals. We would like to close by trying to define a potential design space for social robots whose distinctive characteristic is to be experience mediators. To do so we evoke concepts typical of comics, an art that plays with different ways of representing characters and situations, and has a fundamental role in attracting the reader's sympathy, empathy, and involvement.

Scott McCloud (1993) defined the "picture plane" by a figure (see Fig. 5.2) with three vertices (Picture-Plane, "Reality," and Meaning) linked by edges. The edges define a three-dimensional taxonomic universe in which all comic-book communicative modes can be located. The Picture-Plane-Reality edge is "retinal," the Picture-Plane-Meaning edge "conceptual," and the Reality-Meaning edge "representational"; thus, any mode is defined by its degree of "retinality," "conceptuality," or "representationality." This means that "abstract" modes cluster toward the Picture-Plane vertex, "naturalistic" modes toward the Reality vertex, and "iconic" modes toward the Meaning vertex. This universe is then divided near its "conceptual" edge by a "language border," separating the realm of images from that of verbal language. On the Reality-Meaning edge, it is possible to project the methods of representation of an object, going from photographic realism to iconic representation. The upper vertex refers to the extreme abstractness of pictures not present in nature, therefore the one in which Language and Reality spaces

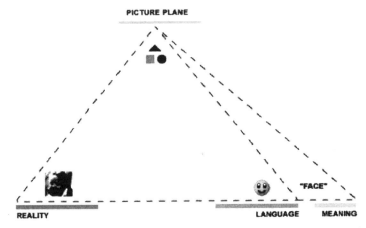

FIG. 5.2. Scott McCloud's picture plane (adapted, with permission, from McCloud, 1993, pp. 52–53).

contract and converge. The vertex of Meaning outlines the space in which the reader fills with meaning the representation played on the meeting point of Reality, figurative components (Picture), and Language.

According to this design space, it is possible to orientate social robots design from extreme realism (humanoid robots or *life-like robots*) to more abstract and iconic solutions (Sony Aibo®). In this respect, Kerstin Dautenhahn (2002) stated that a simple design, one therefore more essential and closer to iconic traits, should be preferred to a realistic, anthropomorphic, or zoomorphic design. She also stated that a "new" design, inspired by fantasy, and not by any living organism, could better support human users in shaping a correct mental idea of how the robot works and behaves. Although one could share this idea, the motivations for adopting a design style may be far more complex.

Actually, to consider the physical and perceptive characteristics of the robot, those connected to the tactile impression, for example, and not only the morphological ones, we could create a "configuration plane of the tactile stimulus" similar to McCloud's "picture plane." In this case, the figure (see Fig. 5.3) would be composed of the following vertices: the Reality of the physical and perceptive properties of the tactile experience (corresponding to Reality in the McCloud picture), the Material Properties (the equivalent of the Abstract Picture), the Mimetic Representations of the tactile stimulus (similar to that of Language), and the Meaning. In this figure, Reality refers to what we physically and emotionally perceive when touching a surface, and that can clearly be associated with a real tool or living entity (we can clearly recognize a cat from touching its body and perceiving its warmth); Material Properties correspond to the abstract tactile attributes of a mate-

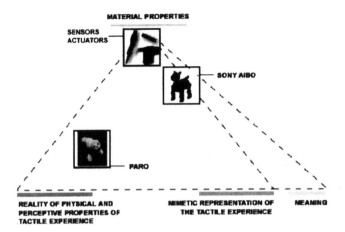

FIG. 5.3. The configuration plane of the tactile stimulus (adapted, with permission, from McCloud, 1993, pp. 52–53).

rial; Mimetic Representations are alternative codings or interpretations of a tactile stimulus (an example may be the Braille code); and Meaning is the significance we attribute to a tactile stimulus.

Although our culture has not provided us with many examples of mimetic representations of the tactile stimulus completely separate from the visual stimulus, this figure may have an interesting heuristic power. For example, we could place a text written in Braille within the lower space defined by the edges connecting Material Properties, Mimetic Representations of the tactile stimulus, and Meaning. We could also place sensors and actuators used to build the reactive behavior of a robot close to the Material Properties vertex. Indeed, if we imagine touching a flex sensor or sound actuator, their material properties do not allow us to connect the tactile experience to any specific tool or artificial or real pet. Similarly, we could place the tactile perception relating to Sony Aibo® close to the vertex of abstractness of materials, but still with some iconic properties related to its morphology (it would be difficult to recognize "a dog" from the tactile impression given by just the robot's body, but not by touching an animal toy which had a head, tail, and four legs). We could place Paro in an intermediate position close to the side that connects the Reality of the physical and perceptive properties of tactile experience with that of the Mimetic Representations of the tactile stimulus; the tactile impression provided by the soft fur, the morphology and purring of the robot body provide a quite realistic impression of touching something like a real baby seal.

The experience of Sony Aibo® and Paro teaches us that the robot's morphology itself is not necessarily a key to the success of social interaction, whatever the method of representation. Rather, we believe that a minimalist and iconic design should be preferred in those cases in which the morphological characteristics are not integrated with perceptual characteristics, like those conveyed by tactile experience. This way, users' expectations will not be let down by a realistic resemblance of the robot not matched by as much resemblance from the point of view of perception and touch. However, should morphology be strictly connected to the robot's perceptual and tactile characteristic, as with Paro, this connection would strengthen the role of significance, characterizing the robot as an experience mediator.

The design of emotional expression and its various approaches, as an end in itself or as a tool of mediation, are now evident.

In the first kind of approach, design aims to model certain personality traits and the control of emotions. For example, the *tool-like* personality is used in robots that work as *smart appliances*: machines that reliably and accurately carry out the tasks for which they have been designed: the robot vacuum cleaner is a clear and familiar instance. Then there is the personality of the *pet-like* robots with which we have largely dealt, but also of those inspired by *cartoons*, usually caricature personalities used to exemplify interaction to nonexperts. *Artificial* and *human-like* personalities are more general and less informative in themselves (they potentially represent a wider variety) and thus are almost always represented in a stereotyped way: the shy robot, the friendly robot, and so on.

The perspective on social robots we present here is instead connected to the quality of interaction and to the personal significance that everyone creates by getting involved with, and involving their own life experiences in, interaction with the robot. That interaction opportunity develops when the object mediates and catalyzes people's creative ability to construct meaning. In this sense the space of design is similar to a learning space: Human–robot interaction is the element that mediates the building of knowledge, a creation of significance that depends on not just the machine's physical and functional characteristics but also, and mostly, the specific context of interaction—on the personal history that every interlocutor calls into play and on the perception of mutual affordances, some of which come from the stimulus given by touching, hearing, seeing, and moving, others from psychological processes that mediate empathic response.

The space of design is defined by the challenge of designing robotic agents that allow the projection of the self and the assignment of significance, the creation of natural ways for involvement in the activity, the perception of interactive experience at a level not only physical and functional but also aesthetic, perceptual, and emotional, and the acknowledgment of the role of emotional and empathic involvement both in reflexive and rational activity and in direct and nonmediated impressions—of, in short, the contagion of emotions.

Chapter 6

Designers and Users: Two Perspectives on Emotion and Design

Donald A. Norman
Andrew Ortony
Northwestern University, Evanston

DESIGNERS, USERS, AND AFFECTIVE REACTIONS

This chapter derives from some of our prior publications that examine emotion theory and its application to design, in particular two volumes (Norman, 2004; Ortony, Clore, & Collins, 1988) and a research article on affect and levels of processing (Ortony, Norman, & Revelle, 2005).

We take as our starting point the distinction between two perspectives on products: designer and user. There is often a mismatch between these two perspectives, but both matches and mismatches constitute a major source of the affective reactions that people have to products and their interactions with them. These reactions extend over a wide range and include not only (relatively short-term) emotions, but also longer-term reactions such as moods, preferences, and attitudes.

The first perspective is that of the designer. The designer works in a space that is constrained by a number of different considerations that, depending on the context, include such things as functionality, physical limitations, appearance, cost, time-to-market, characteristics of market segments, and legacy and brand-identity issues. We focus on two of these in

91

particular—functionality and appearance. We do this not because we think that other considerations are unimportant, but because these two are the most relevant for understanding the relation between emotion and design.

The second perspective is that of the user, and here too, functionality and appearance are important, but for different reasons and in different ways. Specifically, from the perspective of the user, these two aspects of the design space are the principal sources of affective reactions. We focus on three particular kinds of users' emotional reactions to products—reactions that might or might not have been anticipated or intended by the designer. These three kinds relate to what Norman (2004) referred to as Visceral (perceptually based), Behavioral (expectation based), and Reflective (intellectually based) aspects of design. Figure 6.1 shows the relation between the two views.

Differences between designer and user perspectives of the same product are particularly evident with respect to the role of emotions. The designer may intend to induce emotions through the design, but because emotions (which are a special, but particularly salient form of affective reaction) reside in the user of the product rather than in the product itself, the emotions the user experiences are not necessarily the same as those intended by

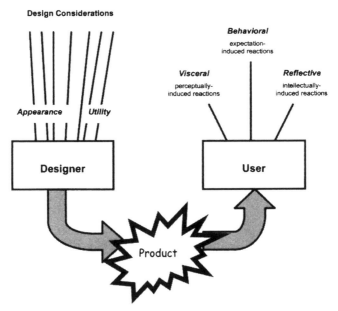

FIG. 6.1. The designer's view of the product differs from that of the user. The designer considers many design aspects; we focus here on two: Apperance and Utility. The user has various reactions to the product; we focus here on three fundamentally different kinds: Visceral (perceptually-induced), Behavioral (expectation-induced), and Reflective (intellectually-induced).

the designer. Certainly, some of the emotions the user might experience might have been intended by the designer, but some might not. And indeed, some might be just the opposite of those intended by the designer. Product-induced emotions are often quite idiosyncratic, depending, for example, on memories the product evokes or on the particular circumstances of use. Yet other emotions result from concerns outside the object, such as the status it might or might not bestow.

Designers have more control over users' Visceral and Behavioral reactions than over Reflective ones, but even here the control is indirect at best. Indeed we characterize the attempts of designers to influence these reactions as attempts to provide *emotional affordances*. (The use of affordances, both physical and emotional, here follows the spirit of the definition introduced into the world of design in Norman, 2002). In other words, designers can do things that provide opportunities for the experience of emotions in users, just as, by building in physical affordances, they can influence the way in which an object is likely to be manipulated and controlled. But whether affordances are actually made use of is beyond the designer's control.

THREE LEVELS OF EMOTIONAL RESPONSE

The user's emotional responses are internally generated. In our theoretical work (Ortony et al., 2005), we suggest that there are three levels of processing that we called *Reactive, Behavioral,* and *Reflective.* For applications of the theoretical work to design, these translate into *Visceral level design* and *Visceral responses, Behavioral level design* and *responses,* and *Reflective level design* and *responses* (Norman, 2004). To understand how the user might develop emotional responses to a product, and to understand the relation between what the designer does and what users feel, we need to do a quick review of the theory.

Visceral Level Design and Responses

When designers attend to the surface features—the appearance—of products, we say that they are engaged in Visceral design. From the user's point of view, Visceral responses involve an automatic evaluation of the perceptual properties of objects, and a quick classification of them as good or bad—safe or dangerous, cold and forbidding or warm and inviting, and so on. In our theoretical work, we argue that these rapid evaluations have evolved as protective mechanisms for animals that must exist within complex and unpredictable environments. Negative assessments flag potentially harmful experiences. Positive ones signal potentially beneficial experiences such as safe situations and places that are ideal for exploration.

Biology has provided people with a vast repertoire of dimensions that are automatically processed and interpreted. Thus, some colors are warm, some cool. Some situations, such as standing at the edge of a cliff, are immediately perceived as dangerous, whereas others, such as experiencing the colorful appearance and sweet taste of fruits, are immediately perceived as nonthreatening and desirable. Designers can exploit these kinds of immediate perceptions.

Note that this level of design relates only to the surface appearance of objects. It is pure style, pure surface. At the immediate, Visceral level, reactions to such features are not based on past experience or deep semantic knowledge and interpretation. There is no comparison with the past, no expectation of the future. All that counts is the current state. These reactions are produced by biologically based pattern-recognition mechanisms driven solely by the here-and-now of perceivable features. This is why we call Visceral level responses "perceptually induced." Because this level is primarily determined by biology, it is generally universal across people and cultures.

Emotion at the Visceral level is very rudimentary: In our scientific work we call it "proto-affect," because we do not wish to invoke the implications of labels such as *emotion* or *affect*. Visceral level emotional reactions are too simple to warrant such labels; they are not conscious, and they are not interpreted. Nevertheless, it is in these reactions that higher-level feelings such as anxiety and concern, and satisfaction and pleasure have their origins.

Behavioral Level Design and Responses

When designers attend to the function and use of a product, we say that they are engaged in Behavioral design. Whereas the Visceral level is innate and biological, Behavioral level responses are learned. The Behavioral level is where skills and routine behavior reside and are controlled. Behavioral level processes are still subconscious and automatic, but because the associated skills and routines are acquired through learning, they also involve past experience and expectations of future states and events.

What we call Behavioral level design includes the general concepts of usability but goes beyond this to include the physical feel of the object as well as the subjective "feeling of control." This is where precise control is essential. It is here that the smooth, viscous feel of a knob (with no backlash) matters so much; it is here that the perfect responsiveness of a well-tuned sports car is felt.

Behavioral responses are intimately connected to predictions of and expectations about the near future. These predictions give rise to affective states akin to fear and hope—primitive forms of recognizable emotions. People frequently become angry at objects that let them down and respond by kicking or hitting them. Such reactions derive from the Behavioral level, where the failure of objects to live up to expectations generates strong emotional responses. It is because of their dependence on how our routine interactions with things ought to feel that we call reactions at the Behavioral level "expectation-induced."

At the Behavioral level, although still automatic and subconscious, there is awareness. Moreover, because Behavioral level routines are learned, they vary from person to person, from culture to culture.

Reflective Level Design and Responses

At the highest level of processing we find reflection, people's self-examination of their own actions, understanding, and monitoring of progress. This is the home of self-image, of metaprocessing, and of the whole range of articulated emotions including emotions such as pride, shame, admiration, and gratitude.

Reflection is the highest level of intellectual functioning in a person, where there can be self-examination, and the assignment of blame (hence emotions such as pride and shame). This level is conscious and self-aware. From our perspective it is the only level at which full-fledged emotions can arise, that is, emotions that incorporate a sense of feeling derived from the affective components from the Visceral and Behavioral levels, along with a conscious interpretation of that feeling.

From the designer's point of view, this is where pride of ownership, quality, and brand play major roles. Here is where people show off or hide their possessions. When designers attend to these components of the use and ownership of a product, we say that they are engaged in Reflective design.

The Reflective level is influenced by experience and culture as well as by one's social group and by the whims of fashion. But Reflective design not only varies from culture to culture, age group to age group, but for some individuals it can even vary from week to week, dependent on the role they play in society. Thus, we all recognize the difference between the clothes one wears to a beach party, to a night out on the town, or to an important business meeting among company executives. These are Reflective level differences: the clothes we wear are often deliberately selected to communicate a message to others about our social status and the role we are playing in the activity. This is why we call Reflective level responses "intellectually induced."

THE DESIGNER'S PERSPECTIVE

To understand the role of emotion from the designer's perspective, we need to ask what the designer is trying to do. Ideally, the designer is concerned with transforming the multiple constraints and dimensions of a product into a single, coherent design. However, for the main theme of this chapter, we restrict our comments to just two aspects of products: the utilitarian aspect and appearance (including "touch and feel," etc.).

Emotion by Accident

In many cases, affective responses to products arise in users without any conscious attention to affect by the designer. This is especially true in crafts, where designers consider themselves skilled craftspeople, making utilitarian objects. In such cases, we say that emotions arise "by accident": They are unintended consequences of the product or a user's interactions with it.

Consider the tyg, a multihandled drinking cup popular in 17th-century England. The three-handled example shown in Fig. 6.2 is clearly utilitarian. The tyg is very practical. It makes it easy to pass a hot cup, for example, from one person to another, where both giver and receiver will have conveniently placed handles to hold, so that neither party will burn themselves. Moreover, if two or three people drink from the same cup, each using a different handle to hold it, their lips would use different portions of the rim— yet another utilitarian aspect of the design.

This cup, predominantly functional from the designer's point of view, might give rise to positive emotions in a user because of the ease of passing it back and forth and perhaps because of the avoidance of another's lip spot, a point to which we return in the next section.

The tyg is an example of a product whose primary design consideration is utilitarian. In other cases, it is can be more difficult to separate function from appearance. One would think that kettles for boiling water should be relatively simple and highly utilitarian. One might think that the primary consideration in the design of a kettle would be that of utility: The thing should be an effective device for boiling and pouring water. However, the basic kettle has several unintended consequences that induce negative emotions in users—

emotions by accident. In particular, there are several ways in which users can burn themselves for a variety of reasons: The handles can become uncomfortably hot, the

FIG. 6.2. The tyg is a cup with three handles, which makes it easy to pass from one person to another. The third handle is invisible in this photograph (photography by D. A. Norman).

steam from the boiling water might scald the user, and the kettle might drip during pouring. Through experiences with unsatisfactory kettles, designers have sought to eliminate these unintended negative consequences and their attendant emotions.

Thus, even when motivated primarily by considerations of utility, the designer may sometimes be concerned about possible undesirable consequences, and therefore about emotions. In such cases, we say the designer's stance vis-à-vis emotions is one of *emotion-prevention*. Consider, for example, the kettle shown in Fig. 6.3. Such a kettle could function perfectly well (and would be easier to construct) if it were made of a single material such as stainless steel. But this would allow the handle to get hot. But notice that the choice of a non-conducting material for the handle can be motivated by either the utility-focused goal of preventing users from burning their hands, or by the emotion-focused goal of preventing the resultant anger. In either case, the end result would be the same.

Note too the shield, which was advertised as preventing the arm from being scalded by steam escaping from the kettle. In fact, the shield was not completely successful, in part because the trigger that pulled the cap off the kettle spout transmitted heat, burning the finger. Whether the shield was motivated by considerations of utility or emotion, it appears to have had the role of emotion-prevention, and although it protected the hand from steam, the result was still unsatisfactory. Accordingly, the company (OXO) brought out yet another version, shown in Fig. 6.4. The advertising copy for this kettle describes it in very utilitarian terms:

Simply lift the OXO GOOD GRIPS Uplift Tea Kettle by its handle and the spout opens automatically! No awkward buttons or levers, just lift and pour!

The soft, non-slip handle is heat-resistant for added safety, and a large lid makes the Uplift easy to fill and easy to clean. (OXO International Ltd., n.d.)

Notice how the text emphasizes the emotion-prevention stance taken by the company: "no awkward buttons or levers," "heat-resistant for added safety," and "easy to fill and easy to clean."

Almost any design stance that minimizes utilitarian difficulties also serves to reduce negative emotions by reducing the undesirable effects that lead to them. As a re-

FIG. 6.3. A utilitarian kettle: the OXO Good Grips tea kettle. The shield is designed to prevent steam from burning the hand.

FIG. 6.4. OXO Good Grips uplift kettle. This kettle, from the manufacturer of the kettle in Fig. 6.3, overcomes its problem. The handle is heat-resistant, and the lid over the spout opens automatically simply by tilting the teapot, so no finger can get burned.

sult, it is not always possible to determine whether the design was motivated by emotion-reduction or by utilitarian goals. Thus, the addition of nonstick surfaces to cookware eliminates or significantly reduces the food that sticks to the cooking surface, thus simplifying cleanup. When designers are motivated to reduce frustration by introducing a nonstick surface, they are motivated by emotion-prevention. However, exactly the same elimination of potential negative affect in users might come about with only the utilitarian motive of maximizing performance (that is, eliminating the sticking), with complete indifference toward the possible affective consequences. The person using the pan, of course, has no way of knowing which of these two motivations lay behind the design of the nonstick pan. In both cases, the net effect is the reduction of potential negative affect. This raises an important point, to be discussed in the next section, concerning the distinction between emotion by accident and emotion by design.

If designers are thinking about user emotions at all when focusing primarily on utilitarian considerations, they are generally adopting an emotion-prevention stance. In many cases, however, designers are motivated by considerations of appearance as much as, if not more than, by considerations of utility. Under these conditions, designers may actively focus on designing the product to cause it to generate affective reactions in users. This is what we call an *emotion-promotion* motivational stance as compared to the kind of emotion-prevention or emotion-indifference cases just discussed. We think of design motivated by emotion-promotion as "emotion by design."

Emotion by Design

Many products are deliberately designed to evoke emotions within the user. Designers have a number of ways of doing this. Most designers have a good intuitive feel for how users are likely to react to a product. Furthermore, although they might not articulate them as we do here, they also have a good intuitive feel for the importance of the three aspects of design that can induce affect in users: appearance (Visceral), behavior and function (Behavioral), and image and brand considerations (Reflective).

The Visceral responses of the user primarily result from the immediate emotional response to the look or feel of the product—a "gut" reaction or what, in positive cases, is often called "the wow effect." For example, many products from the Italian design firm Alessi focus primarily on this aspect of design, making products that many see as "cute," "clever," and "pretty," often to the dismay of professional designers who believe that these reflect *only* surface features of the design—"mere styling," they complain. But this criticism misses the point. The visceral pleasure from the surface styling is indeed enjoyed by many purchasers and users of Alessi products, so in this sense the deliberate attempt to induce positive emotions is successful.

In Behavioral level design the designer wants the user to feel good about the behavioral interaction. This is why a designer might inject viscous oil into knobs to provide a "silky" feel, or put effort into the design of the responsiveness of a vehicle, or feedback in software design. Precision automobiles such as Porsches and Ferraris place high value on this aspect of design. Here, the designer attempts to satisfy or exceed user expectations about the nature of their interactions with the cars.

Reflective level design and responses deal with the prestige and brand components of the product. This is the intellectual side of design. Thus, high-prestige items like designer clothes fit into this category. Indeed, users can experience this level of emotion simply by telling others that they own one of these products ("My Ferrari is in the shop right now"). It is the responsibility of the designer to live up to these beliefs. The designer must maintain the brand identity and image.

Emotion promotion is often focused on generating positive emotions, as in the design of jewelry or ornamental products, where the generation of positive emotions is obviously the goal. Given this, one might be tempted to think that emotion-promotion motivations always target the generation of positive affect, but this is not the case. Sometimes the goal is to create negative emotions. For example, the modern convention of using a skull and crossbones symbol to designate poison, or the use of barbed wire or glass-shredded fences to keep people away, are intended to generate negative affect—fear. And at its root, fear is the currency of deterrence. These are clear instances of deliberate attempts to invoke negative emotions by design.

Although we think that the distinction between emotion by accident and emotion by design is a useful one, there are many cases where both are important. In fact, one of the standard challenges for designers is to deal with concerns over possible conflicts between these two considerations. In the design of the 2004 Chevrolet Corvette, the designers replaced the Corvette's traditional "hide-away" headlights with a more traditional design that incorporated molded lens and small high-intensity headlights. They believed that the new design was both more attractive and more functional than earlier designs (satisfying the Visceral and Behavioral level requirements), but they worried whether they had perhaps lost the distinctive image of Corvette—a Reflective level concern. They were correspondingly relieved to discover that the new design still maintained the brand image.

THE USER'S PERSPECTIVE

We noted at the beginning of this chapter that regardless of what designers might intend, ultimately users' emotions depend on the *emotional affordances* of products. Utility and appearance are central if we are to understand the affective reactions of users to products. Emotional affordances can arise by accident or by design, although they are more likely to arise by accident when designers focus on utility, and by design when they focus on appearance. As users we easily recognize that we are sometimes inspired by the sheer beauty of an artifact—or repelled by its consummate ugliness. We are sometimes delighted at how well some mechanical device such as an automobile performs, or appalled by its ineffectiveness. And we are sometimes impressed by the cleverness of certain design features, whereas on other occasions we are amazed at their apparent stupidity.

In some cases, products induce emotions only very indirectly, often through their conventional or symbolic significance rather than through their physical attributes. A wedding band symbolizes both the institution of marriage and the personal commitments made by the couple. A miniature plastic replica of the Eiffel Tower has little intrinsic merit, but its emotional affordances as a memento are considerable. A very indirect example can be found in the camera phone. The physical affordances that allow one to take a photograph on impulse, to share it with onlookers, and to instantly electronically transmit it to others, also constitute powerful emotional affordances. There are many stages to this process, including activating the camera, taking the picture, sharing it with onlookers, sending it to a recipient, and then, for the remote recipient, perhaps on another cell phone, displaying and at later times retrieving the picture. To the extent that the designer makes all aspects of this process easy the likelihood of deep, reflective emotions is greatly increased. These, and many other examples, demonstrate that both the manner in which products are configured and our interactions with them can generate affective responses and that the physical affordances of products are often also emotional affordances.

However, having said this, we need to remember that what the designer envisions in the design of a product is not always what the user experiences in using it. A predominant focus on the utilitarian considerations often gives rise to unintended and unanticipated emotions in users—both positive and negative. Similarly, a predominant focus on the styling and appearance of Visceral design might very well have unanticipated behavioral side effects.

"Emotion-by-Accident" Products

The key to emotion-by-accident is that it is wholly dependent on the interaction a user has with the product. In this interaction, a user might experience some kind of emotion purely as a result of utilitarian features—a response very different from that intended by the designer.

The most common form of emotion by accident is the frustration or anger that arises from interacting with products that function poorly or that cause other forms of distress in use, such as kettles that burn the hand. However, there are also positive examples. People often have strong affective reactions—for some people negative, for others positive—to the "throaty" sound of a powerful automobile or motorcycle engine. Indeed, many enthusiasts were disappointed when later models of their favorite vehicles changed the sounds to which they had grown accustomed. In one noted case, the Harley-Davidson motorcycle, the sound was so attractive and distinctive that the company attempted to trademark it and ever since has worked hard to ensure that all their motorcycles maintain their distinctive, iconic sound. Here is a case where the Visceral appeal of an accidental by-product of the engine—the sound—became transformed into a Reflective level icon of the company that was purposely recreated in all models.

Because the behavioral level is to a large extent expectation driven, emotions arise when products fail to meet—or perhaps exceed—expectations. Many theories of emotion acknowledge that deviations from norms and expectations are a major source of emotions (see, for example, Lazarus, 1966, 1991; Mandler, 1984; Ortony et al., 1988). Most of our routine interactions with everyday objects carry with them a wide range of expectations of normal performance and regular outcomes. We expect our car to start when we turn on the ignition, we expect food not to stick to the pan, and in general we expect our products to work as they should. When these expectations are violated, we tend to feel bad—we might feel frustrated or angry. When the expectations are surpassed, we tend to feel good—we might feel proud, vindicated, or elated.

"Emotion-by-Design" Products

Many products are intentionally shaped by designers to create some kind of emotional responses in their users, a point already discussed in the last section.

People often express affection, appreciation, or admiration through small gifts. The point, of course, is to generate emotions in the recipient, both through the pleasure of the gift itself and also through the indication that the gift shows that the sender remembers or cares. Numerous services have arisen that make things like this easy to do spontaneously, just when the thought arises. Gift shops at airports or in well-frequented neighborhoods simplify the purchase and delivery of gifts, and services that enable the delivery of flowers to anyone in the world simply by visiting a local florist, a Web site, or by telephone are widespread.

The design of services that satisfy (or even create) spontaneous needs to generate emotions in others requires the integration of numerous components. No single component is necessarily new or unique, but the final combination enables a person to easily (from anywhere at any time) do something that is intended to generate an emotional response in someone else. These services are another way of designing for emotional affordances. An interesting proposal for a service of this kind is the Telekatessen project (Rosella, Slocombe, Sunesson, & Torstensson, 2003) from Interactive Design Institute Ivrea. Here, the goal is to provide a simple way for someone to personally inscribe a surprise gift for another. This service would allow a person to spontaneously decide to surprise someone with a gift of pastry or chocolate, with a short message inscribed with icing on the top. A simple text message to the pastry shop sets the operation in motion. The shop sends a gift certificate and announcement to the recipient, again by SMS. When the recipient arrives at the shop and displays the cell phone message, the shop provides the pastry, with the message already embossed in icing. Here is how it is described on Interaction Design Institute Ivrea's Web site:

> Imagine receiving a message on your mobile phone telling you that somebody you love has arranged for you to collect a surprise at a local pasticceria. When you show the message to the shop attendant, she hands you a beautiful chocolate with a sweet message from your friend crafted on it. (Interaction Design Institute Ivrea, 2003)

The service is, of course, hypothetical, but it could be real. The design in this case is in the service, not the physical product. The impact can be large. Here, the point is to convey directly one person's emotional feeling to another via the intermediary of the cell phone service and the pastry shop. To be sure, the design is of a service rather than a physical product, although, of course, the end result is of something edible and sweet.

SUMMARY

The perspective of the designer is, of necessity, different from that of the user. This is especially true in the realm of the emotional responses a person might have for the use, ownership, or outcomes of a product or service.

Designers can hope to shape the emotional responses of their users through the development of emotional affordances, but in fact, they have no direct control over the emotions that might result.

Designers, of course, must work within a complex realm of multidimensional requirements and constraints. Here, we have examined only two aspects: utility and appearance, showing how even these two play different roles for designers than for users. Designers can attempt to control the users' Visceral, Behavioral, and Reflective responses through the different features of their design and through the affordances they provide.

Part

III

SITUATEDNESS

Chapter 7

Situatedness Revisited: The Role of Cognition and Emotion

Claudio Ciborra
London School of Economics

A POPULAR WORD AND ITS ROOTS

These days the adjective *situated,* the noun *situation,* the Latin expression *in situ,* and the abstract concept of *situatedness,* are liberally employed by those researchers and scholars who want to take and articulate alternative approaches to the study of organizations, the analysis of knowledge and change, the design of sophisticated technical systems, and in general the understanding of the complex interactions between people and technologies.

A few examples taken from the recent organization theory and information systems literature exemplify the "situated" perspectives in the theory of organizational change triggered by the introduction of technological innovations. In relation to knowledge management, Schultze and Leidner (2002) illustrated an interpretive discourse that highlights "the dynamic and situated nature of knowledge" (p. 224). In a more postmodern perspective, Haraway (1991) urged the abandonment of the study of formalized knowledge in favor of "situated knowledges." In her seminal research on planning in relation to expert systems and human communication, Suchman (1987) contrasted "planned versus situated action," suggesting that designers ought to develop systems and programs apt to take into account the emerging circumstances of action. Orr (1996) used the situated action perspective to study how repairmen actually fix photocopier breakdowns during their maintenance interventions. Bricolage and improvisation (Brown & Duguid, 1991) are forms of situated action that are important in organizational breakdowns and emergencies, and when operating in those turbulent environments that high-tech companies routinely face (Ciborra, 1996). Lave and Wenger (1991) unveiled the characteristics of learning as a "situated process" and the importance of situatedness of experience in communities of practice (Wenger, 1998). The concept also crosses other disciplinary boundaries, from Artificial Intelligence (AI; Clancey, 1997) to media studies, where scholars write about the "situated culture."

A common denominator in the discourses that introduce or use the notion of situatedness is their explicit, but more often cursory or implicit, reference to phenomenology (usually via ethnomethodology) as the original source of the concept. Thus, Suchman (p. 39) recalled Dreyfus's (1991) introductory work to Heidegger to highlight the "transparency" of the situated character of action. When Winograd and Flores (1986) wanted to illustrate the importance of being "thrown in the situation" for grasping the more abstract notions of decision and information, they referred to Heidegger (1927/1962) and Dreyfus (1991). Wenger (1988) also saw one of the sources for the social theory of learning and communities of practice as the theories of situated experience based on the phenomenological philosophy of Heidegger. In technology design, Robertson (2002) noted that phenomenological approaches related to situated action have "played a major role in the shaping and progress of CSCW [computer supported cooperative work] research" (p. 302). Finally, in a recent monograph on action, systems, and their embodiment in social practices, Dourish (2001, p. 121) paid his debt to the importance of the "situated" perspective, indicating that "Suchman's work can be related directly to the work of the phenomenologists."

Unfortunately, if our tribute to the original roots of the concept of situatedness expresses an authentic commitment, and not just lip service to the founding fathers, we are bound to encounter a crisis caused by a major oblivion. To wit, German is the original language of phenomenology. "Situated" is the translation of the German *befindlich;* "situatedness" is *Befindlichkeit*. The latter, discussed in Section 29 of Heidegger's *Being and Time*, has been infelicitously translated as "state of mind" (Heidegger, 1927/1962, p. 172; see also Dreyfus, 1991, p. 168). In any event, "Wie ist Ihre Befindlichkeit?" is a courtesy form in German for asking: "How are you?" Hence, the original term *befindlich* not only refers to the circumstances in which one finds himself or herself, but also to his or her "inner situation," disposition, mood, affectedness, and emotion. In particular, Heidegger (1927/1962, p. 182) stated that understanding (i.e., cognition) is always situated, meaning that "it always has its mood." In other words, situatedness refers in its original meaning to both the ongoing or emerging circumstances of the surrounding world and the inner situation of the actor. Surprisingly, concern for the inner situation, or even the "state of mind" of the actor, can hardly be found in any of the current texts that make liberal use of the idea of situatedness.

The main purpose of this chapter is to start an exploration of such a troubling "situation" head on and counter the oblivion of the original roots that the notion of situatedness sports in current discourses.

The organization of the argument is as follows. First, the meaning of situatedness in the recent literature on information technology and organizations needs to be assessed in greater detail. What is really meant by situated action, learning, change, and so on? Second, because such an investigation still leads to fairly vague definitions and statements, one also needs to look at the few controversies in the literature generated by opponents to the situatedness perspectives, in particular in organizational theory and cognitive science (specifically, AI). Concluding remarks summarize the new tasks that lie ahead if one wants to take situatedness more seriously and comprehensively, possibly within the scope of a renewed, authentic, phenomenological tradition.

A CLOSER EXAMINATION

Critics may have a point when expressing the impossibility of identifying a well-delineated research position on situatedness, and dealing instead with a whole congeries of closely related views (Vera & Simon, 1993). For it is difficult to establish the precise contours of the current use of the term in social and organizational analysis, in particular when one wants to establish what are its actual links with phenomenology. The following is our attempt to come to grips with what the most quoted and influential authors had in mind when launching and using the term in the 1980s. In the fields of AI and management information systems, Winograd and Flores (1986) briefly mentioned the importance of the situation in which understanding takes place, and to support their line of argument, they quoted a passage from Gadamer, a student of Heidegger, which evoked the difficulty of capturing (let alone modeling) the subtleties of a situation: "To acquire an awareness of a situation is, however, always a task of particular difficulty. The very idea of a situation means that we are not standing outside it and hence are unable to have any objective knowledge of it. We are always within the situation and to throw light on it is a task that is never entirely completed" (Gadamer, quoted in Winograd & Flores, 1986, p. 29).

In her work of the same period, Suchman (1987) defined situated actions simply as "actions taken in the context of particular, concrete circumstances ... the circumstances of our actions are never fully anticipated and are continuously changing around us ... situated actions are essentially ad hoc" (pp. viii–ix). The notion of situatedness is crucial for Suchman to show that the foundation of actions is not plans, but "local interactions with our environment, more or less informed by reference to abstract representation of situations and of actions" (p. 188).

Soon after, Lave and Wenger (1991) expressed their dissatisfaction with the vagueness of the definitions of situatedness in the literature, also feeling uneasy about their narrowness: "On some occasions 'situated' seemed to mean merely that some of people's thoughts and actions were located in space and time. On other occasions, it seemed to mean that thought and action were social ..., or that they were immediately dependent for meaning on the social setting that occasioned them" (pp. 32–33). To discuss the features of situated learning that takes place within communities of practice, they suggested going beyond the notion of situatedness as an empirical attribute of everyday activity. Instead, they proposed to look at it as an overarching theoretical perspective; that is, as "the basis of claims about the relational character of knowledge and learning, about the negotiated character of meaning and about the concerned ... nature of learning activity for the people involved" (p. 33). This more general perspective implies that there is no activity that is not situated and underlies the need for a "comprehensive understanding involving the whole person" (p. 33). Unfortunately, the latter idea is not developed any further, and their association between situated understanding and the whole person remains unexplored both theoretically and empirically. Ten years later, these ideas and concepts are very much in use within the broadly defined interpretivist accounts of learning, knowledge management, and technical innovation, but little has been added toward their deepening and clarification: A sort of taken-for-granted adoption prevails, possibly becoming a new (alternative) management fad. On the same account, references to phenomenology are often made, but never quite fully explored and exploited. Collateral aspects are mentioned, such as transparency and readiness-at-hand (see, e.g., Suchman, 1987, p. 39), but nobody actually seems to quote Section 29 of *Being and Time* where Heidegger (1927/1962, pp. 172–182) introduced the notion of situatedness (*Befindlichkeit*), contrasting it with the privileged role attributed then (and now) to understanding, cognition, and the purely mental. Such an account can be found in Dreyfus (1991, Chapter 10 on Affectedness), but only within the boundaries of a specialized philosophical study of the first Division of *Being and Time*. Winograd and Flores, as well as Suchman, have been influenced by Dreyfus's work, hence the interesting cross-fertilization from philosophy into computer science, anthropology, and organizational theory taking place on the West Coast at the end of the 1970s. Still, the definitions these authors deploy seem to remain on the surface, not exploiting all their ramifications: In particular, factors such as emotions or moods (part of the comprehensive understanding of the "whole person") are not picked up, neither then nor more recently. Lack of proper references to phenomenology, although using its ascendance, may also induce the reader not versed in philosophy to believe that what these authors say about situatedness is indeed all that phenomenology has had to say on the subject. The consequence is a nonproblematic use of the terms *situated* and *situatedness* by scholars and practitioners who embrace interpretivist or radical perspectives in management and organization studies. Those terms end up meaning just *context* or *emerging circumstances* of action and knowledge.

DEBATES AND CONTROVERSIES

Another means to spell out the definition of situatedness in the current debate is to examine the unfolding, and the temporary outcomes, of recent controversies around the scope and utilization of the concept in both organization theory and AI-cognitive science.

Within the former, Contu and Willmott (2003) addressed the situated learning theory critically, especially its scope. They did not examine situatedness per se, but the way they looked at the dissemination and practice of situated learning elicits some features of the interpretivist and phenomenological approaches as they are being applied today especially in the corporate world. The authors acknowledged first that situated learning is a conceptualization that offers an alternative to cognitive theories of learning. Although in the latter learning is portrayed as a cognitive process involving a selective transmission of abstract elements of knowledge from one context (that is, the classroom) to the site of their application (the workplace), Lave and Wenger (1991) saw learning as integral to everyday practices, to the lived-in world. Looking at learning as a situated process means appreciating that it is embodied (lived-in), historically and culturally embedded. Furthermore, Contu and Willmott (2003) noted that the new perspective does include a due attention to the exercise of power and control in organizations. They pointed out, however, that the subsequent popularization of the idea of situated learning, carried out for example by Brown and Duguid (1991) and Orr (1996) with the notion of communities of practice, has adopted a more conservative approach, casting situated learning as "a medium, and even as a technology, of consensus and stability" (Contu & Willmott, 2003, p. 284). This is the outcome of a subtle cleansing process, whereby the organizational context of learning is looked at "in terms of a transparent background rather than a contested history" (p. 293). In particular, the situation of learning (typically, the community of practice) is conceived as "unified and consensual, with minimal attention being paid to how learning practices are conditioned by history, power, and language."

Without going further into the detail of this critique, and accepting that the focus on power and historicity has been lost in the popularization of the

concept of situated learning (but were these dimensions adequately emphasized by Lave and Wenger (1991) to begin with, or yet again just mentioned
in passing, like their quick reference, not subsequently elaborated, to the
"whole person?"), one should not be surprised if most of the "situatedness
studies" end up supporting a consensus and stability framework for organizational analysis and design. This is often the way any interpretive discourse
has been portrayed lately: as a discourse that "acknowledges the multi-vocal
fragmented, and conflicted nature of society, yet also focuses on the inte-
grative values that allow organizations and communities to function in harmony" (Schultze & Leidner, 2002). Note that the accusation of a consensus
and stability bias is then extended to the root disciplines of the interpretive
discourse—that is, ethnomethodology and phenomenology—so that authors ranging from Burrell and Morgan (1979) to Deetz (1996) have felt it
necessary to put next to the interpretive perspectives others that highlight
radical change and transformation, radical humanism, critical and dialogic
discourses, and so on.

A more pointed critique has been put forward within the AI-cognitive science fields by Vera and Simon (1993) in a special issue of *Cognitive Science*
dedicated to the situated action paradigm. Recall that researchers like
Suchman (1987) took the view that "plans as such neither determine the actual course of situated action nor adequately reconstruct it" (p. 3). But plans
are precisely the main form of symbolic representation on which AI systems
designed to interact with the environment are usually grounded: In this way
the situated action idea is aimed at undermining those efforts in AI and robotic research based on planning programs. The counterargument put forward by Vera and Simon (1993) stated that symbolic systems are able to
interact with the situation by receiving and processing sensory stimuli from
the world. Such systems can deal with the reckoning of local circumstances,
perceive and represent social relations if they have an impact on the system,
and produce appropriate responses even for temporally demanding tasks
embedded in complex environments. For example, if Suchman's concern is
the mutual intelligibility between people and machines in a situation of
technology use, (see the case study later) then Vera and Simon indicate that
such an understanding, and the correlated situated actions, cannot obtain
without internal symbolic representations, learning, planning, and problem-solving programs that feed on them. In summary, "the term situated
action can best serve as name for those symbolic systems that are specifically
designated to operate adaptively in real time to complex environments....
It in no sense implies a repudiation of the hypothesis that intelligence is
fundamentally a property of appropriately programmed symbol systems"
(p. 47). Situated action is here regarded as an approach homogenous, although in competition, with the ones of cognitive science and AI. It differs
only by a matter of degree: Can a symbol-based system be sufficiently so-

phisticated to capture emerging circumstances? Can it be rich enough to represent embedding networks of social relations? Can it be fast enough to perform meaningful actions on the fly? In principle yes, reply the cognitive scientists: When reconstructing situated decision-making, symbolic representations of the ongoing problem space can be drawn, algorithms can be identified, problem-solving programs can be written, which include the stuff of which AI applications are made: plans, if-then-elses, means–ends chains, and so on. In such a view, a physical symbol system interacts with the external environment by receiving sensory stimuli that it converts into symbol structures to be stored in a memory device, and it acts on the environment in ways determined by the newly acquired symbol structures. The memory is an indexed encyclopedia, where representations of external situations are stored as they come in. Stimuli coming from the environment invoke the appropriate index entries, and so on. In other words, cognitive scientists argue that one can design and build symbol systems "that continually revise their description of the problem space and the alternatives available to them" (Vera & Simon, 1993, p. 13). This mimics one of the key ideas of the situated action literature: the importance of moment-by-moment capture of the full situation of action. To be sure, the controversy lies in whether highly adaptive symbolic systems can actually be built. Suchman reviews the developments of AI and the various attempts at representing situations, through scripts for instance, and sees the task of reconstructing a meaningful knowledge background and context of action as infinitely long, hence unachievable in practice. In the end, Suchman insisted, the situation cannot be fully translated into a series of symbols, or a mental state. It is something "outside our heads that, precisely because it is non-problematically there, we do not think about" (p. 47). It is not just knowledge about the world, it is the world as an inexhaustibly rich resource for action.

COMPARISON AND DISCUSSION

The stage is now set to compare the current perspectives on situated action, and their claim to constitute an alternative to the positivist and cognitive views of learning, knowledge, change, and organizations, with the original thinking tried out early on within phenomenology. The recent availability of the teaching notes and lecture transcripts of the courses that Heidegger gave between 1919 and 1926 (before the publication of *Being and Time*) offers fresh material comprising discussions, investigations, and applications of the emerging phenomenological method to a range of domains, also suitable for this purpose, in particular the investigation of what is a situation, why it should be studied, and how to analyze it. In his early courses Heidegger addressed precisely these questions (to be sure, among many other directions of inquiry), and on the basis of his answers we can try to trace in what direction our understanding of situatedness has evolved when jumping from Continental philosophy around World War I to the recent, mostly Californian renditions imported into anthropology, organization theory, and information systems, and ascertain what went lost, or what was gained over the decades, by translating *befindlich* into the current "situated" (see also, Ciborra, 2002).

When contemplating the gap between the initial and the endpoints, one cannot help but feel a slight sense of vertigo. Heidegger had the programmatic vision of founding phenomenology by steering clear of the mind, cognition, psychology, and in general any regional science, while staying closer to everyday "factical" life. He put forward a rich and multiple notion of situation, in which inner life is as important as surrounding circumstances, where the pretheoretical is preserved by giving space to the moods, emotions, and dispositions not linked to thinking. In this respect, *Befindlichkeit* captures the multiplicity of meanings of being in an (inner and outer simultaneously) situation. In comparison, the current renditions of situatedness are much narrower, and deprived of the inner dimension. As mentioned earlier, authors like Suchman pointed to the importance of fleeting circumstances or "the world that stays out of our head." But the heart is also out of our head: It remains consistently ignored by those discourses committed to an alternative approach. The difference gets even more apparent at the methodological level. Current studies bend to the dictates of the scientific method: they strive to keep the observer separate from the situation to be studied. They set up experiments; they record them through technical means so as to obtain an evidence that is objective, can be evaluated independently, and can be shared. Things and people in the situation become objects, and events present themselves as processes occurring in objective time. But when the situation is decomposed in this way, "the self comes to appear as a detached spectator making observations—one item among others in the space-time coordinate system.... The world is 'dis-worlded' and the stream of life is robbed of its character as living ... it gives us a misleading picture of reality and our own selves" (Guignon, 2002, p. 86). Heidegger, instead, is interested in enacting, reenacting, or at least evoking situations. The sense of a situation can be grasped by going beyond objectification or a semantic analysis: It needs execution here and now, and full participation in such an execution.

Then there are more subtle differences, where the claims of present-day scholars are still the same as Heidegger's, but their implementation goes in another (usually scientific and cognitive) direction. Thus, for example, take the notion of time. The study of the temporality of the situation has a far-reaching importance in Heidegger's thinking and will lead in *Being and Time* to the exploration of the time we ourselves are. In Suchman's and others' works, reference to the importance of fleeting circumstances, of moment-by-moment unfolding of action, never challenges the ubiquity of clock time. The empirical study of situated action never hints at a different or problematic notion of temporality. The clock regulates the video and audio recorders.

Also, the distancing from cognitive science shared by the two agendas is handled differently. Heidegger makes every sort of effort to stay away from epistemology and cognition. Although evoking a situation colored by moods, by emotions in facing uncertainty, the objectifying study of situated action reports instead the mismatches between the plans embedded in an expert system and the reasoning of novice users. It identifies sequences of instructions, communication failures, and misunderstandings between users and the expert system. Note how the latter portrait of situated action may become an easy target for the symbol representationists: Preconceptions are symbols stored into "memory" and interpretations of evidence get translated into perceptions of stimuli and their symbolic processing. Despite repeated denials, the mind and the focus on the cognitive level of ad hoc problem solving still prevail. On the other hand, the leaning toward the heart in Heidegger's phenomenological method is quite clear. In the loose sheets for his course from 1919 to 1920, dedicated to the phenomenological intuition, Heidegger reminded his students that phenomenological understanding—as intuition—goes along with and into the fullness of a situation, being intimate, "loving" it (Heidegger, 1993, pp. 185, 262). To be sure, Heidegger is aware of the difficulty of carrying out such a task, and espousing such a method.

CONCLUDING REMARKS

Our going back to the roots of phenomenology to restore the original notion of situatedness and compare it with the current debate on situated action leaves us with three main research agendas. The first, leaning toward AI and cognitive science, states that situated action can be implemented through computer programs interacting with the environment and processing symbolic representations of what happens in the environment. The second, which claims to be an alternative grounded in the social sciences (phenomenology via ethnomethodology), is based on a social ecology of the mind: goals and plans are a vague guide to action. They must be complemented by the ad hoc improvisations of humans exploiting the circumstances and what the world offers at the moment of action. The heart is totally missing from the first agenda. Emotions and moods are sometimes referenced in a footnote, but do not seem to fit the second agenda either, despite Suchman's later claim that her original study fell into "a classical humanist trap." Finally, we have Heidegger's research program where the notion of situation includes at all moments the inner life of the actor, his or her mind *and* heart, and where any form of understanding is situated, meaning "affected." It is the pathos that characterizes the whole person in his or her situatedness in the world. Empirically, the first agenda seeks the construction of expert systems able to interact with worlds of limited variety. The second is validated by studying routine activities within stable organizations and tasks: Microbreakdowns reveal those improvisations that members are able to sport and confirm the situated nature of whatever plan or procedure they are supposed to follow. Phenomenology is interested in studying situations of radical transformation, because what it means to "find oneself" in a situation, to live, here comes to the fore in a sharper way. Overcoming the current state of oblivion and neglect, due consideration of the third, original, agenda should remind contemporary scholars and practitioners that articulating situatedness in organizational analysis and interactive systems design is still going to be a challenging task, and indeed a "touchy" one.

Chapter 8

More Than One Way of Knowing

Gillian Crampton Smith
Philip Tabor
Interaction Design Institute Ivrea

DESIGN METHODS

In 1965 we attended a lecture in the Cambridge School of Architecture in which Alex Reid confidently predicted the death, because of computing, of the design professions. They are not dead yet. But in the 1960s, encouraged by a positivist optimism about the potential "intelligence" of computers, Britain saw several attempts to construct an overarching theory of design and thus to systematize the act of designing into a partly or wholly automatic "design method." Christopher Alexander (1964) and John Christopher Jones (1970), for example, enthusiastically published these attempts, but, soon after their books emerged, rejected these methods. Henrik Gedenryd (1998, pp. 59–60) reported that in 1966 Alexander published an essay explaining why his method did not work, and totally repudiated design methods as suitable only for "incredibly mundane" problems, adding that "until those people who talk about design methods are actually ... trying to create buildings, I wouldn't give a penny for their efforts" (Alexander, 1971); whereas Jones (1970) admitted that "there is not much evidence that they have been used with success, even by their inventors.... The usual difficulty is that of losing control of the design situation once one is committed to a systematic procedure which seems to fit the problem less and less as designing proceeds" (see also, Jones, 1977).

117

The main weakness of the 1960s design methods movement was that it gave the impression that one could apply purely analytic methods to design, whereas designers use both rationality and intuition—the preconscious capacity of everyone to apprehend things or know how to do things without being able to say precisely how or why.

MOODS AND EMOTIONS

One reason why the act of design needs intuition (and other subjective modes of thought) as well as rationality lies in the term *situatedness.* In Chapter 7 of this volume Claudio Ciborra complains about the wide diffusion of this word to mean something different from its avowed source, the state of *Befindlichkeit* described by Heidegger (1962) in *Being and Time.* There is some irony in this accusation, doubtless intended, because Heidegger was hardly reluctant to assign new meanings to existing words. But Ciborra is correct to lament the semantic impoverishment in this case. "The original term *befindlich,*" he writes, "not only refers to the circumstances in which one finds himself or herself, but also to his or her 'inner situation,' disposition, mood, affectedness, and emotion." Much organization theory and information systems literature, he says, retains the idea of situatedness as a characteristic of an action's context (its external situation), but not as a characteristic of the actor's mental state (his or her internal situation).

The idea that actions are always located in an external situation—in a context, geographical, cultural, social, and so on—seems to be common sense. But it deserves restating, and perhaps even dignifying it with this coined word, situatedness, as a reminder that most human actions and decisions take place in a "here and now" which is usually complex, dynamic, and always unprecedented. To stress the external situatedness of designing is thus a useful defense against those who, as shown earlier, theorize the act of designing only from outside, dreaming that it might be abstracted into a realm of generality and predictability in which codified knowledge might somehow be "applied" to a problem so as to generate solutions, which in turn are evaluated by some sort of rational calculus. Given what we indicated in the Introduction to this book—the inherent ambiguity, instability, specificity, and multidimensionality of all but the most trivial design problems—this is a forlorn hope.

This is not, of course, to say that codified knowledge and technical rationality play no essential part in the design process; their role is indispensable—but as ingredients in what Bill Moggridge calls the mental "soup" which nourishes invention. And it is not to suggest that esthetic and ethical qualities cannot or should not be represented by quantities (financial, for example). Rather, that to translate human values or meanings into quantities always involves a significant loss of data—it is often necessary but always reductive.

Heidegger (1962) defined inner situatedness as a predisposition toward the world, prior to objective knowledge, which is made evident in "moods." Heideggerian mood is an "attunement" of individuals into their existential life situation (the arbitrariness of their birth, the certainty of their death, and so on), to which painting, poetry, and music can sometimes refer. But although the designed artifact may undoubtedly aspire to the condition of art, or more modestly use its techniques—the jolting juxtapositions of Surrealism, say, the semantic economy of poetry, or the controlled scan of pictorial composition—we doubt whether objects of use normally induce or echo mood in this sense.

The same cannot be said of emotions. Whereas a mood lacks intention (it is not directed toward an object), it is the foundation of emotions, which do indeed have objects (the complexity of this word processing program irritates me), and each emotion entails a corresponding desire (I want something simpler). Desires imply purpose, and Heidegger particularly honored purposeful objects, "use-objects," because our stance toward them is our primary and most authentic way of viewing the world—that is, its "readiness-to-hand" for our projected endeavors. We see the world from inside, in short, by involving ourselves in its action. However, when, for example, use-objects fail to function, they appear "present-to-hand": the world then seems to disintegrate into "objects of science," seen from outside and disconnected from ourselves.

DESIGN AS REFLECTIVE CONVERSATION

When designing, designers' mental stance is directed toward design's core activity, imagination. This requires them to "get inside" the problem and to "see through" whatever they are designing—just as the hammer, when in use, is "transparent," all attention being on the hammering. An excessively calculating stance will cloud the issue, just as excessive thinking about the hammering will disrupt its rhythm and aim. This "acting in the world" rather than seeing it from the outside is fundamental to the act of designing.

It has been described with much insight by Donald Schön (1983), whose ideas have been developed by Gedenryd (1998), as a "reflective conversation" with a unique and uncertain situation—a participatory act rather than an external observation. "In this reflective conversation," Schön (1983, p. 132) wrote, "the practitioner's effort to solve the reframed problem yields new discoveries that call for new reflection in action. The process spirals through stages of appreciation, action, and reappreciation. The unique and uncertain situation comes to be understood through the attempt to change it, and changed through the attempt to understand it."

As an example, Bonnie Nardi and James Miller (1991, quoted in Gedenryd, 1998, p. 92), described the case of an accountant who, to brief a programmer, found it easier to develop a complex spreadsheet model himself rather than define the problem verbally. Only by developing the solution piece by piece, or move by move, as Schön would say, did he arrive at a full understanding of the problem.

Gedenryd described what he called *interactive* (rather than distributed) *cognition:* the world augments or complements the mind, filling in the richness of reality which is impossible to hold in the head; citing Daniel Reisberg (1987), he pointed out that we can easily bring to mind an image of a tiger, but it is impossible to count its stripes. If you watch designers working, they do not first think things out in their mind, then represent what they have thought: They think as they draw, interrogating the world through their drawing or model-making. Representation is not just to record ideas but a way of experimenting in the world, to both understand the situation better and test hypothetical solutions; as Anton Ehrenzweig (1971, p. 45) wrote "The diagrammatic representation ... can serve two purposes: It can serve as the exposition of a finished design, or as a visual aid in search of a not yet existing solution."

The designer, then, must facilitate the process of interactive cognition while inhabiting the inner world of "disposition, mood, affectedness, and emotion." So the rest of this chapter describes three tools for doing this: the design repertoire, syncretistic attention, and the scenario. We refer mostly to design in general, but where possible draw examples from interaction design.

THE DESIGN REPERTOIRE

Heidegger stressed that people are each the sum of their past, present situation, and future potential. Ciborra therefore recommends that systems be designed to take into account the situatedness of the user. But what can we say of the internal situatedness of the designer? Designers too are each situated in their own history and experience, as both a designer and a human (the two roles are surprisingly compatible). Their experience as *designers* includes what Schön (1983, p. 138) called their *repertoire* of in-the-world "examples, images, understandings and actions" which they compare with the unique problem at hand—a repertoire enriched by each new experience of reflection-in-action:

> It is our capacity to see unfamiliar situations as familiar ones, and to do in the former what we have done in the latter, that enables us to bring our past experience to bear on the unique case. It is our capacity to *see-as* and *do-as* that allows us to have a feel for problems that do not fit existing rules.... Reflection-in-action in a unique case may be generalized to other cases, *not by giving rise to general principles* [our italics], but by contributing to the practitio-

ner's repertoire of exemplary themes from which, in the subsequent cases of his practice, he may compose new variations.

Their experience in practice allows designers in general to build on the general shape of past precedents, then, without defining precisely the relation between the present situation and past exemplars, using metaphor or analogy to draw parallels, direct or indirect, between one unique situation and another. In comparison with architecture or even relative new discourses like that of industrial design, however, it must be admitted that interaction design's repertoire is sparse. One reason is that if we date its birth to the advent of cheap high-resolution bit-mapped color screens in the early 1980s, interaction design has had little time to accumulate its models. Another reason is that the technology with which it deals has been changing fast, reducing the relevance of many of the examples and techniques deriving from obsolete usage. A third reason, finally, is that books, the traditional medium for diffusing most design repertoires, are inadequate for demonstrating interaction design. The repertoire is becoming more accessible (Wendy Mackay of the INRIA research laboratory, Paris, for instance, is planning an interaction design museum where examples can be accessed and explored online) and has grown appreciably since we started teaching interaction design in 1990. But for some time interaction design will need to draw at least partly on the repertoire of practices, notably those of art and design generally.

SYNCRETISTIC ATTENTION

Ehrenzweig (1971) distinguished between two kinds of attention necessary in creative acts: analytic (focused and detailed), which can impede creativity, and syncretistic (unconscious and undifferentiated). As an example, he showed (pp. 25–28) how a musician must be able to switch between analytic attention on individual voices or instruments, and syncretistic, where conscious attention is de-selected to hold in the mind the complexity of the polyphonic whole. He argued (p. 46) that "syncretistic ... techniques rather than analytical clarity of detail are needed for the creative thinker to control the vast complexities of his work," and particularly noted the mathematician Jacques Hadamard's praise of vagueness:

> The student [of mathematics] cannot start by blotting out his conscious attention. He has first to learn the conscious rules governing mathematical transformations and he will check each step according to the rules. But at a certain point ... he has to abandon precise visualization. Instead of each single step he has to reach out and grasp the total structure of the argument as compared with any other possible structure. He has to visualize syncretistically the total structure though he cannot look sufficiently far enough ahead to see clearly

the detailed choices and decisions awaiting him. In Wittgenstein's words: his view must be comprehensive but not clear in detail. (pp. 37–38)

Hubert Dreyfus (1993) also showed that beginners need to think through everything they do whereas experts have typically internalized subconsciously the thought modes behind their expertise. Paolo Legrenzi (2005, p. 110) noted, too, regarding the tacit knowledge informing the performance of experts ranging from Casanova to the tennis player Martina Hingis, that "its codifiability is very low and its transfer requires much interactive learning" (whatever that might mean in Casanova's case).

Given this opposition between creative imagination and the intellect, designers too have traditionally developed techniques for stimulating the former by outwitting the latter. Legrenzi (2005, pp. 66–79) defined *defocus* as to stop concentrating on the problem at hand for a while and let the subconscious mind get to work; how many of us have gone to bed with a problem unsolved, only to wake the next day with the solution? He also recommended *defixating*. Picasso supposedly said there is nothing so dangerous as a good idea because one is reluctant to throw it away; fixation on an idea, by fixing the frame within which one views the problem, prevents one generating others. This is especially relevant in interaction design, typically a multidisciplinary endeavor, where the apparently good ideas which team members bring from their home disciplines come laden with those disciplines' unconscious assumptions.

THE SCENARIO

The experience of designers, as designers, contributes to their mental potential as designers. Their experience as humans also contributes of course, but vicariously, by allowing them to project themselves into the inner situation of the prospective users of a design.

In most fields of design, the notional space in which a problem might be defined or solutions emerge can be summarized by a relatively simple representation. Although the computer has taken over some representational tasks, graphic and product designers and architects still tend to think through the pencil, which becomes a transparent extension of the brain. Whether to work out the shape of a device or ideas for a metaphor, or to diagram a structure or the path through a hierarchy, the pencil allows the designer to keep in what Mihaly Csikszentmihalyi (1990) has called "flow," where nothing interferes with the "creative wave." Even the making of rough physical models, although it takes more time, becomes an intuitive, transparent process.

But in interaction design it is more difficult to act in the world in this ready-to-hand manner because interactive systems are complex, happen over time, and, due to the complexity of possible behaviors, are difficult to represent in summary form. So what can interaction designers do to apprehend, through cumulative experiment, the situation for which they are designing?

A technique particularly appropriate to interaction design is the "scenario": the imagining of a fictional situation and its representation as a written narrative, say, or a storyboard, performance, or video, within which possible interactive behaviors between users and systems emerge. In its purest form (developed by Brenda Laurel in her time at Interval Research and more recently by Nathan Waterhouse at Interaction Design Institute Ivrea) a scenario develops through theatrical improvisation, "improv," in which the players must continually react, instinctively and (one hopes) revealingly, to a situation, only partly defined, as it unfolds. John Carroll (1995, p. 4, quoted in Gedenryd, 1998) wrote:

> The defining property of a scenario is that it projects a concrete description of activity that the user engages in when performing a specific task, a description sufficiently detailed so that design implications can be inferred and reasoned about. Using scenarios in system development helps keep the future use of the envisioned system in view as the system is designed and implemented; it makes use concrete—which makes it easier to discuss use and to design use.

By establishing the scenario and allowing its unpredictable logic to play out, interaction designers imagine themselves into the potential participants' situation; into their "situatedness," indeed, because this self-projection is into not only the physical, institutional, and utilitarian context, but also, inevitably, the whole life-world of the imagined users—their epistemology, esthetic preferences, ethical priorities, emotional repertoire, and so on.

Such self-projection is fraught with danger because another's inner situatedness is profound and ultimately unfathomable. But it is necessary, and, defined as *empathy*, is a human faculty especially prized in design discourse. As early as the first century BC the architectural theorist Vitruvius wrote that a building should exhibit not only *firmitas* and *commoditas* but *venustas*, which last we might interpret as the attributes of Venus, goddess of love: the power, that is, to move the heart, the inner desires and beliefs of the user or observer. The education of designers, and the discourse that arises from their practice, take for granted that artifacts operate in the inner world of significance and value as well as the world of the present-to-hand. Alertness to subtle distinctions of meaning is prized, as is the ability to resist the urge to resolve and oversimplify the contradictions and complexities which characterize mental life.

CONCLUSION

Despite Ciborra's lament, the idea of external situatedness was a useful concept to shift engineers developing productivity software from a narrow focus on the human–machine interface to a broader focus on the physical and social situation in which the user operates. But designers act directly in the world, rather than as if observers from outside, in a reflective conversation with both the materials they work with and their emerging understanding of the situation. Only by this dialectical picturing can they fully define, and in this sense "create," the design problem they are addressing—to inhabit, in short, the situatedness for which they are designing.

There are many understandable reasons for the focus thus far on the pragmatic rather than affective aspects of interactive technologies: the empiricist traditions of the United States, where the major developments have taken place; the predominance of the positivist worldview in engineering discourse, privileging analysis and reason; the concentration in early development on productivity tools whose efficiency was seen as the paramount criterion; and, not least, the sheer difficulty of making interactive applications work at all, let alone respond to the user's inner situation. Only now, when these technologies are no longer just work tools for professionals but an environment in which we all increasingly live our lives, can we realize the experience-enriching potential of deploying, in the interaction design process, the full range of intelligence, intuition, sensibility, and skill traditionally used in other spheres of design.

We hope we have indicated that the designer's way of working is not a substandard version of the analytic-deductive mode but one which also uses different modes of knowing and thinking, less accessible to the conscious mind, but nevertheless effective and accessible to methodical process. We are reminded of the engineer David Boor, the IBM representative at the MIT Media Lab, who told one of us: "After working with artists and designers as I have at the Media Lab, I have come to understand that there is more than one way of knowing."

Part

IV

COMMUNITY

Chapter 9

Participation and Community

Charles Goodwin
University of California, Los Angeles

AN ECOLOGY OF MULTIPLE SIGN SYSTEMS

There exist several different approaches to the study of participation. A research tradition in fields such as linguistic anthropology uses models proposed by Goffman in works such as *Footing* (1981) as a point of departure for the construction of typologies for different kinds of participants within speech events (for instance, ratified versus unratified participant, hearer or overhearer). Within such a categorical framework, little attention is paid to how parties build action in concert with each other through ongoing analysis of what each other is doing, and how such mutual reflexivity is relevant to the collaborative production of future action. Another approach to participation focuses on how newcomers become competent members of a community through processes such as peripheral participation (Lave & Wenger, 1991). Although this is certainly relevant to what is described here, it pays less attention to the detailed, moment-by-moment organization of specific, temporally unfolding activities.

In this chapter, participation is analyzed as a temporally unfolding process through which separate parties demonstrate their understanding of the events in which they are engaged by building actions that contribute to the further progression of these same events. Such a view of participation links cognition to the interactive organization of action. It requires detailed analysis of the specific activities that parties are participating in as they build courses of action in concert with each other. Through the way in which this perspective on participation encompasses participants' orientation toward each other, the details of language use, tools, documents, and relevant structure in their environment, it provides a way of investigating the distinctive work practices and professional vision of particular social groups, and of describing how the structure of tools being used contributes to this process. Central to all of this is a view of human knowledge and action being organized within an ecology of sign systems, rather than in a single semiotic modality (C. Goodwin, 2000). Within such a framework, individual signs can be partial and incomplete because their relevant sense and use is constituted through the way in which they mutually elaborate other co-occurring signs.

This is vividly illustrated by Chil, a man with severe aphasia who nonetheless manages to act as a powerful speaker in conversation by getting others to say the words that he needs. His competence as a speaker is lodged not within his brain, but rather through his ability to participate in language practices in which the actions of others also play a significant role (C. Goodwin, 2003a).

Because of severe damage to the left hemisphere of his brain, Chil is able to speak only three words: "Yes," "No," and "And." Despite this severely restricted vocabulary he functions as a powerful participant in conversation. He is able both to say a great many different things and to produce complicated action. To accomplish this he does not function as an isolated actor, but instead builds meaning and action in concert with others within a relevant environment. His situation provides an opportunity to investigate how ongoing participation in courses of action with others within a consequential community is central to the organization of human action and cognition. Conceptualizing someone with aphasia as building action within a world that is simultaneously being structured by the actions of others, rather than as an isolated individual faced with the impossible task of constructing rich linguistic structures, has implications for the design of tools that might aid people in such circumstances.

CONVERSATION 1: THE BIRD CALENDAR

As noted previously, participation is investigated here as a temporally unfolding process through which separate parties demonstrate to each other their ongoing understanding of the events in which they are engaged by building actions that contribute to the further progression of these very same events. Parties participate in specific courses of action while taking into account: (a) what each other is doing, (b) the consequences this has for the organization of future action, and (c) the frequently relevant structure in the environment.

The practices used by Chil to construct meaning help make this more clear (for more detailed analysis of this sequence, see Goodwin & Goodwin,

2001). In Fig. 9.1 the participants around the table are admiring a calendar with pictures of birds that one of them has just received. As a new picture is revealed, Pat, the woman on the left, assesses or evaluates it by saying "*Wow!* Those are *great* pictures" (line 2).

Chil, the man with severe aphasia, is seated on the right in Fig. 9.1. Despite his impoverished linguistic abilities, he also assesses the picture, using a string of nonsense syllables ("Dih-dih-dih-dih") to carry an appreciative prosodic contour (line 1). Note, however, that his assessment occurs much later than Pat's, indeed when her talk has almost reached completion.

It might be argued that Chil's delay is a manifestation of his cognitive deficits, for example that he lacks the ability to respond to relevant events with normal timing. However, when his embodied behavior is examined, a quite different picture of what is happening emerges. When Pat begins her "*Wow!*" he is looking down at the food he is eating. To assess something, to judge it in some fashion, an actor must perceive it. Immediately on hearing Pat's "*Wow!*" he raises his head and moves his gaze to the object being assessed. Only when this has been completed, and he is actually looking at the calendar, does he perform his own assessment. Note also that he does not move his gaze toward the source of the sound to which he is reacting, Pat, but instead recognizes that the activity in progress is an assessment and immediately moves to the object being assessed. His understanding of, and contributions to, the events in progress are displayed as much through the precise movements of his body as by his talk.

FIG. 9.1.   Conversation about a bird calendar. Chil, a man with severe aphasia, participates with Pat through vocalization and visible embodied behavior.

Chil's use of visible embodied behavior as well as talk to participate in the assessment, the activity that the parties are currently pursuing together, both displays his understanding of the events in which he is engaged, and contributes to the further shaping of these very same events. If analysis is restricted to his linguistic output, he appears to be a severely impoverished actor, indeed almost an idiot who talks in nonsense syllables. However, focusing on how he *participates* with others in the joint construction of relevant action allows us to recover his cognitive competence, and to demonstrate his ability to engage with precision in speech activities, despite his almost complete inability to speak. Rather than acting as an isolated, self-contained agent, his cognitive abilities are lodged within a community of other actors who participate with him in the construction of the actions and events that make up the lifeworld they inhabit together.

This view of participation has several consequences:

1. Study of participation in this fashion requires analysis of the specific activities in which the parties are engaged in. The notion of a situated activity system is central (Goffman, 1961; C. Goodwin, 1996; M. H. Goodwin, 1990).

2. Rather than being accomplished within a single semiotic modality, such as language, participants build meaning and action by using the resources provided by a larger ecology of sign systems (see also, Hutchins, 1995) that can include talk, a range of different kinds of sign systems displayed by the visible body (gesture, for example, displays of orientation through gaze and posture, or multiparty participation frameworks), and semiotic and other forms of structure in the environment. Within such a framework, any individual sign can be partial and incomplete. Chil's nonsense syllables, prosody, and gaze mutually elaborate each other to create a whole that is not visible in any of its constituent parts.

3. The organization of participation within emerging courses of action has consequences for vision and perception as forms of socially organized practice. The temporally unfolding activity in which Chil is participating systematically leads him to gaze at a particular place within the complex visual environment of the room in Fig. 9.1, and to formulate what he sees there in ways that are relevant to the activity. The multimodal organization of this activity, the way in which it encompasses not only language but also visible displays by the body and orientation to, and formulation of, objects in the environment, allows us to describe with some precision how actors construct relevant events through participation in emerging courses of action.

Chil manages to function as a powerful speaker in conversation by getting others to speak the words that he needs, and also by using structure in his local environment (including relevant objects, the talk of others, and the way in which the spaces that constitute his lifeworld are sedimented with meaning). Timing and sequential positioning are crucial to this process. The practices that he uses may have consequences for the design of tools that could facilitate the communication of people in his position. To oversimplify, much research focuses on the construction of tools that would give someone such as Chil resources for the construction of complex symbolic objects, such as sentences. Of necessity, many of these tools are quite complex, and indeed their construction can probe the boundaries of research in fields such as computer science.

The practices that Chil uses suggest an alternative: the design of rather simple tools that would allow someone with aphasia to invoke structure in the environment in a way that is appropriate to the unfolding organization of the activities in which he or she is engaged. Rather than focusing primarily on construction of complex symbolic objects, such tools might place a premium on timing, the ability to rapidly act in concert with others in ways that are appropriate to the moment-by-moment unfolding of human interaction—to reflexively participate, that is, in the construction of the ongoing events.

Before providing a specific example relevant to such possibilities, let me note a few caveats. First, I am not a designer, and this is being offered simply as data and practices that might stimulate the thinking of others. Second, aphasia and other forms of brain damage are highly variable. Chil's particular mix of strengths and weaknesses should not be taken as typical for all aphasics.

CONVERSATION 2: SAN FRANCISCO OR REDDING?

In Fig. 9.2, Chil is sitting at his kitchen table with his daughter Pat and son Chuck. They have been talking about the births of Pat's two children. Both were born in California, one in San Francisco, and the other in Redding (a city in northern California). The births occurred approximately 20 years ago when Pat lived in California. Chil and his wife, who live near New York City, went to California for the births. Chil has been using gesture and other resources to get Pat to recall incidents about the births which they are telling Chuck.

In line 2, Pat starts to talk about something that happened in San Francisco. Chil immediately intercepts her talk with one of his three words, "No." Pat then changes "San Francisco" to "Redding" (note how the replacement of the first place name with the second is displayed explicitly through the way in which the "I was in X" format is recycled). Because of the injury to his brain, Chil is completely incapable of either saying a word such as "Redding" or of constructing the sentence that encompasses that lexical

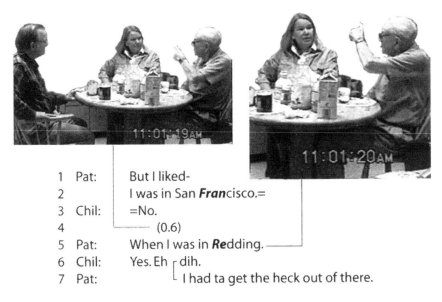

1	Pat:	But I liked-
2		I was in San **Fran**cisco.=
3	Chil:	=No.
4		(0.6)
5	Pat:	When I was in **Re**dding.
6	Chil:	Yes. Eh ⌈dih.
7	Pat:	⌊I had ta get the heck out of there.

FIG. 9.2. Conversation about a birthplace. Chil uses his limited vocabulary (yes, no, and) and directional gestures to shape the conversation.

item. However, in a number of significant ways he is the author of what is said in line 5. Thus, if he had not intervened, Pat would now be talking about something quite different, some event that occurred in San Francisco. Moreover, although the transcript does not fully capture this, as she speaks Pat displays that Chil is the ultimate authority as to the accuracy of what she is saying, as indeed would be the case if she is now trying to provide the correction he signaled was needed with his "No." Thus, she raises her head while gazing intently at him while lifting her eyebrows with a facial expression that seems to indicate that she is checking with him. Chil does in fact treat what Pat says as an action that requires his verification by responding to it with a "Yes." In essence he has gotten Pat to speak words that he can't and, in so doing, to move the conversation in a new direction, one that he has chosen.

What resources enable Chil to function as a consequential speaker in conversation despite his almost complete inability to speak? First, his limited vocabulary ("Yes," "No," and "And") presupposes that he is living and acting in a world already inhabited by others, and structured in fine detail by their semiotic activities. "Yes" and "No" are second pair parts, terms designed not to stand alone, but instead to function as next moves to actions produced by others. They thus have a strong indexical component in that recipients use the semiotic structure of the talk being responded to as a point of departure for understanding an action such as "No" by Chil.

With his "No" here, Chil is not objecting to life in general, or any of the millions of things in the world to which he could be opposed, but instead to something that the prior speaker just said, the most salient possibility being the place name that she just produced. Pat can reasonably infer that he is asking her for a different place name. These possibilities are further constrained by the local history of the discourse in progress where Pat has been talking about two births that occurred in two different cities. She can and does succeed by producing the other city (Redding instead of San Francisco) in response to his objection. The locative character of the solution Chil wants is further suggested by the pointing gesture that co-occurs with his "No." Indeed, he is actually pointing in the direction (west) that is at issue. Note also how his actions presuppose a cognitively complex coparticipant, one who is not simply decoding what he says, but using that talk as the point of departure for structured inferences. One pervasive model of a speaker's competence focuses on mental processes within an isolated individual. Here Chil functions as a consequential speaker through his ability to participate in public, socially organized language practices.

POSSIBLE TOOLS

Much research into the design of tools that could help someone such as Chil communicate focuses on tools that would enable a speaker to produce complex symbolic structures, such as sentences. The computer program through which the physicist Stephen Hawking (whose speech problems result from something other than aphasia) is able to talk is one example. Such tools, and the research that makes them possible, are important and can help many people who have difficulty producing speech. In essence, such research tries to recreate the complex symbolic processes of the prototypical competent speaker. By contrast, Chil can use very simple tools, a word consisting of only a single syllable, to say something novel and complex. He does this by tying to and invoking relevant structure in his environment. He is not an isolated monological speaker, but instead an actor operating within a world inhabited by others and structured in fine and relevant detail by their activities.

This may have the following relevance to the design of tools for someone such as Chil. Instead of trying to produce complex symbols, and treating an actor such as him as an entity required to produce sentences from scratch in isolation, it might be possible to design simple tools that could rapidly and reflexively intervene in unfolding courses of action by tying to semiotic structure produced by others. Something like a simple buzzer, although with a more pleasing sound, comes to mind, perhaps one that could include relevant intonation contours. Although not described here, Chil's use of intonation for both action and the display of emotion is crucially important (see Goodwin, Goodwin, & Olsher, 2002).

Looking at this from a slightly different perspective, some aphasic speakers are able haltingly and slowly to construct far more vocabulary items than Chil. Although their aphasia is considered less severe, the onward movement of the conversation in progress can be severely delayed, as the construction of each word becomes a task in its own right. This situation can become difficult for interlocutors. By way of contrast, what would be preserved by a simple tool that tied to structure in the ongoing talk of others, and what was preserved in Chil's way of participating in the talk of others, was the rapid, reflexive *timing* of typical interaction. It has been suggested that the very severity of his aphasia paradoxically helped him function as an engaging and effective conversational partner, by eliminating futile efforts to produce relevant vocabulary.

I raise the possibility of trying to design very simple tools that invoke structure in their environment in part because of a conversation I had with a new PhD student in computer science at a conference recently. I was interested in talking with her because she had just given a paper on aphasic speech. I suggested that she look at the actual interaction of people with aphasia, but she said that for her research it was adequate to focus on transcripts of the talk they produced. Consider what transcripts of Chil's talk, in isolation from that of his interlocutors, would look like. I also suggested that very simple tools might be extremely powerful. She told me that she could never get tenure unless she designed complex computer programs. Moreover, it helped her lab, and her standing at her new university, to require expensive equipment for her research.

In brief, despite his catastrophically limited ability to produce language, Chil is able to function as a powerful speaker in conversation. This is possible because he does not act as an isolated speaker, the prototypical locus for the study of language in contemporary formal linguistics, but instead constructs meaning and action by participating in talk-in-interaction with others.

There is not space here to investigate how participation in activities can encompass not only talk and different kinds of embodied displays, but also tools, documents, situated writing practices, and various kinds of structure in the environment. For example, the tools used by archeologists, chemists, and oceanographers provide architectures of perception that entrain the embodied participation of different actors in specific ways, structure cognition and provide historically shaped solutions to the distinctive tasks posed by the work of particular communities (C. Goodwin, 1994, 1995, 2003b; Hutchins, 1995).

CONCLUSION

By participating together in courses of action, separate parties both display their understanding of the events in which they are engaged, and build meaning and action in concert with each other. Through this process, a community is constituted in a number of different ways. Chil's aphasia provides a particularly clear example. Not only his social, but also his cognitive life depends on the way in which talk is embedded within the activities of a small local community, those who are interacting with him. He is able to build consequential meaning and action only by participating in courses of action with others.

That participation has a moral dimension. Despite his impairment, those who share Chil's lifeworld with him treat him as a cognitively alert human being, someone who can understand others, and who has intelligent, relevant things of his own to say. Indeed, they invest considerable effort to figure out just what he wants to tell them. This situation could be very different: It would be quite possible for others to assume that someone who can barely speak is an idiot and exclude him from participation in those discourse practices that constitute him as a fully fledged human being. Yet in most central ways, the community that encompasses Chil is brought into being and structured through the ways in which members of that community participate in relevant courses of action together.

Chapter 10

Participation in Interaction Design: Actors and Artifacts in Interaction

Pelle Ehn
Malmö University

WHAT INTERACTION DESIGN IS NOT

> *Real artefacts are always part of institutions, trembling in their mixed status as media-tors, mobilizing faraway lands and people, ready to become people or things, not know-ing if they are composed of one or many, of a black box counting for one or of a labyrinth concealing multitudes.*
>
> —(Bruno Latour, *Pandora's Hope*, 1999, p. 193)

Interaction design is not computer science, or even human–computer in-teraction (HCI), even if it deals with humans, computers, and interaction. Interaction design is not graphic design, even if it is both visual and com-municative. Neither is interaction design another computer study like com-puter-supported cooperative work (CSCW) or participatory design (PD), nor another design discipline like product design, architecture, or media studies, even if all of these disciplines and practices and many others in the background give shape to interaction design.

Interaction design is neither art nor technology in isolation, but maybe a social and esthetic synthesis. Interaction design is not statements about facts, or even propositions of what ought to be, but maybe it could be design as an anxious act of political love, as a rethinking of the Aristotelian intellectual virtues of *techne* and *phronesis*, and the reunion of art, technology, and politics in the era of ubiquitous computing.

We are designing interaction design!

POSITIONING THE FIELD

To me the idea of bringing design to software, and information technology to design, represents the birth of a new and challenging design discipline— interaction design. This discipline has a design-oriented focus on human interaction and communication mediated by artifacts. The identity and actuality of this emerging design discipline is emphasized by the convergence of digital artifacts with physical space and the objects surrounding us (ubiquitous computing) as well as the convergence between different media (multimedia, new media). This discipline represents social, technical, and esthetic challenges to the interaction designer.

Another way to position interaction design would be with reference to *embodied interaction*. The term was coined by Paul Dourish (2001) for the creation, manipulation, and sharing of meaning through engaged interaction with artifacts. Embodied interaction starts from the observation that computing is getting both more tangible and more social. Design of information technology is becoming more tangible in the sense that emerging, radically new kinds of digital artifacts, beyond the desktop computer, deliberately amalgamate the interaction qualities of physical objects with computational qualities—augmenting papers, pens, toys, and all kinds of everyday objects. Computers are becoming more and more embodied as embedded aspects of our experience of our everyday environment. There is also more embodied interaction in the sense of the embeddedness of artifacts in social practice, community, place, and situatedness, beyond the disembodied human–computer interface. The embodiment of our experiences in the world is coming more and more into focus.

Interaction design has grown out of a merger between the traditional design fields (especially product design and graphic design) and socially oriented computer studies (especially HCI, but also CSCW and PD). Other important initial contributions come from architecture, art, sociology, and media studies. *Bringing Design to Software*, edited by Terry Winograd (1996) and with contributions from many design fields as well as computer science, was an early manifestation of this trend. The design of computer artifacts is here understood as an activity that is conscious, is a conversation with the material, is creative, is communication, and is a social activity that keeps human concerns in the center and has social consequences.

However, one might ask, just how new is the design brought in, in practice, by interaction design? What we meet is in many ways closely related to the modern design tradition from the Bauhaus. As Lev Manovich observed in *The Language of New Media* (2001), we clearly see the traces of "new vision" (Moholy-Nagy), "new typography" (Tischold), and "new architecture" (Le Corbusier).

The real challenge in interaction design is maybe the other way around. Really new is the bringing of computational technology to design, and the phenomenon of ubiquitous computing (Weiser, 1991). The design materials for ubiquitous computing and the appearances of computational artifacts are both spatial and temporal (Redström, 2001). With computational technology we can build temporal structures and behavior. However, to design these temporal structures into interactive artifacts, almost any material can be of use in the spatial configuration. Interaction design deals with a new kind of combined interactive narrative and architecture, a kind of mixed object and places (Binder et al., 2004).

Another challenge of particular significance to interaction design—because it is physical, digital, and social—is the paradox of *demassification*, an expression used by John S. Brown and Paul Duguid in their early article (1994). They pointed at how information technology and new media introduce new material and social conditions for the design of artifacts. Demassification concerns the *physical* or material change—artifacts literally lose mass and can be distributed and accessed globally: Think of a digital book or a library. But there is also *social* or contextual demassification. This concerns the ability to customize and make individual copies of digital artifacts—a loss of mass as in "mass medium": Again, think of a personalized version of the book or the digital library. Why is this a design problem? Is it not just great to have totally mobile and individualized artifacts? As Brown and Duguid suggested with their paradox of demassification, this is achieved at the price of losing the intertwined physical and social *experiences* of the artifacts. Physical demassification deprives the artifact of what they called its "border resources"—the aspects which, although possibly peripheral to its content, remain unchanged in a physical object, thus signaling its commonly held significance, but which are lost or changed when transmitted digitally. The cover of the book may not be decisive for the content, but its shape, texture, weight, wear and tear, and so on may still be an important aspect of its "bookness" and how we experience it as a book. It is these border resources that are lost when every digital copy gets its own form, and hence a relatively "immutable" source for interpretation dissolves.

Entangled with this, and adding to the problem of lost physical mass, is social demassification. The individualized versions of a digital artifact, reaching only a few people, underline the loss of shared border resources by jeopardizing their function as relatively immutable contextual sources for shared interpretations within a community.

These dilemmas in interaction design reach far beyond the digital-physical and are fundamentally concerned with the embeddedness of artifacts in social practice, community, and place.

In this chapter I relate this emerging field of interaction design to my own "moves" in this direction with a focus on participation and community. I focus on three such moves. The first concerns the *work-oriented design of computer artifacts* where questions of community and participation were centered around users, democracy, and the design process. The second move deals with *design-oriented studies of information technology in context* and the attempt to form a framework for uniting the technical, functional, and subjective knowledge interests in the design of computational artifacts. The third move has the form of a *manifesto for a digital Bauhaus* and the attempt to create an arena, a meeting place, a school, and a research center for creative and socially useful meetings between "art" and "technology." Finally, I suggest a fourth move toward the games people play when *appropriating places* out of community, artifacts, and architecture.

MOVE 1: WORK-ORIENTED DESIGN OF COMPUTER ARTIFACTS

A "movement" toward *participation and skill* in the design and use of computer artifacts evolved at Scandinavian workplaces and in academia in the 1970s and the 1980s. This movement was based on two design ideals:

1. *Industrial democracy* and the attempt to extend political democracy by also democratizing the workplace.
2. *Quality of work and product* and the attempt to design skill-enhancing tools for skilled workers to produce highly useful quality products and services.

Personally, I was heavily involved in this movement as both an active researcher and a reflective academic. In my *Work-Oriented Design of Computer Artifacts* (Ehn, 1988) I tried to give a comprehensive view of the theory and practice of this movement. What was suggested was an approach that from an emancipatory perspective deals with both the inner everyday life of skill-based participatory design and the societal and cultural conditions regulating this activity.

This kind of locally anchored, trade union-based, politically significant, interdisciplinary and action-oriented research on resources and control in the processes of design and the use of computational artifacts, has contributed to what abroad has often been seen as a specifically Scandinavian approach to computer systems design.

Participation As a Fundamental Epistemological Category

There was, however, also a complementary focus to this labor process approach to skill and participation. This was the focus on the role of skill and participation in design as an everyday practical activity, and participation as a fundamental epistemological category. This concern grew out of a dissatisfaction with traditional theories and methods for systems design, not only with how systems design had been politically applied to de-skill workers, but more fundamentally with the theoretical reduction of skills to what can be formally described. Hence, one can say that the proactive critique of the political rationality of the design process pointed at a transcending critique of the scientific rationality of methods for design of computational artifacts.

An alternative foundation for the practice of a skill-based participatory design approach was outlined, based in the tradition of the language-game philosophy of Ludwig Wittgenstein and the "paradox of rule-following behavior" (Wittgenstein, 1953). To be able to follow a rule is to have learned how in practice to continue an example we have been given. Mastery of the rules puts us in a position to invent new ways of carrying on. There is always the possibility that we can follow a rule in a wholly unforeseen way. Tradition and transcendence, this is the dialectical foundation of design. This more epistemological foundation for participatory design, reflecting on the design process, design skills, and design artifacts like mock-ups and games, was described as a number of lessons learned:

(General lessons on the language-games of work-oriented design)

1. By understanding design as a process of creating new language-games that have a family resemblance with the language-games of both users and designers, we have an orientation for really doing work-oriented design as skill-based participation, a way of doing design that may help us transcend some of the limits of formalization. To set up these design language-games is a new role for the designer.

2. Traditional "systems descriptions" are not sufficient in a skill-based PD approach. Design artifacts should not be seen primarily as means for creating true "pictures of reality" but as means to help users and designers discuss and experience current situations and envision future ones.

3. "Design-by-doing" design approaches like the use of mock-ups and other prototyping design artifacts make it possible for ordinary users to use their practical skill when participating in the design process.

(Lessons on skill in the design of computer artifacts)

4. Participatory design is a learning process where designers and users learn from each other.

5. Besides propositional knowledge, practical understanding is a type of skill that should be taken seriously in a design language-game, because the most important rules we follow in skillful performance are embedded in that practice and defy formalization.

6. Creativity depends on the open-textured character of rule-following behavior, hence focus on traditional skill is not at the cost of creative transcendence, but a necessary condition. To support the dialectics between tradition and transcendence is at the heart of design.

(Lessons on participation in the design of computer artifacts)

7. Really participatory design requires a shared form of life—a shared social and cultural background and a shared language. Hence, participatory design is a question not only of users participating in design, but also of designers participating in use. The professional designer will try to share practice with the users.

8. To make real user participation possible, a design language-game must be set up in such a way that it has a family resemblance with language-games the users have participated in before. Hence, the creative designer is concerned with the practice of the users in organizing the design process, understanding that every new design language-game is a uniquely situated design experience. There is, however, paradoxical as it may sound, no requirement that the design language-game make the same sense to users and designers, only that the designer sets the stage for a design language-game so that participation makes sense to all participants (see Fig. 10.1).

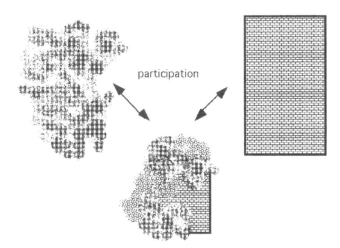

FIG. 10.1. The participatory design approach. A communicative and participative language-game view of the design process: two or more language-games are fundamentally related via shared experiences in a common design language-game (the design community-of-practice) which resembles the ordinary language-games of both users and professional designers. A fundamental competence of the designer is the ability to set the stage for a shared design language-game that makes sense to all participants.

(A final lesson on the boredom of design)

9. Formal democratic and participatory procedures for designing computational artifacts for democracy at work are not sufficient. The design process must also be organized in a way that makes it possible for ordinary users not only to utilize their practical skill in the design work but also to have fun doing this.

In retrospect, this view on design, learning, and participation comes close to approaches inspired by how we become legitimate participants in communities-of-practice (Lave & Wenger, 1991).

I would like to claim the relevance of this kind of understanding design of computer artifacts when laying the theoretical foundations for the field of interaction design, especially concerning the design process and the use of design artifacts in that process. However, interaction design is in no way limited to work, and given other settings, participation can take very different shapes. Furthermore, PD is very weak on understanding the computer as an interactive design material that has to be formed not only technically and functionally but also esthetically. As my introduction suggested, this material is both spatial and temporal, a kind of mixed object. With computational technology we can build temporal structures and behavior. However, to design these temporal structures into interactive artifacts, almost any material can be of use in the spatial configuration. This is a challenge for interaction design that goes far beyond the work-oriented design of computer artifacts, also with regard to the design process.

In the epilogue to my *Work-Oriented Design of Computer Artifacts* (Ehn, 1988) the future of participatory design was outlined as "back to Bauhaus and beyond postmodernism." This also became my route toward interaction design.

MOVE 2: DESIGN-ORIENTED STUDIES OF INFORMATION TECHNOLOGY IN CONTEXT

The next move for me on this route, however, was when in 1991 I took up the chair in Informatics (Information and Computer Science) at Lund University in Sweden with a program to push the discipline in the direction of design-oriented studies of information technology in context. It was clear from the outset that such a design orientation would go beyond the "technical" and focus on the "social and functional," on participation and usability. But as a design discipline, it would also have to deal with "experience and esthetics," and the general design orientation was toward "quality-in-use" (Ehn, 1995).

A general framework for such contextual design inquiries in the use of computational artifacts was outlined with reference to the social philosopher Jürgen Habermas and the concepts of "communicative actions" and "knowledge interests" (Habermas, 1968, 1985). As designers we can be said to have relations to three "worlds": the objective, the social, and the subjective. The languages of these worlds are very different. The objective world has to do with rationalistic design: Quality here is a question of prediction and control. The social world concerns understanding, interpretation, and communication: Quality becomes ultimately a question of ethics. In the subjective world, we deal with emotional experiences and creativity: Quality is a question of esthetics. We relate to these worlds and their language both in design as product (*artifact-in-use*) and process (*design process*).

Artifacts-in-Use

Looking at artifacts-in-use, what was needed was a way to address significant aspects of the control, ethics, and esthetics of computational artifacts. This was, by the way, the approach to "design" taken by the architect Vitruvius about 2,000 years ago, and when architects assess buildings in terms of their *structure, function,* and *form,* this goes back to his *de Architectura* in which he divided the study of buildings into *firmitas, utilitas,* and *venustas* (firmness, commodity, and delight). These are exactly ways to assess the objective, social, and subjective quality of artifacts. However, the ability of informatics and other computer science-oriented disciplines to judge the quality of artifacts has historically focused on the "objective" structural or technical aspects. Taken alone, no matter how well they are understood, these aspects say very little about quality-in-use. To understand the quality-in-use of computational artifacts, we also have to be concerned with the contextual aspects of function and form. To that end we now have well-elaborated design perspectives with which to judge the "social" functional aspects of an artifact. Finally, when it comes to the "subjective" experience of computational artifacts, we were just beginning to form an esthetic perspective. Without such a perspective, the ability to judge the quality of a computational artifact was severely hampered (Ehn & Löwgren, 1997; see Table 10.1).

Design Process

Shifting from product to process, it is interesting to notice that not until the 16th century did the term *design* emerge in European languages. This coincided with the need to describe the process of design and the profession of designing. The term indicated especially that designing was separated from doing (Cooley, 1988). In modern times the design process has been studied as an academic field since the early 1960s. The field has been dominated by architectural and industrial design.

TABLE 10.1

Artifacts-in-use

'World'	Objective	Social	Subjective
Artifact aspect	Structure • Hardware and software • Other materials • Material or medium	Function • Practical use • Symbolic use	Form • Experience of use
Quality perspective and language	Control • Software metrics • Quality standards	Ethics • Usefulness • Utility • Power • Interests • Values	Aesthetic • Appropriateness • Style • Balance • Resemblance
Vitruvius	Firmitas (firmness)	Utilitas (commodity)	Venustas (delight)

Note. Artifacts in use can be analyzed according to their (a) physical and functional aspects, (b) qualitative and linguistic aspects, (c) the Vitruvian triad, and (d) overall usability and style.

The development of design approaches can be described in three generations corresponding to each of the three design worlds (Cross, 1984; Ehn, 1995). The "first generation" design approach focused on *engineering*. It addressed the "objective world" and the answer had to do with control—with the correct representation and manipulation of objects, facts, and data. The second one focused on *participation*. It addressed our "social world" and the answer had to do with ethics—with democracy and appropriate social interaction. The third one focused on *design ability*. It addressed our "subjective world" and may be described as having to do with esthetics—with the expressive and creative competence of designers and users. In retrospect, the design approaches seem complementary rather than mutually exclusive (see Table 10.2).

Three "Worlds" of Design and Seven Questions

This left us with six related questions about how to design computational artifacts and a seventh holistic question that has to do with our ability to relate these questions to each other in a proper way in the practice of designing computational artifacts for quality-in-use (see Table 10.3).

Struggling to come up with useful practices to address these questions in education and research, informatics as a discipline has gotten a stronger design orientation. It is especially interesting to note that there seems to be a

TABLE 10.2

Design Process

'World'	Objective	Social	Subjective
Design process aspect	Engineering • Manipulation of objects • Formalisation • Refinement of specifications • Logic	Participation • Communication • Learning • Politics	Design ability • Creativity of environments • Emotional expressions • Artistic expressions
Quality perspective/ language	Control • Correctness of descriptions • Predictability	Ethics • Democracy • Appropriate social interaction	Aesthetics • Creativity • Innovation
"Usability" design style?	Appropriateness: A "proper balance" between engineering, participation, and design ability		

Note. The design process can be analyzed according to its (a) process aspects, (b) qualitative and linguistic aspects, and (c) overall usability and style.

TABLE 10.3

Three 'Worlds' of Design and Seven Questions

'World' Aspect	Objective	Social	Subjective
Artifact-in-use	Structure How do we make sure that an artifact is made of the right materials?	Function How do we make sure that an artifact is useful in its context?	Form How do we make sure that an artifact supports appropriate experiences?
Design process	Engineering How do we control the technical development of an artifact?	Participation How do we support appropriate interaction in the design process?	Design ability How do we support creativity in the design process?
Quality-in-use	Appropriateness: How do we find a proper balance among our responses to the questions above in our design practice?		

Note. The three "worlds" of design, analyzed according to function, process, and quality in use, raise seven questions.

direct connection, worth investigating, between the focus on user experience and artifacts-in-use (as opposed to objects in isolation) and the current interest in embodied interaction and interaction design in socially inhabited places (as opposed to abstract spaces). However, the discipline had not really become a design culture and a proper home for interaction design, even if it had developed a design philosophy and had also integrated HCI with participatory design. Esthetic practices were still poorly developed and integrated at that time. To play along it seemed more useful to turn back to the original design tradition and the meeting of art and technology, and to make a move toward a "digital Bauhaus."

MOVE 3: A MANIFESTO FOR A DIGITAL BAUHAUS

In the history of the modern society, several grand projects have been launched in attempts to unite the two sides of the Enlightenment: the hard (technology and natural sciences) with the soft (values, democracy, art, and ethics). One remarkable such project was the Bauhaus and the birth of modern design in the 1920s. It was a great project with a background in the radical and revolutionary movements of that time in Europe. The Bauhaus designers were collective designers and their design manifestos envisaged a new unit of art and technology in the service of the people. However, like all utopias, the Bauhaus was also full of contradictions. Transformed into modernism and functionalism, it produced rational living contexts of regular geometric shapes far from the dreams of the people that had to occupy them.

Today, in a digital age, we can witness new attempts to make creative and socially useful meetings between "art" and "technology"—an emerging "third culture," as C. P. Snow's *The Two Cultures and the Scientific Revolution* (1959) called the overcoming of the division between the arts and the sciences.

In 1997 I was offered the opportunity to participate in such a "third-culture" Bauhaus project, building a new university and a school of design, focusing on the meeting between art, media, and information technology. As director of research for the School of Arts and Communication at Malmö University (K3) I was given the opportunity to participate in trying to create a proper environment for the emerging discipline of interaction design. This became my third move in the direction of interaction design:

> What is needed in design and use of the most post-modern media and technologies—the information and communication technology—is not a modernism caught in a solidified objectivity in the design of modern objects in steel, glass and concrete, but a comprehensive sensuality in the design of meaningful interactive and virtual stories and environments.

What is needed is not the modern praise of new technology, but a critical and creative aesthetic-technical production orientation that unites modern information and communication technology with design, art, culture and society, and at the same time places the development of the new mediating technologies in their real everyday context of changes in lifestyle, work and leisure.

What is needed in the development of the aesthetics of the information and communication technology society is a Scandinavian design that unites a democratic perspective emphasizing open dialogue and active user participation, with the development of edifying cultural experiences and the production of useful, interesting, functional and maybe even beautiful and amusing every day things for ordinary people.

What is needed is humanistic and user-oriented education and research that will develop both a critical stance to information and communication technology, and at the same time competence to design, compose, and tell stories using the new mediating technologies.

This bold quotation comes from the founding vision document for the school, *Manifesto for a Digital Bauhaus* (Ehn, 1998, p. 210). At this neo-modernist "Digital Bauhaus," bachelors programs in material and virtual design, interaction design, media and communication studies, and performing arts technology, masters programs in interaction design, imagineering (art and technology) and for creative producers, as well as a doctorate program in interaction design—are all combined with studio-based interdisciplinary and interart research.

At that time, pioneering schools and institutes in the field of interaction design included Stanford University's Computer Science department, the Computer Related Design (now Interaction Design) department at London's Royal College of Art, the Domus Academy in Milan, and the Interactive Telecommunication program at Tisch School of the Arts. Today several universities and institutes also have masters programs in interaction design, and at least the Carnegie Mellon University School of Design and the Malmö University School of Arts and Communication (K3) have doctorate programs in the discipline. MIT Media Lab is one of the major research contributors to the field, but newcomers like the Interactive Institute in Sweden and not least Interaction Design Institute Ivrea illustrate that the field is rapidly growing.

Educating the Interaction Designer

The explicit but ambivalent neo-modernist references to the Bauhaus express fundamental challenges to, and contradictions in, the field of the design of interactive computational artifacts and to interaction design as a new discipline.

Can and should interaction design studies be shaped into an academic subject? What is the relation between the professional and the scientific? Is information technology in context a technical, a social, or even an artistic field of study? And what about ethics and esthetics if we try to cross borders between engineering and social sciences as well as between art and technology?

Let me use our graduate program in interaction design at K3 in a provisional attempt to deal with these questions in practice. The program has an orientation toward a design doctorate rather than a traditional PhD degree. Examples indicating this orientation are as follows:

- The program in general is oriented toward coaching, learning by doing, and reflection-on-action.
- Focus is on a design-oriented synthesis of constructive action and critical reflection, and on a synthesis of art and technology.
- Students have a commitment to design studies but come from very varied backgrounds; for instance, computer science, engineering, architecture, product design, interaction design, set design, music, visual art, and literature.
- Research work is carried out in a production-oriented, studio-based, interdisciplinary and interart environment.
- The thesis may take the form of a portfolio of work and a reflective summary.
- Form is allowed to follow content, hence the form of the thesis may be an interactive multimedia production.

Such a design orientation of the discipline challenges the traditional role of theories in academic life as explicit, abstract, universal, and context-independent carriers of knowledge. Instead it draws attention to the practice of knowing, to politics, sensuousness, embodiment, and particularity.

The kind of support expected from "design theory" is not so much in terms of "scientific theories" for predicting the results of an activity independent of context and situation, but rather support for reflections about the conditions for changed human activity. Such "design theories" are instead practical instruments to support designers as reflective practitioners (Löwgren & Stolterman, 2004; Schön, 1987) in improving their competence or design ability to make ethical and esthetic judgments appropriate to their context.

The epistemological basis for such an orientation of graduate interaction design studies may be found in rethinking the intellectual virtues Aristotle (1985) named *techne* and *phronesis* (see, e.g., Macintyre, 1981).

Techne may well be a cornerstone in building a firm platform for a teachable doctrine of interaction design. In *techne*, art and technology are not yet separated. *Techne* does not separate the methods and theories of science and technology from the creativity and freedom of art, but focuses on pragmatic, concrete, context-dependent, means-end, knowledge-oriented toward production. In practice, studies are carried out as shared projects in the interdisciplinary, studio-based research environment, focusing on learning by doing, coaching rather than teaching, and a dialog of reciprocal reflection-in-action in what Donald Schön (1987) called a "reflective practicum." Students learn to reflect on their own theories-in-action in the presence of patterns of phenomena of practice (theories on action) to build up their own repertoire of paradigmatic exemplars.

Because interaction design is both a scientific technological practice and a design-oriented artistic practice, the relation between text and artifact in the thesis work becomes potentially contradictory. Is an artifact new knowledge? Are texts the ultimate form for thesis arguments? That artifacts can illustrate textual arguments is not the question, but what are the demands on artifacts if they are to be valuable arguments in their own right? This is an essential question for higher education in interaction design, addressed in Durling and Friedman's *Doctoral Education in Design: Foundations for the Future* (2000), for example.

Another related question has to do with the lack of canonical texts and exemplary artifacts. As I discussed in my introduction, interaction design is a very young and highly interdisciplinary design field, with contributions from not only several traditional design fields and computer science, including HCI, but also media studies and performing arts. Beyond introductory readings like Winograd's (1996) *Bringing Design to Software*, mentioned earlier, it is hard to point to an agreed body of canonical texts to which all interaction design students should develop a stance. In a similar way, as opposed for example to the situation in architecture, it is not obvious which exemplary artifacts to point to in interaction design. There certainly exist well-known interactive artifacts, but there is not really a discourse around this repertoire of exemplars.

The booklet *Searching Voices: Towards a Canon for Interaction Design* (Ehn & Löwgren, 2003) addresses this lack of canonical texts. In our doctorate program, we asked the students to write essays about interaction design focusing on texts from their native fields in set design, architecture, engineering, literature, media studies, and so on. This kind of eclectic exchange with other fields is one small step in finding a theoretical home for the discipline. Texts like *The Language of New Media* (Manovich, 2001), *Hertzian Tales: Electronic Products, Aesthetic Experience and Critical Design* (Dunne, 1999), and *Where the Action Is: The Foundations of Embodied Interaction* (Dourish, 2001), and more general texts by Bruno Latour and Donna Haraway, are becoming central points of reference as well as contested terrain in their interaction design discourse.

Phronesis

In the Aristotelian virtue of *phronesis*, wisdom and artistry as well as art and politics are one. *Phronesis* concerns the competence to know how to exercise judgment in particular cases. It is oriented toward the analysis of values and interests in practice, based on a practical-value rationality that is pragmatic and context-dependent. *Phronesis* is experience-based ethics oriented toward action. Hence, it is fundamentally concerned with neither statements of fact nor prescriptions of what ought to be, but speculative propositions enacted as "anxious acts of political love"—an expression I have borrowed from J. M. Bernstein's *The Fate of Art* (1992).

Students are encouraged to focus on their own hidden politics-in-practice rather than on espoused design philosophy (sociotechnical methods, human relations theory, or participatory design procedures, for example). No one is seen as a naïve neutral technician, independent free artist, or simple manipulator in the service of power, but as a designer with a humanistic stance recognizing the collective and political character of the design process in real cases and facing questions like: How do they "get things done their way?" What tactics and strategies are enacted? How are interventions legitimized technically, ethically, and esthetically (Ehn & Badham, 2002)?

However, the virtue of *phronesis* in interaction design studies is not necessarily a question about user-centered interaction design. In the current hybrid networks of mind and matter, the most important purpose might be to contribute not yet another useful modern digital product but critical interpretations taking the form of tangible design proposals. This is the approach to interaction design taken by Anthony Dunne and Fiona Raby in *Design Noir* (2001). For example, by investigating the secret life of electronic products they hope to stimulate debate about the dominant perspective in pervasive or ubiquitous computing where the consumer or the user is the hero who needs to do everything as fast and as easily as possible. Instead, they want to provoke reflections about "design noir," where the user or customer is a kind of antihero as in *film noir*, where things do not always work out happily. Hence, design noir claims not to solve human needs but to suggest dilemmas, conflicts, and ambivalence and to provide narratives where these darker feelings are expressed, explored, and acted out. Design noir is not glamorous, with great utopias and modern heroes as in the Bauhaus, but it still has a humanistic stance and a consciousness about political dilemmas that can move us beyond modern design, and challenge both the traditional Bauhäusler and the postmodern designer as the interaction designer of tomorrow.

WHERE INTERACTION DESIGN TAKES PLACE: THE NEXT MOVE?

So what could be a reasonable next move, a move that continues the outlined participatory and collective rule in a creative and sensible way? In the beginning of this chapter I positioned interaction design in the convergence of digital artifacts with physical space and the objects surrounding us (ubiquitous computing) as well as the convergence between different media. I also tried to position interaction design in relation to what Dourish (2001) has called *embodied interaction*, including the observation that artifacts are getting more and more embedded in social practice, community, place, and situatedness, beyond the disembodied human–computer interface.

In Europe such an understanding of places for interaction design was central to many research projects within the "Disappearing Computer" research program, including projects like ATELIER in which I had the privilege of participating (Binder et al., in press). So my final move in this chapter is in the direction inspired by these projects.

Interaction design cannot be limited to artifacts like handheld or desktop appliances. In spatial arrangements, scale is an important aspect. However, space is inherently a physical concept as opposed to place that cannot be thought of without including social activity. With the perspective of embodied interaction, both the social dimension and our bodily experiences come into focus. *Place* reflects the emergence of practice as the shared experience of people in space and over time. The interaction design challenge is not to design space, but to design technology and architecture for the *appropriation* of space for the activities that take place among a particular set of people embodying that place.

One might say that the approach to skill-based PD inspired by Wittgensteinian language-games, which I outlined at the beginning of this chapter as a first move, was a reflection of the "pragmatic turn" in philosophy making its way to design and computers. Design became a question of intertwined language-games between designers and users, of interacting communities of practice. Now we are in the midst of a general complementary "move" that has been called a "spatial turn" in social and cultural theory (see, for instance, Crang & Thrift, 2000). Again, the philosophy of language-games developed by Wittgenstein is an interesting approach to reflecting on how our social and tangible practices take place.

As the first move toward PD, it was suggested that users and designers may be seen as fundamentally related via shared experiences in a common design language-game which has a family resemblance to the ordinary language-games of both users and professional designers. A fundamental competence of the designer became the ability to set the stage and make props for this shared design language-game that makes sense to all participants, allowing interaction and mediation between different language-games.

However, with the task of designing for the appropriation of artifacts to take place in practice, there is a shift in focus toward the kind of language-game which Anders Hedman (2004) has called "place-making games." Studying an exhibition with mixed digital and physical objects designed by the researchers within SHAPE, one of the Disappearing Computer projects, he observed how visitors shifted between different games during a single visit. Moreover, the kind of place games that occurred constituted an open-ended set of activities. The designer's task moves toward being one of dealing with conditions for these place-making language-games.

In the ATELIER project, focusing on places for education in interaction design and architecture, this challenge, to design for the place-making processes of unfolding social negotiations and the appropriation of space, was met by the concept of *configurability* of architecture and artifacts. Hence our wish to support the design students in organizing space and artifacts into assemblies according to the situation at hand, playing with foreground and background, juxtaposition of narrative connections between objects, or improvizational movements between private and public. One instance of this is how design students decomposed a mixed-object "tangible project archive," built earlier in their studio, to support semipublic place-making games at a railway station. They configured for place-making games and appropriation around refrigerator poetry, reusing exactly the same physical building blocks and technological components as in the tangible archive in the studio (see Figs. 10.2, 10.3).

Hence, the designer's role shifted toward regarding whatever space is available as a potential place for meaningful interactions and place-making games, and architecture and technology as ways to support people in appropriating that space. The challenge we face in interaction design is not necessarily to better support user participation in design language-games, but the more demanding one of designing configurations that can make sense to people in the multiple intertwined place-making language-games that they make up as they go along.

FIG. 10.2. A "tangibile project archive." Improvising in response to the social and built situation, design students configured their objective archive, previously assembled in their studio (see Fig. 16.3), in a railway station. This research project, Atelier, focused on place-making for education in interaction design and architecture.

FIG. 10.3. Place-making games around refrigerator poetry. Another view of the Atelier railway station installation.

How such design for place-making in use may be supported by pattern languages based on ethnographic analysis is another of the really interesting moves to explore as we play along in the making of interaction design (Alexander et al., 1977; Crabtree, Hemmings, & Rodden, 2002).

Part

V

CONVERSATION

From Conversation to Interaction Via Behavioral Communication. For a Semiotic Design of Objects, Environments, and Behaviors

Cristiano Castelfranchi
Institute of Cognitive Sciences and Technologies–CNR[1]

Interaction design implies communication design; not only because communication is a form of interaction, but because interaction is frequently based on communication. However, while speaking of communication, we immediately have in mind linguistic exchange (in particular, conversation) or other forms of explicit and conventional signs and messages.

The message of this work, on the contrary, is the following one:

1. Conversation in strict sense is not the right model for interaction; it is misleading: too intention based and too cooperative.
2. The focus of attention should not only be on natural language, special languages, and communication protocols, but also on (a) communica-

[1]OMLL-ESF Project. I'm grateful to Anna Galli and to Francesca Giardini for their contribution to the exploration of this theory. Thanks to Luca Tummolini, Emiliano Lorini, other students of the Siena PhD in Cognitive Science, and to my colleagues at ISTC for precious comments and discussions. This paper is part of a long term research activity; see Castelfranchi C. (2001–2005, in progress), When Doing is Saying. The theory of Behavioral Implicit Communication. Retrieved from http://www.istc.cnr.it/search.php?searchauthors=castelfranchi&searchtitle=&searchtext=type0=other&lang=&hits=10&offset=0&submit=Search

tion by action without sending special messages, and (b) communication by leaving traces in a shared environment.

In this perspective I introduce the theory of silent and physical communication, which has a fundamental role in any human interaction and in any object mediation, and which will play a crucial role in future technologies for domotic (home automation), human–robot, robot–robot, human–computer interaction.

AGAINST THE LINGUISTIC COMMUNICATION PARADIGM

Conversation and dialogue are very important, and typically human, forms of communication, and they must be understood in their principles (we discuss cooperativeness and the fact that the participants build a common plan [Parisi & Castelfranchi, 1981]). However, they are also very peculiar forms of interaction and of communication. They provide a misleading model of both forms.

First of all, they are highly cooperative activities that shouldn't be generalized to "interaction" (covering also competition, interference, and conflicts; Castelfranchi, 1998) or to communication. In current use of human language there are two goals:

1. The speaker's goal: X's behavior has the goal or function that Y recognizes the act, understands the meaning, comes to believe the fact p.
2. The interpreter's goal: Y has the goal of interpreting X's (speech) act in order to give it a meaning, to understand what X means.

Thus they have a common goal (Conte & Castelfranchi, 1995) which can also presuppose mutual knowledge:[2] The goal of "Y understanding what X means by …." This is a common goal on which X and Y usually cooperate for a successful communication (Meijers, 2002). The communication has been successful only when X and Y achieve this goal: that Y understands precisely what X intends to communicate to him (Parisi & Castelfranchi, 1981). This is also because what characterizes linguistic communication is a metamessage "I'm communicating; I intend that you …" (Grice, 1957).

Moreover, linguistic communication is characterized by special cooperative principles: altruism in knowledge sharing. These are the famous Grice (1957) principles that I prefer to present as default rules of the speaker and of the hearer:

1. The speaker's rule: *Provided that you do not have specific motives and reasons for deceiving, by default, tell the speaker the truth, relevant for her or him.*

---

[2]X and Y also cooperate in building and maintaining a "common ground" (Clark & Shaefer, 1989).

2. The hearer's rule: *Provided that you do not have specific motives and reasons for being suspicious of him or her, by default, believe what the speaker is saying.*

In conversation—which is a coordinated exchange of speech acts—X and Y cooperate in a stronger way: They are building a common goal structure. The act of responding, in fact, consists of adopting some of the goals of the act of the speaker. Let's give three examples of different possible responses in conversation (Poggi, Castelfranchi, & Parisi, 1981).

1. In a *direct answer*, the goal of Y's speech act is to satisfy directly the main and explicit goal of X's question. Suppose the question is "What time is it?" A direct answer is "It is five o'clock," and in this case there is a common goal motivating both plans: That X knows what time it is.

2. In *overanswering*—and in general in what we call overhelp in collaborative relationships—the goal of Y's speech act is not to directly respond to the question, but to satisfy the higher goal motivating X's question. For example, X is anxious and would like to know whether it is late or not. The common goal, adopted by Y, is that X be quiet: It is not late. Y is not literally answering X's question; Y is helping X to do something different from what has been requested (Figure 11.1). Overhelp is a very important aspect of human collaboration: Y is really helpful because he does not limit himself to doing as requested, but he does more than is expected or even something different. He is able to go beyond our literal request to satisfy our needs or interests. This is true and intelligent help. In fact our request might be wrong for our own needs, or (currently) impossible, or not the best solution, and so forth, while Y provides

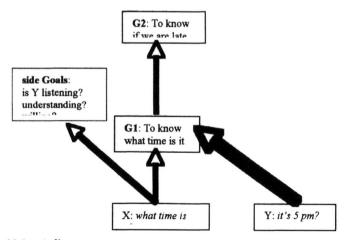

FIG. 11.1. A direct answer.

us a contextual and intelligent solution to our problems (Falcone et al., 2000). How much interactive technologies—on the contrary—are able to anticipate our needs and to over-help us?

3. In the *side answer,* examples of side answers to the question such as "What time is it?" might be "What?," "I don't know," "I do not speak English," and so on. Any question (and any speech act), in fact, presupposes that the other understands, attends to our utterance, is able to answer, and so forth, and there is a side or control goal in X's plan to check, to know whether this is, in fact, true. In side answers Y adopts just one of these sides; control goals of X: to know whether Y is listening and understanding what X is saying (in our example Y listens to but did not understand X).

Usually, in conversation there is goal adoption and thus some common goal on which the agents cooperate with their linguistic actions. However, to be true, the cooperative perspective and the conversation prototype are misleading, not only for a correct view of communication in general, but even for a complete view of linguistic communication, which is not always cooperative (as assumed, e.g., by Clark & Schaefer, 1989; Clark, 1996). Consider, for example, exchanges of reciprocal insults, where it is not necessarily a goal of Y to listen to, to understand, to dialogue with, to maintain a common ground with X. Cooperation—if any—is just accidental. (Castelfranchi, 1992). Analogously, a normal use of linguistic communication, but noncooperative at the Gricean layer, is deception, where we have (a) hidden intentions and manipulation[3] of the other; and (b) violation of the cooperative principles, or better exception to the default rule.

Communication is just a kind of social action—not necessarily bilateral, not necessarily social interaction, but necessarily interactive, because its outcome, its success, depends on another agent's activity.

I do not discuss here those general notions for answering the questions "What is Communication?" And what about its relationships with interaction? I only answer these questions indirectly—and provide a different idea of communication—while exploring the theory of the most basic form of communication: *behavioral communication,* when simple behavior is or is not communication. This is very important—I believe—for the theory of meaningful interaction, and thus also for interaction design.

I discuss coordination and interaction without communication at all, the fundamental notion of "signification" (the very basic role of the receiver), the crucial theory of *behavioral implicit communication* (BIC), how BIC is important in social interaction, and the role of the physical environment in it (*stigmergy*).

---

[3]By "manipulation" I do not mean any form of influencing and persuading, but only the intention of influencing the other hiding the intention of influencing her.

I will explain why this will be very important in designing domotic interaction, and human–agent and human–robot interaction.

BEHAVIORAL IMPLICIT COMMUNICATION THEORY

Usual, practical, even nonsocial behaviors can be used contextually as messages for communicating. Behavior can be communication without any modification or any additional signal or mark. We will call this form of communication without specialized symbols *behavioral implicit communication* (BIC). Behavioral, because it is just simple noncodified behavior; implicit, because—not being specialized and codified—its communicative character is unmarked, undisclosed, not manifest, and thus deniable. In BIC, communication is just a use and, at most, a destination, not the shaping function (Conte & Castelfranchi, 1995). Normally, communication actions are, on the contrary, special and specialized behaviors (like speech acts, gestures, signals, etc.).

BIC is a very important notion, never clearly focused, and very frequently mixed up with other forms of communication (typically the so called nonverbal, expressive, extralinguistic, or visual communication). It has been source of a number of misunderstandings and bad definitions.

This ill-treated notion is crucial for the whole theory of social behavior: coordination, control, social order creation, norms keeping, identity and membership recognition, social conventions building, cultural transmission, deception, and so forth. A lot of social control, collaboration monitoring, and coordination are, in fact, based on this form of communication and not on special and explicit messages (communication protocols).

Even for the theory of linguistic communication, BIC theory is fundamental:

- It is the basis of several pragmatic inferences ("What does X mean by saying this?")
- It is the origin of the Gricean metacommunicative character of linguistic communication (discussed earlier).
- It is the basis of meaning and linguistic conventions negotiation, and so forth.
- Moreover, this form of emerging and spontaneous communication is one of the forerunners and a premise for the evolution and acquisition of language.

Against Watzlawick: Are We Damned to Communicate?

A famous thesis of the Palo Alto psychotherapy school was that in the social domain "It is impossible not to communicate, ... any behavior is communication" (Watzlawick et al., 1967). In this view, a non-communicative behavior is nonsense.

This claim is too strong. It gives us a notion of communication that is useless because is nondiscriminative. Is simple understanding already communication? Is it possible to clarify when behavior is communication and when it is not?

In order to have communication, having a recipient who attributes some meaning to a certain sign is a nonsufficient condition. We cannot consider as communication any information or sign arriving from X to Y unless it is aimed at informing Y. A *teleological* (intentional or functional) sending action by the source is needed. The source has to perform its behavior with the aim that the other agent interprets it in a certain way, receives the message and its meaning. Is, for example, an escaping prey "communicating" to its predator or enemy its position and movement? Watzlavick's overgeneralization cannot avoid considering as communication to the enemy the fact that a predator can observe the movement of the prey. Although this information is certainly very relevant and informative for the enemy or predator, it is not communication. Receiving the information is functional (adaptive) for the predator and for that species which has developed such ability, but it is not functional at all; it is not adaptive for the prey. Thus, sending that sign is not a functional (evolutionary) goal of the prey; that is what matters for having communication.

Analogously, is a pilferer informing or communicating to the guard his presence and moves? The pilferer does not notice that there is a working TV camera surveillance system and thus he does not know that there is a guard who is following him on a screen! Or when a pilferer, while escaping from the police, is leaving prints and traces of his direction on the ground, are those signs (very meaningful for the police) messages to them?

We should not confuse mere *signification* with *communication*. Following Eco (1976) prints on the ground are signs for the hunter of the passage of a deer; smoke is the sign of a fire; some spots can mean "It is raining" (they are, for Y, signs of the fact that it is raining). We have here the simple processes of signification.

Notice that meanings are not conventional but are simply based on natural perceptual experience and inference. Notice also that the signal, the vehicle, has not been manufactured on purpose for conveying this meaning; it doesn't need to be encoded and decoded via some conventional artificial rule.

The Goal of Communicating: Functional Versus Intentional Communication

The crucial component for the notion of communication is the teleonomic nature of the act of sending the message. Should we ascribe intentions and mental states to any animal (like insects) for accounting for animal communication? Or should we renounce to a general notion of communication?

We want to have a general notion of and to coherently use the notion of animal communication. In order to do this we need two kinds of finalism of teleology. In fact, we distinguish between goal-governed and goal-oriented agents and between *intentions* and *functions*.

There are two kinds of goal-oriented systems and behaviors: (a) the cognitive, intentional ones (goal-governed), and (b) the merely goal-oriented ones, without any internal anticipatory representation of the goal of the action, where the teleonomic character of the behavior is merely in its adaptive function.

The teleonomic notion we need has two different meanings: (a) either the message is sent on purpose, intentionally, by X, which should be a cognitive purposive system, an intentional agent (in this case, X believes and intends the result—discuss later), or (b) the message is not intentional but simply functional. The sending behavior is not deliberated but is merely goal-oriented behavior, either designed by some engineer, selected by natural or artificial evolution, or selected by some learning.

Thus, we have two basic kinds of communication: (a) intentional (or goal-governed) communication, and (b) functional (or goal-oriented) communication.

Functional communication has several sub-types: by evolution–selection, by design, or by reinforcement learning based on the effects.

Fully intentional BIC presupposes an intentional stance and, more precisely, a theory of mind in the interpreter, since the message brought by the action is about the mind of the source: his intention, or emotion, or motives, or assumptions, and so forth.

The definition of BIC at the intentional level (in this volume we analyze intentional BIC) is as follows:

> In BIC, the agent (source) is performing a usual practical action a, but he also knows and lets or makes the other agent (addressee) observe and understand such a behavior a; that is, to capture some meaning m from that message, because this is part of his (motivating or non motivating) goals in performing a.

In summary, BIC is a practical action primarily aimed to reach a practical goal, which is also aimed at achieving a communicative goal, without any predetermined (conventional or innate) specialized meaning.

Why BIC is not Nonverbal, Extralinguistic Communication

BIC is not the same as and has little to do with the so-called "nonverbal" or "extralinguistic" communication (NVC), although NVC is communication through some behavior or behavioral features, and BIC is certainly nonverbal and extralinguistic. The few of BIC that have been identified have actually been confused with the never well-defined notion of nonverbal behavior (e.g., Porter, 1969).

Nonverbal and extralinguistic communication refers to specific and specialized communication systems and codes based on facial expressions and postures, specific gestures, supersegmental features of voice (intonation, pitch, etc.), and so forth, that communicate specific meanings by specialized, recognizable signals (either conventional, such as a policeman regulating traffic; or universal, such as emotional signals). BIC, on the contrary, is not a language. Any verbal (or nonverbal) language has some sort of lexicon, that is, a list of (learned or inborn) perceptual patterns specialized as signs (Givens, 2003), where "specialized" means either conventional and learned as sign, or built in, designed just for such a purpose (function) by natural selection or engineering. BIC does not require specific learning or training, or transmission; it simply exploits perceptual patterns of usual behavior and their recognition. BIC is an observation-based, non-special-message-based, unconventional communication, exploiting simple side effects of acts and the natural disposition of agents to observe and interpret the behavior of the interfering others. BIC gestures are just gestures; they are not symbolic but practical: to drink, to walk, to scratch oneself, to chew. They represent and mean themselves and what is unconventionally inferable from them (like the agent's intentions and beliefs).

The Stigmergic Overgeneralization

The notion of stigmergy comes from biological studies on social insects, and, more precisely, the term has been introduced to characterize how termites (unintentionally) coordinate themselves in the reconstruction of their nest, without sending direct messages to each other. Stigmergy essentially is the production of a certain behaviour in agents as a consequence of the effects produced in the local environment by previous behaviour (Beckers, Holland, & Deneuborg, 1996).

This characterization of stigmergy is not able to discriminate between simple signification and true communication,[4] or between prosocial and antisocial behavior. It would, for example, cover prey—predator coordination and a pilfer (unintentionally) leaving very precious footprints for the police. In order to have communication, it is not enough that an agent coordinates its behavior with the behavior the traces of the behavior of another agent. Also, this is an overgeneralization. Stigmergy is defined as indirect

---

[4]For example "… Coordination of the agents' movements is achieved through stigmergy. The principle, initially developed for the description of termite building behavior, allows indirect communication between agents through sensing and modification of the local environment which determines the agents' behaviour" (Beckers, et al., 1996, p. 181).

communication through the environment.[5] To us, such a definition is rather weak and unprincipled.

Here, clearly, the authors, while attempting to generalize the stigmergic (trace-based) notion arrive rather close to the BIC concept. But this is the wrong way: from insect stigmergy to broader low level forms of behavioral communication. What we need is a principled and general theory of behavioral communication where stigmergy represents a specific case, not the other way around.

First, a lot of usual communication and even linguistic messages are directed toward unknown or unspecified addressees. Second, and more important, isn't an utterance propagated through the environment as energy? Isn't a letter or a book a physical environmental sign? Any kind of communication exploits some environmental channel and some physical outcome of the act; any communication is through the environment! This cannot be the right distinction.

The real difference is that in stigmergic communication we do not have specialized communicative actions, specialized messages (that unambiguously would be direct messages because they would be just messages); we just have practical behaviors (like nest building actions) and objects that are also endowed with communicative functions. In this sense, communication is not direct (special communicative acts or objects) but is via the environment (i.e., via actions aimed at a physical and practical transformation of the environment).

From our perspective, stigmergy is communication via long term traces, physical practical outcomes, and useful environment modifications, not mere signals. Moreover, in insects (and in some simple artificial agents) stigmergy is a functional form of behavioral communication, where the communicative end cannot be represented in the agent's mind (intention), but it is a functional effect selected by evolution or built in by a designer. This means that stigmergy is a sort of innate and trace-based behavioral communication (discussed later) and, in our point of view, is very effective, especially for subcognitive agents.

Stigmergy is just a subcase of BIC, since, in fact, any BIC is based on the perception of an action that necessarily means the perception of some trace of that action in the environment (e.g., air vibrations). We restrict stigmergy to a special form of BIC where the addressee does not perceive the behavior (during its performance) but perceives other post hoc traces and outcomes of it. To be true, perceiving behavior always means perceiving traces and environmental modifications due to it; the distinction is just a matter of per-

---

[5]Holland and Beckers generalize this notion and distinguish between two forms: (a) cue-based stigmergy, where the agents coordinate with each other basically observing their behaviors and patterns (like in birds flocking), and (b) sign-based stigmergy, where the agents create and leave special physical signs in the environment (like in termites).

ception of time and of duration of the trace. The only difference is that when we refer to communicating via traces we have in mind more long term traces that persist also when the authors are no longer there: The receiver observes the trace, although he could not observe the authors performing the action. But for a trace-based communication to be stigmergy, it is necessary that the perceived object also be a practical one and that the originating action also be for practical purposes (like nest building).

Stigmergy is not only for insects, birds, or noncognitive agents. There are also very close examples in human behavior. In animals, stigmergy is nonintentional, but intentional forms of it are possible. Consider a sergeant who—while crossing a mined ground—says to his soldiers: "Walk on my prints!" From that very moment, any print is a mere consequence of a step, plus a stigmergic (descriptive "Here I put my foot" and prescriptive "Put your foot here!") message to the followers. Consider also the double function of guardrails: On the one side, they physically prevent cars from invading the other lane and physically constrain their way; on the other side they also communicate that "It is forbidden to go there" and also normatively prevent that behavior. This is what law theorists call materialization of the norm: The norm is hardwired, because either the external or the internal conditions for the agent's doing differently are excluded; there is no possible decision to violate it. In fact, the communicative function is a parasitic effect of the practical act and of its long-term physical products. It is just BIC and, specifically, stigmergy.

Intentional Behavioral Communication Step by Step

There are several steps in the evolution from mere practical behavior to BIC and to a conventional sign. Let's examine this transition.

1. Just behavior: An agent X is acting in a deserted world; no other agent or intelligent creature is there, nobody observes, understands or ascribes any meaning to this behavior α.[6] Neither signification nor—a fortiori—communication is there.

2. Signification: An agent X is acting by its own in a world, but there is another agent Y observing it which ascribes some meaning μ to this behavior α.

There is, in this case, signification (X's behavior has some meaning for Y, informs Y that p), but there is not necessarily communication.

By signification, we mean that the behavior of X is a sign of something, means something else, for Y. For example, p can be "X is moving," "X is eating," "X is going there."

---

[6]Although sometimes we use BIC and stigmergic messages with ourselves.

As we know, to have communication, the signification effect must be on purpose; but this presupposes that X is aware of it. Thus, in signification we have two possible circumstances: (a) X does not know (consider the pilferer example where he is not aware of being monitored), or (b) X's awareness: weak BIC.

Consider now that X knows he is being monitored by a guard, but that he does not care, because he knows that the guard cannot do anything at all.

Y's understanding here is among the known but unintended effects of X's behavior. Although, perhaps, being an anticipated result of the action, it is not intended by the agent. Not only can indifferent or negative expected results be nonmotivating, nonintended, but positive (goal-realizing) expected results can also be nonintended in the sense of not motivating the action, neither sufficient nor necessary for the action. In our example the pilferer might be happy and laughing about the guard being alerted and powerless and angry.

3. True or strong BIC: The fact that Y knows that p is comotivating the action of X. The behavior is both a practical action for pragmatic ends (breaking the door and entering, etc.) and a message. We call this *strong* or *true* behavioral communication, where the pragmatic behavior which maintains its motivation and functionality acquires an additional purpose: to let or make the other know or understand that p.

The important point for fully understanding BIC (and the difference with the following meta-BIC) is that we have here a fully intentional communication act, but without the aim (intention) that the other understands that X intends to communicate (by this act). The intention of communicating and communicating (this) intention are not one and the same. Given the well-consolidated (and fundamental) Grice-inspired view of linguistic communication—that frequently is generalized to the notion of communication itself—these two different things are usually confused, and it is difficult to disentangle them; but they are clearly different, both at the logical and at the practical level.

With a BIC message, X intends that the other recognizes her action, and perhaps recognizes and understands her practical intention motivating the action (eating, having the door closed, knowing what time is it, etc.), but X does not necessarily have (at this communicative stage) the intention that the other realizes her higher intention that Y understand this, that it is her intention to communicate something to Y through that practical action: I want Y to understand that I intend to go, but not that I intend that he understands that I intend to go.

4. Meta-BIC: In meta-BIC, there is a metacommunication, typical of higher forms of communication like language. A BIC metamessage is as

follows: "This is communication, this is a message, not just behavior; it is aimed at informing you."

Frequently, BIC has such a high-level (Grice's way) nature. For example, the act of giving or handing is not only a practical one, but is a meta-communicative act where X intends that Y understands that she is putting something closer to Y so that Y (understanding that she intends so) takes it.

5. Beyond BIC: Actions for communication only: The behavior α is intended and performed by X only for its meaning μ, only for making Y believe that p. There are no longer practical purposes. The act is usually performed either out of its practical context or in an incomplete and ineffective way.

6. Simulation: Notice that in the pilferer's scenario, that fact that the α has only a communicative goal means that it is a fake action! In fact, if α has no other goals apart from communicating to Y, Y will be deceived, and the information he will derive from observing α will be false (and a is precisely aimed at this result). It is just a bluff.

7. Ritualization: The practical effect becomes irrelevant: The behavior is ready for ritualization, especially if is not for deception but for explicit communication. Ritualization means that a can lose all its features that are no longer useful (that were pertinent for its pragmatic function) while preserving or emphasizing those features that are pertinent for its perception, recognition and signification. After ritualization, the behavior will obviously be a specialized communicative act, a specialized and artificial signal (generated by learning and conventions, or even selection). This is the ontogenetic and evolutionary origin of several gestures and expressive movements.

Given our previous discussion about cooperation in linguistic interaction, let us notice that in BIC there are two goals or functions meeting each other:

1. The communicator's goal: X's behavior has the goal or function that Y understands, recognizes, and comes to believe that p (and this holds from step 3).
2. The interpreter's goal: Y has the goal or function of interpreting X's behavior to give it a meaning (and this holds from step 2).

However, those goals in the initial forms of BIC are simply independent from one the other. Cooperation is just accidental. X and Y do not really have a common goal.

Since, in step 2 (signification), X does not know that Y wants to understand her behavior; while in step 3 (true or strong BIC) Y does not know that

X is communicating to him through behavior α. Thus, Y has not the goal of understanding what X means by α; that is the true common goal of higher forms of communication (like linguistic communication) in which X and Y usually cooperate for a successful communication.

In meta-BIC, on the contrary, Y knows that X is communicating. Therefore he has a special form of goal 2, the goal of understanding what X is communicating; the interpreter's goal is to understand what X intends to communicate, to understand the meaning in X's mind.

The agents arrive in such a way to cooperate in strict sense (like in linguistic exchange), and the two goals become complementary, convergent, and functional to each other; that is, X and Y have the same goal and they know the goal of each other.

UBIQUITOUS BIC

We are so used to BIC and it is such an implicit form of communication that we do not realize how ubiquitous it is in social life and how many different meanings it can convey. It is useful to give an idea of these uses and meanings—even risking being a bit anecdotal—first of all just for understanding the phenomenon, second, because several of these uses can be exploited in Human–Computer Interaction (HCI), in computer mediated collaboration, in agent–agent interaction.

BIC acts can convey quite different meanings and messages. Let's examine some of the most important of them for human social life (also applicable to agents and robots).

"I'm Able" or "I'm Willing." The most frequent message sent by a normal behavior is very obvious (inferentially very simple, given an intentional stance in the addressee) but incredibly relevant: (as you can see) I'm able to do, and/or I'm willing to do; since I actually did it (I'm doing it) and on purpose.

There are several different uses of this crucial BIC message.

Skills Demonstration in Learning, Examinations, and Tests. When Y is teaching something to X via examples and observes X's behavior or product to see whether X has learned or not, then X's performance is not only aimed at producing a given practical result but is (also or mainly) aimed at showing the acquired abilities to Y.

More generally, doing the same action of a model, imitating, is the basis for a possible tacit BIC message of X, for the possible use of the action α as a message to Y: "I'm doing the same". But for this, specific additional conditions are needed:

1. X performs α (imitates Y)

2. Y observes and recognizes 1, and forms the meaning μ "X is doing α, like I am"
3. X knows that 2
4. X intends that 2
5. X performs α also so that 2 (that is because of 4 & 3).

In this case α is a real (successful) message to Y.

When and why should X inform Y about imitating him? Especially when Y has the goal that 1.

Also, the behavior of the teacher is a BIC; its message is: "Look, this is how you should do it." Usually this is also joined with expressive faces and gestures (and with words) but this is not the message we are focusing on.

In general, if showing, displaying, and exhibiting are intentional acts, they are always communication acts.

Warnings Without Words.   This is a peculiar use of exhibition of power that deserves special attention: the Mafia's warning, monition. The act (e.g., burning, biting, destroying, killing) is a true act and the harm is a very real harm, but the real aim of this behavior (burning, killing, etc.) is communicative. It is aimed at intimidating, terrifying, via a specific meaning or threat: "I can do this again; I could do this to you; I'm powerful and ready to act; I can do even worse than this." This meaning—the promise implicit in the practical act—is what really matters and what induces the addressee (who is not necessarily already the victim) to give up. The practical act is a show down of power and intentions; a message to be understood. The message is "If you do not learn, if you will do this again, I will do even worse."

Nations do the same: Consider, for example, the repeated reactions of Sharon after terrorist attacks in Israel; it is not only a revenge (the military outcome is marginal), it is a message: "Do this again and I will do this (bombing) again," the same holds for terrorists' bombs. This is a horrible way of communicating. Perhaps it would be better to communicate via words and diplomacy.

Is all this expressive—nonverbal—communication? Bombing is bombing (not particularly "expressive"), and can be unintentional (by mistake and accident), or intentional just for destruction and/or for mere revenge or material prevention, but it can (also) be a message, possibly without any different features at all.

"I Did It," "I'm Doing It."   Another typical meaning of BIC is simply "I did it; I did so." This is very relevant in several human interactions where a given behavior of X is expected by Y. Consider, for example, a child showing the mother that he is eating a given food, or a psychiatric patient showing the nurse that he is drinking his drug.

This message is particularly important in the satisfaction of social commitments and obligations (discussed later), but it has other uses.

For example, for serial synchronization in coordination and collaboration, if the action of Y in a common plan presupposes the previous accomplishment of the act of X, and the coordinate is based on observation, then the act of X means: "done! It's your turn."

For example, when being invited to dinner, finishing the food and cleaning the plate means "I finished it, I liked it," as the guest wishes and expects.

"I conform; I agree." Imitation-BIC as Convention Establishment and Memetic Agreement. Imitation (i.e., repeating the observed behavior of Y—the model) has several possible BIC valences (we already saw one of them; Castelfranchi, 2003). The condition is that Y (the model) can observe (be informed about) the imitative behavior of X.

We can consider, at least, the following communicative goals:

1. In learning or teaching via imitation, X communicates to Y "I'm trying to do like you; check it: is it correct?"
2. In convention establishment and propagation, "I use the same behavior as you, I accept (and spread) it as convention; I conform to it."
3. In imitation as emulation and identification, "I'm trying to do like you; I want to be and to behave like you; you are my model, my ideal."
4. In imitation as membership, "I'm trying to do like you; I want to be and to behave like you; since I'm one of you; I want to be accepted by you; I accept and conform to your uses (see 2)."

Let me focus on the second BIC use of imitation, which is really important and probably the first form of memetic propagation through communication.

X interprets the fact that Y repeats its innovation as a confirmation of its validity (good solution) and as an agreement about doing so. Then X will expect that Y will understand again its behavior next time, and that Y will use it again and again, at least in the same context and interaction.

An Example: Linguistic Negotiation (Terms, Meaning, Rules, etc.)

Very rarely do we explicitly negotiate and discuss the new terms that we introduce, the use and meanings of words, the linguistic rule and conventions. We just try to understand and to be understood and to understand whether the other understood us. Linguistic conventions are just a particular case of social conventions of which we live. Our claim is that tacit agreement or consent (*Qui tacet consentire videtur*) is the way social conventions and informal norms emerge. Let's simplify this complex domain with a simple example from linguistic negotiation about the creation and establishment of a new name (as for social norms see the section [this chapter] on BIC of social order).

To name X, I use the new term *bbb* (for example, to call my friend Amedeo, I introduce the name Amed) with my hearer H:

1. My hearer understands (I infer this from her answer or reaction).
2. My hearer does not protest or discuss.

I interpret H's not protesting or discussing as an implicit acceptance (at least passively and for the moment) of my use; and—more than this—of an implicit behavioral communication of such an acceptance (in not reacting, H is communicating me "OK, I let you use this term").

This is some sort of weak implicit acceptance of my use of *bbb* by H. When I use *bbb* again with H, I will expect (believe and want) that

- H understands again.
- H will not protest or discuss.
- H knows about my expectations.

In strong implicit acceptance, H reuses the term *bbb* (in the same occasion or later). In doing so, H expects that:

- I understand.
- I do not protest or discuss.
- I know about these expectations.

There is now a true, implicit convention, a tacit agreement about using *bbb* (at least between us and in similar contexts). If somebody else hears us using *bbb*, or H uses *bbb* with other people, the new term is spread around and a diffuse collective linguistic convention is established.

We can distinguishing two phases (see Fig. 11.2).

1. One is a tacit negotiation and produces weak implicit acceptance.
2. The other is active reuse and produces a true convention.

Stigmergy in Humans: Some Nice Examples With Deontic Components

Let's now mix up several possible meanings of BIC messages while focusing on various interesting uses of them, like human stigmergy.

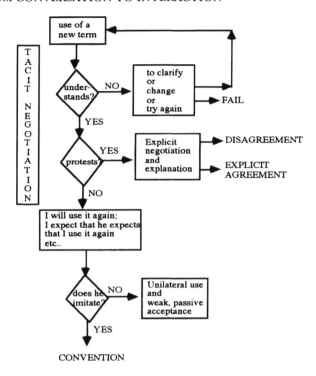

FIG. 11.2. Overanswering (overhelp).

Leaving the Coat on the Seat.   This means "Already taken, not free." This is a sign, deliberately used to mean (signal) this. It's communication. But for communicating we simply use a usual object in its usual practice: putting a peg on a seat. Since people derive from this trace the fact that this seat is already in use by somebody that is momentary absent but will be back, and we know that, we purposely use this as a BIC message. We didn't need to establish this meaning arbitrarily. Let's notice that this diffuse social practice has later become a convention, this sign or message starts to be conventionalized and specialized.

Bestsellers.   While buying a book (for your own pleasure)—we in fact leave a strange trace in the environment: We modify the number of copies sold. This changes the position of the book in the bestsellers list, and this is information (intentionally sent by the publisher or by the booksellers to the potential clients) that will be taken into account by other persons. This is communication, although your act just remains the practical act of buying a book, with its practical intended effect for you. You do not intend, in this case, to communicate anything at all, but, in fact, in that market your behavior has acquired such a parasitic (exploited) communicative function.

Parking Marks.   A beautiful example of stigmergic communication with normative (prescriptive and permissive) character is the use of painted blue or white lines on the ground for car parking, delimiting the car area and indicating their disposition: either in form of a comb, or parallel to the side-walk. Those lines not are just signs and instructions: "You are allowed to park and should park in this position," but they also have a practical and physical function.

They are not merely messages; in fact, they cannot be replaced by a simple poster illustrating the prescribed car disposition in that street. They also have the practical function of a visual referen ce point in the maneuver to be used during the act of parking.

So we put into our physical environment—for coordinating our actions—physical objects that are at the same time messages: precisely like termites, but with an additional deontic character.

BIC Soccer: BIC Actions Plus Stigmergic Communication Through the Ball

In soccer the players of the same team usually communicate with one the other—in order to coordinate their actions—simply through their movements and through the ball itself ("Look, I'm going on the right!"; "Look, I'm passing you the ball! Take it!"). The same moves are, for their opponents, just signs, not messages, except when they are faked or, in a few circumstances, where I intend that my adversary understands what I'm doing (see Fig. 11.3).

Silence as Communication

It is very well known that silence can be very eloquent. In general, doing nothing, abstaining from an action, is an action (when it is the result of a decision or of a reactive mechanism), thus it can be—as any behavior—aimed at communicating via BIC.

The meanings of silence or passivity are innumerable, depending on the context and on the reasons for keeping silence (or doing nothing) that the ad-

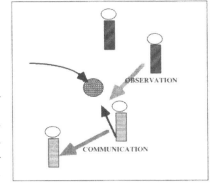

dressee can ascribe to the sender. We can, for example, mean indifference ("I'm not involved, I do not care," "I do not rebel," "I do not know," etc.), or respect and obedience, or stoicism, and so forth. The most important social use, however, as we already saw, is for tacit agreements that, by definition, are BIC-based. It is opportune that we discuss BIC and social order.

BIC BASEMENT OF SOCIAL ORDER

BIC has a privileged role in social order, in establishing commitments, in negotiating rules, in monitoring correct behaviors, in enforcing laws, in letting conventions and rules of behaviors spontaneously emerge. If there is a "social contract" at the basement of society this social contract has been established by BIC and is just tacitly signed and renewed.

Fulfilling Social Commitments and Obeying Norms as BIC

This is another kind of demonstrative act, not basically aimed at showing power and abilities, or good disposition, but primarily intended to show that one has done the expected action. Thus the performance of the act is also aimed at informing that it has been performed! This is especially important when the expectation of X's act is based on obligations impinging on X, and Y is monitoring X's nonviolation of his duty. Either X is respecting a prohibition, or executing an order, or keeping a promise. A second-order meaning of the act can also be: "I'm a respectful guy; I'm obedient; I'm trustworthy," but this inferential meaning is reached through the first meaning "I'm respecting, obeying, keeping promises."

A social commitment of X to Y of doing the act, in order to be truly (socially) fulfilled, requires not only that agent X performs the promised action, but also that the agent Y knows this. Thus, when X is performing the act in order to keep his promise and fulfill his commitment to Y, he also intends that Y knows this.

> If there are no explicit and specific messages, any act of social commitment fulfillment is also an implicit communication act about that fulfillment.

Notice that what is important for exchange relationships or for social conformity, is not that X really performed the act, but that Y (or the group) believes so.

One of the functions of norm obedience is the confirmation of the norm itself, of the normative authority of the group, and of conformity in general. Thus, one of the functions of norm-obeying behaviors is that of informing the others about norm obedience. At least at the functional level, X's behavior is implicit behavioral communication.

Frequently, X either is aware of this function and collaborates on this (thus, he intends to inform the others about his respect of norms) or he is worrying about social monitoring and sanctions or seeking social approval, and he wants the others see and realize that he is obeying the norms. In both cases, his conforming behavior is also an intentionally implicit communication to the others. Of course, X can also simulate his respect of the norms, while secretly violating them.

At the collective level, when I respect a norm I pay some costs for the commons and immediately I move from my mental attitude of norm addressee (which recognized and acknowledged the norm and its authority and decided to conform to it), while adopting the mental set of the norm issuer and controller (Conte & Castelfranchi, 1995):

> I want the others to respect the norm, pay their own costs and contribution to the commons.

While doing so, I'm reissuing the norm, prescribing a behavior to others, and checking their behavior (expectation). Thus, the meaning of my act is twofold: "I obey, you have not to sanction me," "Do as I do, norms must be respected."

As an example of the second behavioral message, let me propose Socrates' drinking the poison. Although his friends and fellows were pushing him to escape, Socrates wanted to drink in order to teach us and his fellows that norms (although iniquitous) must be respected: the content of the message, the conveyed meaning of the act is its motivation, its reason. Which sermon could be more eloquent than his act?

Also, the act of violating a norm can be a communicative act, either intentionally or even functionally. This is, for example, the case of the "provocative" attitudes of adolescents.

DESIGNING FOR BEHAVIORAL AND TRACE-BASED COMMUNICATION

While designing interaction, we have to design the following:

• Visibility and monitoring—We should wonder whether the actors in a given context can see each other (glasses, windows, etc.), or who sees whom and for what, and whether monitoring the other's behavior is possible or not (consider the rooms of certain offices with transparent walls). This was the idea of the celebrated *Panopticon* of Bentham.

• Self-explanatory artifacts and behaviors—The very famous capability that tools and artifacts should have for signifying their function, for making intuitive the comprehension of their use, when designed on purpose, would be a form of communication (Norman, 1997). Is the tool designed in order to signify (make understandable) and thus communicate its function and proper use (e. g., how to handle it, what to press, which part is what, etc.)?

• Stigmergy in the shared environment—We should wonder whether the environment is apt for leaving traces in it so that the others can recognize and use them (or for avoiding this). For example, having additive traces of multiple uses (like in garden paths) and exploiting this information for making more accessible the most used objects.

• Messages "written" in environment, objects, dresses, acts, ...

• Coordination channels and artifacts should be foreseen for:
  1. Coordination without any communication.
  2. Coordination via BIC.
  3. The possibility for an object to work as a coordination device.
  4. Specialized coordination artifacts (Omicini, Ricci, Viroli, Castelfranchi, & Tummolini, 2004).
  5. Coordination via communication protocols and devices.
  6. Coordination via natural language.

• Environments and devices both for cooperation, for isolation and privacy, for conflicts—Different environments afford different levels/power of observability, which agents can exploit by issuing suitable epistemic actions to realize different forms of observation-based coordination on top of it, such as the BIC. And we know that coordination does not mean only cooperating and acting together, but also avoiding interference, maintaining privacy, or even competing and fighting. Environment should be differently conceived for these very different social interaction forms.

BIC in Domotic, Robotic, and Human–agent Interaction

A wrong, or at least limited, paradigm is currently dominating the approaches to human–computer and human–robot interaction. HCI is in fact moving toward a new paradigm: from the *interactive* paradigm to what we call the collaborative paradigm. In the interactive perspective, the priority was about dependability, comprehensibility, the right feedback and interface, the possibility to intervene actively by the side of the user, the personalization and presentation of information, and so forth, while in the collaborative perspective, there is a "mixed initiative" approach. The computer (especially in terms of some agent in the role of personal assistant, or mediator, or representative, or mobile agent, etc.) is supposed to have some initiative, to learn or understand the current user's intentions, plans, or needs and to anticipate her request or overhelp her, for example, by providing more information or operations than explicitly requested. A real collaborator is expected to be able to do more and better than planned by us, and to go beyond the request for really helping us (Falcone & Castelfranchi, 2000).

In this collaborative paradigm based on mutual understanding, anticipation, and initiative, BIC communication (the possibility to rely upon the other perceiving the results of our action, and understanding what we did or our plan and aim) is crucial. This holds in our interaction with our personal computer that is supposed to understand and anticipate what we are doing or with our personal assistant agent.

This should also apply to Computer Supported Cooperative Work (CSCW) and computer-mediated collaboration: We cannot send boring messages to inform partners that we did our job and kept our commitments, especially when we did them on and through the computer or the net. Some form of observation by the partners should be allowed and exploited; or some machine-understanding of our acts.

When humans cooperate, the product of the action of X passed to Y is the message "I did; this is for you; now its your turn." We do not want be obliged to unnatural and explicit coordination messages. For example, after making a commitment with Y, after doing what was promised—possibly on the same computer—I'm obliged to send an explicit message to Y to inform Y that "I did." In human collaborative work, usually, our action, or its product, or the transmission of the results, per se, is also a message "I did; this is for you; now its your turn."

Agents might also relieve the users from unnatural and boring practices like this, recognizing the tacit message and automatically sending an explicit message to Y.[7]

The same holds in human–robot and Robot–robot interaction (Fong, Nourbaksh, & Dautenhahn, 2003; Beckers, Holland, & Deneoborg, 1996). It is not simply a matter of specialized and artificial messages (words or gestures). Also, expressive NVC signals (faces, emotions; Breazeal, 1998) are not enough. Even before this, one should provide the robot—for example, for coordination with humans in a physical environment—the ability to interpret human movements, understand them and react appropriately. At that point the human action in presence of the robot will be performed also for its understanding, that is, as a BIC message to it. Analogously, the human should be in condition to monitor what the robot is doing and to intervene on it by adjusting its autonomy. At this point, the robot behaviors—in front of the human—might become a message for approval, help, coordination, and so on.

Who would like to interact with her domestic robot or with her smart and animated house always by verbal orders or special gestures? First, one

---

[7]This idea emerged during a nice discussion with Francis Brazier.

would expect that the robot or the house "observes" (perceives) us, understands us and what we are attempting to do, and coordinates with our behavior, for example—as for the robot—by letting us pass in the corridor or by imitating us or by following us; as for the house, by anticipating (adjusting light, conditioning, etc.; opening and closing doors, windows, etc.) our movements and activities (Giardini & Castelfranchi, 2003).

CONCLUDING REMARKS

Designers should not have in mind conversation and dialog as the prototypes of communication and interaction. Linguistic communication (and, even more, conversation) biases our view of human interaction. I have illustrated the importance of mere behavior and its environmental traces as communication, and explained how important it is in human coordination, cooperation, social order, cultural transmission, and so forth.

My suggestion—on that basis—is that, while designing an object, one should take into account as primary properties not only the practical, so called "functional"properties of it and the esthetic ones (from mere formal beauty to affective impact; see Ortony & Norman, this volume), but also its semiotic and communicative properties and its functionality as a coordination or communication artifact.

Objects are repositories of knowledge, of practices, of memories, of culture. Are they able to store this, to preserve this, to make this accessible, readable? Or are they apt for or resistant to distributed cognition and memory, to semiosis? Are they able to pass messages, to inform the user, or to recall their appropriate use (Norman, 1997), to facilitate learning, participation; to favor and not to obstruct interpersonal coordination?

In particular, I have stressed how objects and environments are fundamental vehicles and mediators of interaction and of communication, not in terms of explicit and symbolic messages (like blackboards, answering machines, or traffic signals), but in terms of tacit, although very eloquent, behaviors and traces.

Designers should take into account this prominent function of objects, environments, and behaviors as vehicles of meanings more or less extemporary or conventionalized. Otherwise it would be like designing money simply as strange disks of metal or pieces of paper with figures.

In addition, I have claimed that silent and physical communication will play a relevant role in human–computer, human–robot, human–smart environment (domotic), robot–robot interaction, and this should be taken into account while design these technologies.

Chapter 12

From Function to Dialog

Mario Mattioda
Interaction Design Institute Ivrea

Federico Vercellone
Università degli Studi di Udine

> The absolute nature of scientific statements was for long considered symbolic of universal rationality. We, on the other hand, think that science will open up to the universal when it stops denying, wanting to have nothing to do with the preoccupations and demands of the societies within which it is rooted, when it will be capable of conducting a dialog with nature, whose many charms are at last appreciated, and with men of all cultures, whose problems are at last respected.
> —Ilya Prigogine and Isabelle Stengers (1984/1986, translated from the French by the authors of this chapter)

PREMISE

This chapter argues that the interaction between the interactive tool (product or service) and its user is best seen as a kind of conversation. A successful conversation depends on its participants sharing to some extent a language and its grammar, but these need not be rigorously codified. More important for mutual comprehension is to share both a wider cultural context and a local context to whose development the conversation itself will partly contribute. In the same way, a successful user–tool interaction, and thus its design, depends on not following universal procedural rules but responding instead to each unique context of use—a context which constantly shapes, and is in return shaped by, the user and the interaction.

The reflections developed in this chapter stem from the idea of function in its commonly accepted sense. We aim to renew this idea by setting it against the concept of interaction used in the world of interaction design. The interaction we have in mind is mainly in the form of dialog, for the model of the dialog both relieves the friction inherent in the interrelations between the user and the interactive tool (the product or service) and expresses the complexity and the influence of context which applies to all fields of communication.

Using the dialog model, function can also be freed from its logical-methodological bonds and lead directly to the user's requirements. What does this kind of scheme imply? First, it means embedding the design in the supportive domain of the user's real expectations and habits, developing and renewing them while never losing sight of the original need or desire and its context. Second, it means replacing mechanical logic with a search for the functional (which might be called "organic") so as to achieve an almost natural exchange between the user and the interactive tool.

By "natural" we mean here an approximation to forms of communication and hence to almost immediate responses and uses—to something, that is, not far removed from the mental sphere of individuals and their languages. It is well known that a design modeled on psychological and cultural premises, deeply rooted in the subjects concerned, leads to a more familiar and hence more truly functional interaction—one, that is, less entropic than strongly rational procedures.

By introducing the model of the dialog we are led to reexamine the technological imagination, no longer characterized by the forms of rigid mechanical necessity but based on flexible design procedures which leave the designer wide margins of freedom and creative scope. By adopting the method of the dialog, finally, we arrive at a (kind of) science modeled on the living individual, which must thus take into account the irrational element always involved in individuals (their own special, one might say idiosyncratic, features), which are an integral part of their "heritage" of individual requirements, and to which the interaction design must respond and correspond.

TOWARD ORGANIC FUNCTION

Interaction design must be based on the user's way of life, for a purely rational design cannot guarantee that the interactive tool will be usable. In designing products and services the intrinsic complexity of the context in which they are used must not be interpreted in a reductively banal way, or rigidified within a logical cage. On the contrary, it must be raised to a threshold of linguistic-communicative familiarity to function enjoyably and easily.

In interaction design, indeed, a merely rational solution may even be harmful: abstract, neglectful of the users and their environment, and derived exclusively from general principles which may turn out to be inadequate or unrelated to the context, given the constant shifting of meaning characteristic of actual life and work experience.

In fact, users' work procedures are concerned less with logical concepts and rational schemes than with a living symbolic plane that produces understandable, if contextualized, meanings. In this creation of meanings, it is primarily the context which lays down the rules by which users recognize their needs—rules which designer must not ignore, inasmuch as they define the user's uniqueness, and from which are derived, from languages and usages not strictly classifiable, fundamentally important ways of working.

Basing interaction design on real situations of use, face to face with the user, overcomes the difficulty of managing a complex system. Thus a tool or service can be located in a set of relationships made up of emotional and evocative moments as well as functions. All this questions the very idea of the subject, for interaction design deals with not a pure, transcendental subject but a subject seen in its feeling, emotional nature, partly comprising automatic reactions rising from the unconscious.

Precisely this awareness, that cold, unevocative functionalism may well lead to functional failure, makes interaction design seek new forms of the functional that tend toward dialog and are linked to aesthetic pleasure, to enjoyment of the use of the thing designed—which, of course, enhances its functional effectiveness. For it is assumed that a user will feel a greater urge to use a tool or service if it offers pleasurable, easy-to-grasp ways of interacting with it. The idea that aesthetic pleasure can be aroused at the point where aesthetics and technology meet, incidentally, goes back at least as far as the mid-19th century and the Arts and Crafts movement associated with William Morris (1992).

In this way, as Morris might have wished, the relationship between the user and the interactive object is humanized, indicating a new awareness that technology cannot exist outside the scope of "natural" communication. The design becomes an organic "prosthesis" that can only work in conditions that guarantee its vitality. Hence interaction design, placing design at the service of the quality of life (Harrison, 1986), emphasizes a more complex premise than that of traditional industrial design. The design of this technology tends toward an aesthetic experience achieved by guaranteeing communication with users and responding to their wishes and real needs.

Besides the pursuit of taste, inherited from industrial design, interaction design proposes its own special design orientation. It seeks to knit together the functional and the pleasurable, rendering communication models natural; indeed, its distinctive features seem to lie in its merging of aesthetics, functional values, and the field of communication. Its axiological position, its value system, distances itself from the alienated approach of traditional technology, rooted in a claim to be independent of the individual user.

In the end, the feature that seemed to proclaim the universal character of modern technology (Heidegger, 1977) is found to be its limit. The idea that the universality of technology might win worldwide favor, irrespective of the context in which it is applied, turns out to be that technology's intrinsic limit, contradicting its functional pretensions. Insofar as it is itself a culture, technology cannot ignore culture. Only through this awareness can it avoid the hyperbole that would rebound against it and the environment in which it flourishes.

Although interaction design exemplifies how the pursuit of organically functional values superseded the mechanical paradigm, this pursuit, which surfaces at the moment of communication, means that interaction design reduces the friction of impact with the user. In this connection, one might mention some epistemological considerations, with reference, for example, to "complexity" (of the systemic relation between an entity and its context) and particularly Ilya Prigogine's (Prigogine & Stengers, 1984) "dissipatory structures," an open system which thanks to external input, can move in the direction of order, counteracting the environment's tendency to return it to entropic chaos. Similarly, the aesthetic-communication exchange appropriate for an interactive object also tends—an objective of interaction design—to reduce entropy.

Finally, consideration of functional values cannot be disconnected from that of the transformation of machine models. If machines are really meant to alleviate human labor and replace people, on closer consideration there is no real substitute for the original. Mechanical function, founded on the idea of function in which the relation of the parts is extrinsic, proves insufficient. Perhaps it would be worth going back to Jean-Marie Guyau (1891), a thinker of the positivist age, who conceived the idea of an organic machine whose parts, not being intrinsically related like those of mechanical machines, were endowed by definition with greater power because they were less exposed to entropy. Guyau's organic machine might be presented as a theoretical reference for what might be defined as the ecological adaptation of interaction design.

THE DIALOG MODEL

The user's subjectivity is enhanced by a background of organic use, because a real-life context is sensitive to the elements it encloses. If the context determines the rules and the user forms an active part of it, the user has a dialogic relation with the context that interaction design must absorb. So effective designing lies not in minimizing dependence on the users' capacities but, on the contrary, in enhancing communication with users. Indeed, the interaction design must contain within it their habits and codes of expression, their whole world of meaning. In this way, design procedures involve defining interactive languages that, generated to govern flexible procedures, must translate into dialog the working exchanges between the user and the digital tool.

User-based design therefore involves setting up a field of communication within which information circulates as naturally as possible. To achieve this it is not sufficient to use methods directed solely at technological innovation. The design must equally attend to existing technologies and their hinterland of common dominant languages. It must evolve by adapting and extending already-known relations, and establishing the interactive point that ensures the greatest user-friendliness.

In this connection we would mention a user interface that an insurance company commissioned us to design. The request arose from the need to free the staff of its accidents office from their excessively abstract, rational procedures, returning them to their normal way of working, so that they would agree to work with a computer tool (a workflow management system) recently introduced to deal with accident cases. While the preliminary design was being developed, research on the professional competence of the users, based on actual use of the program, linked the work procedures processed by the program to the staff's usual methods. That is, the design programming consisted of bridging a communication gap between the new procedural treatment of accidents and the traditional business of inspection, adapting the system's interface to the users' work habits. The staff feared their profession would be downgraded to merely repeating fixed tasks. So rather than imposing rigid and bewildering procedures, we sought procedures that, although respecting new criteria, would use the habits and skills of the users. Hence the idea of a technical interface, designed solely to shorten a throughput of cases, was abandoned in favor of one directed toward maximum communication and active dialog between user and computer at all stages of the process.

The dialog model is complex. Analysis of the phases of an ordinary conversation shows a tension in speakers between asserting their own individuality and the opposite impulse that directs them toward the listener. This polarization, however, is not radical: Speakers neither impose their will absolutely nor completely surrender to listeners. So dialog involves a constant search for compromise and integration, a continual shifting from one state of meaning to another.

The same message, moreover, even the same word, can differ in meaning according to the context and circumstances in which it is used. Communication is not a transparent transmission of information, carried out by means of a neutral vehicle. If it succeeds it is not as a result of previous agreement on the meaning of the words used and on the rules applied to combine them, but partly because the participants share the same context of events, forms, aims, procedures, and so on. Put another way, we understand, or are understood, partly because we express ourselves within a set of interests and aims shared by our listener, who can fully grasp the message's meaning only with reference to him or her.

Thus communication is possible if the speakers belong to the same cultural background, whose principal expression is language. When we start from this premise the concepts of context and conditions of use become more relevant. No act of communication exists without a context and every context presupposes an exchange of experiences that identify certain features of the message rather than others as necessary for understanding. The unexpressed reference is to the "language-games" defined in Wittgenstein's *Philosophical Investigations* (1953), which bring meaning back into use and are expressed and rooted in "forms of life."

To return now to dialog, the model proposed here is useful for interaction design inasmuch as taking part in a dialog, as Jürgen Habermas (1981) said with reference to Hans-Georg Gadamer, is a bridging, a search for mediation as opposed to rigid alternatives, an exchange within which messages define their own field of meaning, partly through their evocative power, with the intent of restoring conventional codes to the domain of natural languages.

Here philosophical hermeneutics, which includes the theory of interpretation and dialog as an interpretative activity, can be an important reference for the theory and practice of interaction design. In this framework, hermeneutics is not antimethodological (Gadamer, 1982), although it limits method's autonomy (and hence technology's), bringing it back to the fundamental question from which, and in response to which, it arose. It is not a matter of creating a humanist methodology but of identifying from case to case the user's characteristics, without which the method becomes empty and ineffective or, as we have seen, can even seem hostile and confusing to the user.

The dialog model, last, shows the inadequacy of formalized languages. The extent to which logical languages, unamenable to intuitive understanding, often appear unfamiliar and unclear to the user, is well known. The solution to this limitation can be found in interaction designs sensitive to the user's working context and able to relate flexibly to the user's habits and mental pathways, and to transfer into the user's familiar world the potential offered by the interactive tool.

GUIDELINES AND CREATIVE FREEDOM

What guidelines to interaction design might assist designers in their work, bearing in mind that humans think linguistically in structured ways? As is well known, the linguistic structure of thought and the development of cognitive processes in identifiable and reproducible sequences enable us to construct models that describe the mental pathways that guide human action. Recognition, memorization, and learning are just a few of the mental functions that cognitive psychology has been able to formulate.

Hence cognitive models might be proposed as the "grammar" of the exchanges between the user and the digital tool, making an important contribution to the designer's work, a body of theory guaranteeing a successful interaction for every kind of project. One point, however, must be made: Generic formulas cannot be applied to design decisions with certain success.

Only in this sense can it be claimed that a design is primarily the result of a set of individual creative acts. The dominance of the context does not detract from the project's originality. On the contrary, by focusing on specific situations, designers draw from them the direction in which they can evolve a design that is not abstract but responds to real-life activities. For the contexts of use, by definition infinite in number, are what define the extent of design freedom, the sphere of the possible, and so can modify the rules rather than freeze them into some codified methodology.

Regarding the three basic elements of design—context, user, creative freedom—we find that the designer, by linking users to their context, becomes the stabilizing factor in contingent circumstances. Interaction design appears as the result of interactions between the designer, the user, and the user's context (Fig. 12.1).

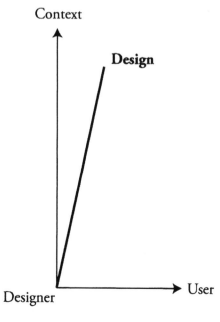

Let us return to the parallel between cognitive modeling and language. Natural languages are not just vocabulary plus grammar: Even if we acquired all the vocabulary and rules of a language, their combination would be of little practical use. Similarly, no general theory can lead the designer from the infinite interactive relations between the user and the tool to an optimal design. Like language, a

FIG. 12.1. Interaction design as the result of interactions between the designer, the user, and the context. The designer is the stablizing factor in contingent circumstances.

cognitive model, however fully formulated, must be expressed in specific and live situations. The cognitive models applied to interactive design do not constitute an arbitrary field but establish a discipline in which rules and grammar fix the inevitable point of departure for every design. Yet to stop here would mean to claiming to give precise content to a mere list of context-less theories. Hence the designer's work emerges as a mixture of rules, context, and creative freedom.

We have seen that to start from the interaction means going back to the very origins of the communication act itself. The context of which the user is a part, and on which depend his or her operating decisions, will be inspired, case by case, by design criteria developed according to cognitive principles that depend on the context and the designer's choices.

If the design follows an individual pattern (not anarchic, that is, but responding to the logic of the individual rather than some a priori generalization), it is somewhat contradictory to base it on some previous classification of contexts, which are by nature infinitely various. And even if contextual data could be satisfactorily categorized, one could not rigidly classify the forms of design intervention, which always spring from juxtaposing the designer's power of expression and the user's personal approach to dialog, and integrating them.

In interaction design the mistake is not to propose a method, which may indeed work in some contexts, but to attribute universal validity to it. When developing ideas, in other words, an a priori method that claims to solve diverse design problems will not work. For no guideline can tell in advance which way of looking at the facts under consideration is best.

Designers may, however, operate by defining the same project in different ways. There is in fact a range of approaches from which one might draw, without having to pronounce one absolutely more effective than the others and thus fall back into normative procedures. Indeed, we know that interactive design is to be measured in terms not of its intrinsic value but in its relation to the user's experience, because its aim is the translation, or rather the regulated transformation, of one language (the tool's) into another (the user's).

Consequently, interaction design emerges as an activity directed toward identifying and strengthening new ways of communicating. If it is not to crystallize into rigid theoretical and logical positions, it must continue defining its own field of operation, contextually and in relation to the user.

A METHOD FOR REVEALING COMPLEXITY

We have seen that the aim of placing the user and the interactive tool at the center of the design involves an effort to define a very complex interrelational field. The dialog model has shown that the relation's two terms must be grasped and developed without flattering one at the expense of the other. It is not a question of putting the user's habits first at all costs, or of conceiving the interactive object as the basis for all solutions, but of combining the tool's potential and specific features with the user's own experience.

On what methods, if they can be called that, can designers draw when tackling interaction problems, in the terms in which they have been outlined so far? Edgar Morin's (1992) thinking might help. Starting from the relation among order, disorder, and organization, he developed a method to give access to the complexity of real-life situations by framing apparent incommensurables as each other's contexts, as complementaries linked by a "virtual circle" of reciprocal definition through which alone their true complexity is constantly rerevealed.

We can apply Morin's ideas and terminology exactly to interaction design. Relating the real scope of the interactive tool to that of the user involves a never-ending effort. The two terms, in their irreducibility and continual reference to each other, have a circular relation evident in the reciprocal influence that grows between the interactive object and the user's tasks. Although the new tools modify people's work procedures and behavior, designs in their turn become ever more like incarnations of theories in people. So to pose the interaction question does not mean searching for a single all-inclusive theory to transcend the relation's intrinsic circularity. The choice is not between, on one hand, the specificity, narrowness, absolute precision, and rationality of contemporary computer science knowledge, and, on the other, a universal method, a general and abstract system. It is rather a matter of outlining a way to articulate what is separate and to connect what is disconnected: a procedure that does not hide but reveals the links, implications, and interdependencies of the complex relationship between the user and the interactive object.

The circular nature of the interaction between human and mechanical components therefore provides a framework for interaction design, in which each participant in the relationship offers something lacking in the other. Although defining the link between a mind and an interaction design cannot be straightforward, the path of the design must be placed within a virtuous circle. Maintaining this circularity means maintaining the association between two entities, user and tool, which, taken separately, are both recognized as true but, placed in contact, risk negating each other. It means, moreover, to conceive these two truths as two sides of an articulated situation, revealing the interdependence between two ideas that, when separated, become isolated and opposed.

In this way a kind of mutual belonging comes to link the user and the interactive tool, analogous with that created between two speakers in a dialog who constantly depend on a third contextual element, the dialog itself. Only by referring to communication links, to unconnected areas of interaction understood in their structural elements and dynamic tensions, can the circularity intrinsic to human–machine interaction become virtuous and begin a practice of design which creates an ever more complementary relationship between them.

The methodological revolution developed by Morin, applied to interaction design, seems at first to be an antimethod that, in the early design phases, to resist destructive oversimplification, accepts the uncertainty, confusion, and relativism associated with complex phenomena. But taking into account what cannot be simplified is the first step toward better understanding the relationship between user and interactive tool. Thus the systemic ideas behind old analytical methods become essential scaffolding to be taken down or put up in a different order to confront the task of designing user–machine interaction.

It should not be thought, however, in the absence of laws that might offer solutions applicable to specific categories of design problems, that the method outlined here has no guiding rules but only the individual features of a specific interactive relationship. On the contrary, we can establish typologies that go beyond the level of the absolutely specific without blocking the tensions of a complex relationship within universal, generalizable laws. The categories of relationship theorized by Morin—of meaning (a way of interpreting a relationship's complexity), development (the transformations and exchanges that take place within the relationship), structure (how it is organized), and the connection between the whole and its parts (how they combine with and complement each other)—may restore to projects their full interactive character and accept and resolve the diversity and incongruencies inherent in any encounter between user and tool.

DIALOG WITH THE USER

We have seen that an interaction design project can only start properly from an analysis of the user. Being not a tool of logical communication but the satisfying of that user's real-life expectations and intentions, it must establish common ground with the user's deepest roots, overriding the rational constructions frequent in professional stereotypes.

Here we come to another crucial point: The relationship between designer and user does not always proceed in an even, straightforward manner, as the user's expectations often need interpretation. How can these expectations be identified, given that an authentic relationship with the user cannot always be set up in the first few meetings, and the designer often suspects a gap between what users say and what they really do? The rest of this chapter answers this question by paraphrasing an essay by Aldo Giorgio Gargani (1995) on the relationship between master and disciple, and applying it to interaction design.

It could be maintained that the designer must guard against believing what the user says without question, for instance, because it may be shaped by the subordinate position often assumed by the user and thus an over-ready acceptance of the designer's solutions. Users' arguments about what should lead to the best working solution may derive from inauthentic forms of communication in which, reasoning in the abstract, they surrender their role and position as interlocutor. In these cases the users, carried away by their rationalizations, put the designer on the wrong track.

It is useless for them to pile up documents as proof of their ideal position if they misrepresent their situation. When users tend to suppress their own individuality, what they maintain and defend lacks the significance of the expectations that prompt them. But the designer must keep this significance in mind, having observed and listened to them, and having compared what they assert with what they habitually do and the context in which they act.

This design phase is not easy. The conflict that can arise here is caused not by a wish to go against another's will but rather by a recognition of real expectations, despite the hiding and feigning techniques that users may adopt. Hence the designer must show users the problem resolution that they sometimes seem not to identify but that the expert designer can glimpse behind and beyond them. Apparently it is often the designer who waits for users at a point they do not yet realize they have reached.

This does not mean that designers are a sort of adversary who, as a result of the mental transformation they intend to bring about in the users, aim to impose their own will when the users are not prepared to recognize it. Designers want not the users' reluctant consent but to identify their real needs, because only by starting from them can they establish genuine communication. By refusing users' rationalizing resistances and basing their designing on the nature of the user's work, designers achieve a merging of meanings and outline the scope of a dialog that is more than the sum of their and the users' wills but instead a triad (designer, user, dialog) that transcends their reciprocal positions and allows new ways of operating.

In this tension between user and designer is neither the volatility of a sharply polarized discussion nor a conflict of wills but an effort to identify a common outlook. In this case, we get a paradoxical situation in which the designers are not influenced by the users' arguments because they reject them. Whereas the users describe, infer, jump from one idea to another, the designers tenaciously maintain their own understanding of the problem, guarding the design field while the users flounder in their explanations and deductions. Saving the plan for a possible meeting, the designer keeps the users in touch with the reality which preceded the debate. For, as we have seen, discussions sometimes alienate users, preoccupied by the apparent attempt to draw near and communicate. By stirring up users' rigid assumptions (about this technology, that model of communication, that company philosophy, and so on) the designer helps the users go beyond them, assume the role of their interpreter, and start to develop them themselves.

Thus interaction design starts with the users' real situation as its primary value and—by transforming their professional skills, habits, and awareness of the nature of their work—restores users to themselves and their mode of working. This way avoids features that might lead to entropic collapse, and, as the vitalists would put it, generates an increase in energy.

Chapter 13

Interface Design and Persuasive Intelligent User Interfaces

Oliviero Stock
Marco Guerini
Massimo Zancanaro
Istituto per la Ricerca Scientifica e Tecnologica, Trento

INTELLIGENT INTERFACES AND INFORMATION PRESENTATION

What we want are interfaces that understand us, that are nonintrusive, natural, and powerful, that adapt to us, that help us focus our attention and memorize, and that are pleasant and entertaining.

Natural language as a means of communication is an obvious aspiration.

Natural language processing has been a focus of research for many years; it has shown considerable potential in the area of interaction, especially with the development of the field of computational dialog. The last decade has seen a transformation of the field, due mainly to two factors: (a) the availability of a large quantity of linguistic data together with a dramatic increase in computer power and memory which allows them to be processed, and (b) the introduction of short-term competition in the field, basically imported from the speech research tradition. These factors have been the basis of a more engineering-oriented development, as opposed to the ambition to understand cognitive processes, and specifically to the prevailing emphasis on statistical rather than knowledge-based methods. Speech technology, a culturally different area, has meanwhile produced notable results, and speech recognition can be realistically integrated in many interfaces—although it provides limited dialog capabilities and is currently appropriate only for certain applications.

The natural language processing community has also contributed to the emergence of intelligent user interfaces.

Let us focus briefly on intelligent multimodal information presentation (Stock & Zancanaro, 2005). At the root of intelligent information presentation (IIP) are several scientific areas, but at least three are fundamental. Probably the first which should be mentioned is natural language generation, the branch of natural language processing that deals with the automatic production of texts. The field is normally described as investigating communicative goals, the dynamic choice of what to say, the planning of the overall rhetorical structure of the text (sometime called strategic planning), the actual realization of sentences on the basis of grammar and lexicon (sometimes called tactical planning), and so on. With a similar objective but different means, the second field, adaptive hypermedia, combines hypertext (hypermedia) and user modeling. Adaptive hypermedia systems build a model of the goals, preferences, and knowledge of the individual user and uses this throughout the interaction to adapt the hypermedia to the user's needs. By keeping a model of some aspects of each user's characteristics, the system can adapt to individual users and help them navigate and filter information that best suits their goals. A third important field is computer graphics, which experienced a fundamental transition toward the end of the 1980s when it was understood that graphics production should start from internal representations and communicative goals in a way similar to language production. This transition has led to the possibility of developing multimodal systems that in output would consider available modalities, possibly the context and the user's characteristics, and allocate and realize the message in a coordinated way on several media.

Intelligent interactive information presentation has gone further along that line. It relates to the ability of a computer system to produce automatically multimodal information presentations, taking into account the specific attributes of the user, such as needs, interests, and knowledge, and engaging in a collaborative interaction that helps the user retrieve and understand relevant information. It may include dimensions such as entertainment and education, opening important connections to areas unrelated to HCI, such as broadcasting and cinematography. This vision has led to substantial and novel collaborations, evident in a number of projects, in Europe as well as America and Japan, where the teams have included very diverse expertise.

The following sections now introduce a more specific theme we think will be important for interface design: persuasion.

AUTOMATING PERSUASIVE COMMUNICATION

Future intelligent interfaces will have contextual goals. They may aim to induce the user, or the audience in general, to perform certain actions in the real world. They will have to take the "social environment" into account, exploit the situational context, and value the emotional aspects of communication.

Foreseeable scenarios of this kind include dynamic advertisement, preventive medicine, social action, and educational entertainment—all scenarios where rational reasoning is not enough. For "intention adoption," what really matters is often not only the content but the overall impact of the communication.

We want to provide the interface with the capability to reason on the effectiveness of the message as well as on its high-level goals and content (Guerini, Stock, & Zancanaro, 2003). According to Perelman and Olbrechts-Tyteca (1969), persuasion is a skill that humans use to make their peers perform certain actions or collaborate in various activities. Argumentation has often been considered as addressing similar points. But in our view, persuasion is a wider concept: Argumentation can be regarded as a resource for persuasion, whereas negotiation puts the accent on interactivity in argumentation.

In the first place, persuasion is a "superset" of argumentation. Although argumentation aims to make the receiver believe a certain proposition, persuasion aims to make the receiver perform a certain action. The link relies on the fact that, apart from coercion, the only way to make people do something is to change their beliefs. As Castelfranchi (1996) put it, "It is impossible to directly modify the goals [...] of an Autonomous Cognitive Agent. In order to influence him (to modify his goals) another agent is obliged to modify the former's beliefs supporting those goals."

In this prospect, argumentation is a resource for persuasion. The statement that persuasion involves more than argumentation acknowledges that persuasion also involves a-rational elements, such as inducing emotions to obtain a given result or using a specific language for threatening or promising. All can be regarded as resources for inducing the receiver to act in a desired way.

Natural argumentation, a growing research area in AI, comes closer to persuasion, as it also concerns, for example, the adequacy—the effectiveness—of the message. Even in professional settings such as juridical argumentation, extrarational elements can play a major role.

Persuasion mechanisms include the following four aspects:

1. The cognitive state of the participants—The beliefs and goals of both the human and the system.

2. Their social relations—Social power, shared goals, and so on.
3. Their emotional state—Both the emotional state of the human and that expressed by the system.
4. The context in which the interaction takes place.

A brief description of these aspects follows. The beliefs and goals of both the user and the system about the domain of the interaction are prerequisite for persuasive interaction, because persuasion is communication leading to belief adoption, with the overall goal of inducing an action by the user by modifying his or her preexistent goals. In a museum guide system aimed at persuading visitors to see some exhibits, for instance—a theme we have explored in PEACH, a large project devoted to cultural heritage appreciation (Stock, Zancanaro, & Not, 2005)—we can instantiate all these elements.

Social relations exist between the visitor and the system, which plays the role of a competent guide, and between the visitor and other relevant persons such as experts and parents.

Emotional elements can enhance or lower the message's effectiveness. Although our reference framework is similar to that suggested by Gmytrasiewicz and Lisetti (2001), whose article focuses on how the emotions of an agent can change his or her own behavior, we focus on how emotional elements can be used to increase or diminish the persuasiveness of a message. There are four dimensions to be considered:

1. The current emotional state of the persuadee—How it affects the strategy selection of the persuader.
2. The emotional state expressed by the persuader—What emotion the persuader must display to maximize the persuasive force of the message.
3. The emotional state possibly produced in the persuadee by the message—The induced emotional state may not be desirable, but must be taken into consideration for subsequent interaction.
4. The current emotional state of the persuader—How it affects his or her strategy selection.

It is still a matter of debate whether the latter should actually be taken into consideration in persuasive interfaces. The two main standpoints are as follows:

1. A perfect persuasive agent should be emotion-neutral; he or she just has to display the most effective emotion for the current persuasive goal.

2. For the persuader, to feel emotions is a good way to handle unpredicted situations and a resource for responding to the persuadee's moves.

Our current work focuses specifically on the role of the emotional state of the receiver (how this affects the system's selection of a persuasive strategy) and on the emotion the system must express to maximize the message's effectiveness.

Finally, persuasion strategies can use contextual elements. For instance, a reference to a painting the visitor has seen previously ("this painting is by the same person who painted ... ") can enhance the probability that the user will stop in front of the present painting.

The emphasis on modeling persuasion mechanisms, and therefore on performing flexible and context-dependent persuasive actions, goes beyond the current focus of *captology*, the term introduced by Fogg (2002) with reference to persuasive technologies. Most current approaches to persuasive technologies provide hardwired persuasive features. We focus instead on reasoning capabilities to make human–computer interfaces able to provide flexible persuasive communication with their users.

CONCEPTS FOR PERSUASION IN HCI

In the HCI field there is increasing research interest in providing the persuasive features of systems with reasoning capabilities. To build persuasive systems it is necessary to individuate systematically the several dimensions of persuasion at play in the kind of interaction we want to address:

1. Action versus behavior and attitude inducement:
   a. Behavior inducement—Changing, in a stable and persistent manner, the way an agent acts, for example in response to certain events or state of affairs in the world. This is meant to be a long-term effect.
   b. Attitude inducement—Changing, in a stable and persistent manner, the way an agent evaluates events, state of affairs, or objects. This effect is also long term.
   c. Action inducement—Changing a particular planned action of an agent. This effect is short term.
2. Argumentation-specific versus fully persuasive:
   Another distinction between argumentation and persuasion (other than that discussed previously) can be drawn by considering their different foci of attention. Argumentation focuses on the message's correctness, the argument's validity, whereas persua-

sion is more concerned with its effectiveness. The point is that an argument can be valid but not effective, or, on the contrary, effective but not valid.

3. Audience-specific versus universal:
   a. The first definition of persuasion given by Perelman and Olbrechts-Tyteca (1969) claimed that persuasion is characterized by being audience-specific—it can adapt the topic to specific listeners.
   b. Cialdini (1993) took the opposite position: all the strategies he analyzed are meant to be universal, because they use cognitive patterns of the receiver which are common to everybody—an approach specially relevant to broadcast presentations.

4. Monological versus dialogical:
   a. Perelman and Olbrechts-Tyteca's (1969) analysis of persuasion, because concerned with rhetoric (how to create effective discourses), is more involved with monological interaction.
   b. Cialdini's (1993) analysis of persuasion, because concerned mainly with selling agents, is more involved with dialogical interaction—the "foot in the door" or "door in the face" strategy, for example.

5. Domain-specific versus universal:
   Some strategies are typically domain-specific, like "fake discount" strategies to increase sales. Others, like "resorting to fear" with impressionable people, are not.

6. Language-only versus multimodal:
   This distinction is relevant if multimedia is available, in particular in relation to innovative HCI aspects such as the use of embodied conversational agents (ECAs).

Let us review some prototypes that have touched on some of the above aspects. STOP is one of the most well-known systems for behavior inducement that exploit persuasion (Reiter, Sripada, & Robertson, 2003). Mainly a natural language generation system employed for real human settings, it aimed to induce users to stop smoking. Its authors had a hard time evaluating it because of the difficulty of assessing real change in behavior and the particularly awkward nature of smoking addiction.

Some systems, such as the one proposed by Zuckerman, Jinah, McConachy, and George (2000), use argumentation strategies to generate persuasive messages. They are concerned with the abstract form of the argument's unfolding: *reductio ad absurdum*, inference to the best explanation, reasoning by cases, and so on. In general, however, logical reasoning is just one resource to support persuasion.

There are computational models of persuasion (Grasso, Cawsey, & Jones, 2000, for example) based on work developed by linguists, philosophers, and cognitive psychologists—like Toulmin (1958) and Perelman and Olbrechts-Tyteca (1969), to mention only the most important.

Research on these theories and systems showed the limits of purely logical reasoning and the need to consider uncertainty in modeling persuasive dialogs (Zuckerman et al., 2001) and to introduce more sophisticated argumentation schemes (Walton, 2000).

A recent issue in this research area concerns widening the persuasion modes from considering "rational" or "cognitive" arguments to appealing to values, social relations, and emotional states (Grasso et al., 2000; Guerini et al., 2003; Sillince & Minors, 1991). Carofiglio and de Rosis (2003) focused on emotions as a core element of affective message generation in dialogical argumentation, whereas Guerini et al. (2003) focused on the same aspects but from a monological point of view.

Table 13.1 proposes a possible grid for systematizing persuasion prototypes on the basis of the main dimensions of persuasion they address. Rows indicate the types of persuasive goal; columns indicate the type of interaction, audience, and exploited feedback.

ECAs AND OTHER REALIZATIONS

ECAs are synthetic characters, usually human-like, whose physical appearance can display dynamic expressive behavior; Prendinger and Ishizuka (2004) surveyed this research area. How to plan autonomous behavior led by internal beliefs, desires, and intentions remains an open topic, but flexibly persuasive ECAs have a large potential. Synthetic characters allow richer communication with the user because

- They are more natural. Their gaze can be used to display the focus of attention or indicate turn-taking, for instance, as can gestures for deictic (that is, context-dependent) references, for communicating contents, and for expressing emotional states. On the importance of

TABLE 13.1

A Systemization Grid for Pursuasion Prototypes

	Monological single	Monological group	Monological broadcast	Dialogical	Limited feedback single	Limited feedback group
Effect Now (short-term effect)						
Effect Later (long-term effect)						

eye gaze in ECAs to improve the quality of communication see, for example, the study of Garau, Slater, Bee, and Sasse (2001).
- They are more involving. They may display emotion and express personality. There are many studies of the display of behavior and emotion (e.g., Allbeck & Badler, 2002; Poggi, Pelachaud, & De Carolis, 2001).

Taking also into account the natural predisposition of people to treat anthropomorphic agents as human peers, the possibility of ECAs to persuade them, by leveraging social responses, is crucial and critical—for example, by using their realism to increase their believability.

Nevertheless the relation between realism and believability on one side and effectiveness on the other is not totally obvious. In certain situations it is reasonable to use cartoon-like (unrealistic) characters, which can be effective because they do not generate an overattribution (discussed later) that could lead to frustration, and are more suitable for particular kinds of audience, children for example. When talking about realism and effectiveness we are referring to behavior as well as appearance: cartoon-like characters often use behavior to display exaggerated emotion and thus obtain the desired effect. In any case, we should point out that ECAs are not the only resource in persuasive interfaces.

Multimedia yields many features relevant for HCI persuasion, as pointed out in Fogg's (2002) assessment of the relation between Web site appearance and trustability. Other cases are the use of music (Scherer, 1995) or techniques used to animate texts (*kinetic typography* or KT; Forlizzi, Lee, & Hudson, 2003) to emphasize and induce emotion. Used appropriately, it can enhance the emotional impact of the content conveyed by words: For instance, hopping words can emphasize a happy message.

A BIRD'S EYE VIEW OF THE DESIGN OF AN INTELLIGENT PERSUASIVE SYSTEM

In designing the architecture of monological intelligent persuasive systems, there are three levels of processing to be considered (see Fig. 13.1):

1. At the first level the persuasive message's specification is planned. This level selects, on the basis of the participants' cognitive, social, and emotional states, possible persuading strategies. It results in a perlocutionary act (that is, one having an effect on its receiver), possibly accompanied by an illocutionary act specification (that is, its intended effect). It can also specify the emotional state the interface agent must display, and provide other information to pass on to the next level, like the message's rhetorical structure. A markup language like APML (De Carolis, Carofiglio, Bilvi, & Pelachaud, 2002) can be used to define the elements necessary for multimodal generation.

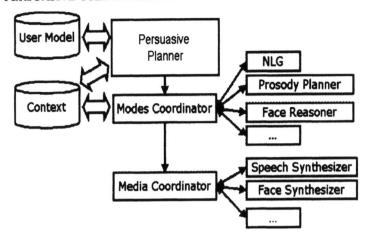

FIG. 13.1. An intelligent persuasive system: a possible overall architecture.

2. At the second level, realization strategies are defined involving the available modes. Specific modules, "realizers," take into account information passed on by the first level. Selection is performed as indicated by the first level, but is independent of it. For example, a severe mood can be conveyed by harsh intonation, by a text whose words are selected for their emotional import ("kick the bucket," for example, instead of "die"), or through perceptual elements such as font selection.

3. At the third level, the different available media are coordinated to integrate and synchronize the various components.

This architecture is intrinsically multimodal. Each component realizes a portion of the message in its own way, including characterizing it emotionally. Coordination takes place at the system's second and third levels. At the second level, content is allocated to the different modes and coordinated according to rhetoric rules. For example, harshly intoned synthetic speech can be accompanied by the serious facial expression of an embodied agent. At the third level, a different kind of coordination takes place, mainly involving temporal constraints, for example, synchronization between certain facial muscular movements and pitch levels in synthesized speech.

An example is the museum scenario, mentioned earlier, in which the visitor has an electronic organizer combined with a locating component connected to the persuasive module. At the beginning of the visit, the system, a dynamic user model, starts with an initial user profile which is then continuously updated on the basis of the visitor's behavior (time spent in front of an exhibit, movement, general attitude in the course of the visit, topic selection on the PDA). Of particular importance is the interest model and the attentional–emotional state (if the visitor is near a painting, the system is describing but not looking at it, for instance, the system can infer that the visitor is bored). Let us consider the visitor passing in front of a painting. Because the system knows the visitor's position, it might assume the goal of directing his or her attention to that particular painting. This communicative goal fires the instantiation of a number of strategies that, in turn, trigger abstract rules and metarules, which are used for the following:

1. Content selection, such as the statements to be made.
2. Content modification, such as the emotion to be displayed in association with the statement. If applicable strategies point out the object's negative features of the object, for instance, the agent can display a "sorry-for" mood in association with them.
3. Content structuring (ordering of statements). If there are applicable strategies on the positive and negative features of the object, for example, the agent can obtain the persuadee's trust by putting the negative features first.

These latter rules instantiate the specifications for the final persuasive message, built in accord with Mann and Thompson's (1987) *rhetorical structure theory* (RST), which proposes that the structure of many texts is a tree built recursively, starting from atomic constituents (such as clauses) connected through specific relations. These *rhetorical relations* (RRs) account for the structure and content-ordering of the text.

The relation between persuasive strategies and RRs can take place at two levels. At the macrolevel, the RRs connect different strategies, whereas at the microlevel (not considered here) the RRs articulate the content of a single strategy (see Guerini, Stock, & Zancanaro, 2004). A persuasive strategy can be seen as an atomic constituent (elementary unit), and, by means of selection theorems, it is possible to identify the RR connecting different segments (adjacent elementary units) composing the structure of the persuasive message. In deciding which RR most appropriately links two particular segments, selection theorems take into account the typology of the strategies involved and their content.

The persuasive message of our example, generated by the metarule illustrated in number 3) earlier, can be a text such as "Although this painting is much degraded, it is one of the most important of the Middle Ages and the only one that represents fox hunting." The text span, as a final modification, is converted to an APML file for multimodal realization so that a talking head can render the APML file as Fig. 13.2—the sorry-for specification for the first part of the message is generated by the example abstract rule reported in number 2).

```
<EXPML>
<affective type="sorry for"> Although this painting is much degraded, </affective>
<affective type="joy"> it is one of the most important of Middle Ages and is the only one that
represents fox hunting </affective>
</EXPML>
```

FIG. 13.2. Text for a museum-guide system, realized in APML (Affective Behavior Markup Language).

We want to stress that such specification does not have to be realized through an ECA: the same text can be rendered with KT, using movement to convey the corresponding emotion. Figure 13.3 shows two screenshots of ECA and KT rendering.

As a final example, Fig. 13.4 shows two APML realizations of the same persuasive message for inducing a pupil to pay attention to the teacher during the museum visit. The difference in the messages is that one conveys a reproachful attitude, the other an empathetic attitude, by changing the display of emotion accompanying part of the message.

ETHICAL ISSUES

As interface agents become more complex (autonomous, proactive, and so on) and more common in everyday life, the need for agents to be designed ethically is becoming more compelling. The need is even greater when agents are intended to persuade—that is, have a very proactive and influence-wielding attitude toward people.

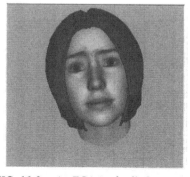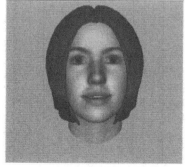

FIG. 13.3. An ECA (embodied conversational agent), a synthetic character displays dynamic expressive behavior. Here "sorry-for" and "happy" expressions accompany, respectively, the first and second phrases shown in Fig. 13.2.

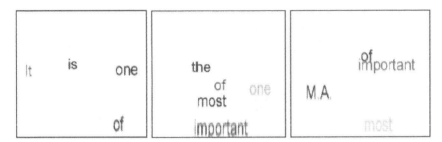

FIG. 13.4.   Kinetic typography. The second phrase in Fig. 13.2 is expressed "joyfully."

Up to now some thought has been given to sketching guidelines for the development of ethical interface agents, mainly considering privacy and attribution. But (except for Berdichevsky & Neuenschwander, 1999) the behavior of the agent itself has not received much attention:

- Privacy—How the information which interface agents may collect about us is used (see, for example, Kobsa, 2002).
- Attribution—The problem of misleading and culpable behavior by the interface (especially human-like ECAs) where the user (over)attributes human traits and capabilities to the system.

The challenge in the long run is to create ethically aware agents: able, that is, to reason about the ethicality of their actions and possibly correct their behavior accordingly. Let us pose some general questions about the ethicality of a persuasive agent or system, which are relevant to any kind of communicative agent, artificial or human:

1. What is the ethical status of a system when it tells the truth but hides important information from the receiver? Or, worse, what if the agent tells a falsehood which nevertheless might benefit the receiver? Is the first unethical and the second ethical?

2. What about persuasion which uses information, conclusions, or values that the system or its developer does not believe in but the receiver does?

3. Related to this, is the overall goal and belief structure of the persuasive message intelligible to the receiver? Is the agent hiding its true intentions? And when and why is this unethical?

4. Is the overall goal of the persuasive message in the interest of the receiver? Is the agent making receivers act against their own best interests? Are there situations in which this is ethical?

5. A "tutorial" goal is where an agent aims to influence a receiver to decide to perform actions that are in the receiver's interest without, however, the receiver being aware of that interest. Is having a tutorial goal with regard to the receiver a sufficient condition for persuasive interaction to be ethical? What if the tutorial goal is not recognized by society? What if the tutorial goal overcomes other interests?

6. When is it ethical to induce extreme emotions in receivers to persuade them?

7. When is it unethical not to try to persuade someone?

Cognitive and emotional notions are central for elaborating answers to these questions. In the long run it will not be just a matter of hardwired ethicality. It will be necessary to endow interface agents with the ability to reason about the ethical consequences of their actions and act accordingly.

EVALUATION METHODOLOGY

Persuasive interfaces need to be evaluated. It is not sufficient that they be theoretically sound: They must also be effective with real users. Evaluation is not at all straightforward, as the STOP experience showed: It is necessary to point out carefully all the variables that can affect the system's effectiveness (context of use, scenario of the interaction, typology of the user, required task, persuasive strategies at hand, and so on) and how they work together. Given the complexity of the task, specific evaluation methodologies have to be defined.

One solution is to use qualitative, theory-driven experiments to measure "indirect aspects" like attention and memorization. Variations in attention and memorization can indicate the impact of the persuasive message. Measuring the memorization of the message's key concepts can help one understand the message's persuasive strength.

Petty and Cacioppo's (1986) *elaboration likelihood model* (ELM) can be used for this kind of experiment—a research issue currently investigated by Shlomo Hareli at the University of Haifa. ELM claims that there are two routes to persuasion: central and peripheral. The central route is most effective when the receiver is motivated to elaborate the message thoroughly; here strong arguments are more appropriate. But when the receiver is not able or motivated to process the information, the peripheral route is more effective: emphasis on the claimed expertise of the persuader, for example, can compensate for the relative weakness of the argument.

Using this model we can infer, for example, that when the receiver is persuaded via the central route, the change of attitude lasts longer and memorization is stronger. This model also offers predictions about how emotions can affect the elaboration level—the amount by which receivers amplify the received argument for themselves and thus increase their belief in it.

CONCLUSIONS

This chapter has proposed a view that emphasizes intelligent communication capabilities in future interfaces. We briefly discussed the state of the art of natural language processing, then introduced the field of IIP, which aims to make computer systems able to produce multimodal information presentations automatically. IIP systems take into account the specifics about the user, such as needs, interests, and knowledge, and engage in a collaborative interaction that helps the user retrieve and understand relevant information. In particular, we think the dimensions of emotion and personality must have a role in individual-oriented, context-aware communication systems.

So we then discussed persuasive interfaces in particular—interfaces that aim to induce the user, or the audience in general, to perform specific actions in the real world. Such interfaces will have to take the "social environment" into account, exploit the situational context, and value emotional aspects in communication. To model persuasion mechanisms and perform flexible, context-dependent persuasive actions are aims much more ambitious than those of most current approaches to persuasive technologies. We focus on reasoning capabilities, as opposed to hardwired persuasive features, to make human–computer interfaces able to provide flexible persuasive communication with their users; foreseeable scenarios are dynamic advertisement, preventive medicine, and social action. Finally, we described the characteristic elements of persuasive systems, proposed the outlines of a basic architecture, and briefly discussed some ethical and evaluation issues.

Building flexible persuasive systems is just at the beginning. It is a complex task and this short discussion has had to leave out many important issues. Let us just mention now, however, the temporal dimension we briefly characterized as a feature of a persuasive system. For emotions provoked by a message may be future-oriented, for instance when people are persuaded to choose a product at a later time. A choice may be influenced by immediate emotions, but one can also consider the expectation of direct emotional experience in the future.

The effect of a persuasive message on memory and the evolution of an emotional state is an interesting theme yet to be investigated. Cognitive studies on the impact of advertisement may have something to teach us, although we have the additional dimension of flexibility and the potential for individual orientation. Another aspect involving time and memory is postaction assessment: recall of all the emotional experiences involved may have a role in completing the assessment of the situation. A successful choice performed by the individual under the influence of a persuasive message produced by an intelligent agent will help establish fidelity toward not only a brand but the system that originally produced the message.

ACKNOWLEDGMENT

The authors of this chapter thank Cristiano Castelfranchi for his contribution to the formation of the ideas underlying it.

Part
VI

MEMORIES

Remembering Together: Some Thoughts on How Direct or Virtual Social Interactions Influence Memory Processes

Giovanna Leone
Università degli Studi, La Sapienza, Roma

Memory is usually studied and commonly represented as an individual faculty, allowing past information (coming from external environment through senses, or deriving from internal activities such as thoughts, fantasies, or mental images) to be recalled when the original stimulus has disappeared from current experience. It may therefore be conceived as a mental link between past perceptions and current states of mind. Referring to this first definition, similar to many ideas of memory that arise spontaneously to mind, the aim of this chapter is twofold.

First, I review some classical contributions, challenging this definition of memory as a purely individual faculty. Starting from different points of view, these pioneering authors studied how social interactions not only may be a context in which memory processes are embedded but may, on the contrary, deeply influence the basic functions of individual remembering—such as information recall, reconstruction of the meaning of past stimuli or events, or monitoring activities that spontaneously follow acts of remembering. These authors showed that a definition of memory as a purely individual activity could be seen as ignoring many important aspects, if not as actually erroneous; their precious new insights therefore al-

211

low us a deeper understanding of the importance of social influence on memory acts. Apart from profound differences in their overall perspectives, in fact, all the authors chosen show how social interactions may change our memory processes.

Second, I try to speculate about how these classical intuitions may be applied to the new perspectives made possible by the virtual interactions that characterize our current technological ambience. This new kind of social reality could not be imagined at the time of these seminal studies, all made in the first half of the last century. Nevertheless, their pioneering intuitions are richly suggestive of applications to our changed and changing social world because of their unsurpassed sensitivity to the way in which historical and cultural conditions may be mentally internalized.

Finally, in the last part of this chapter, I try to suggest, starting from the innovative points of view proposed by these classical authors regarding the social dimensions of remembering, how an interaction design, if appropriately conceived, could be used to change and eventually improve individual acts of memory.

The first question to ask, then, to arrive at these eventual observations, is whether memory is only a capability whose performance depends on whether the individual is more or less endowed by nature, or if it may be socially improved, at least to a certain degree. Although this question is basic to all the classical authors we intend to review, it is essentially the work of Lev Vygotskij that overtly proposes it as a key theoretical issue. (Although I use the Anglo-American "Vygotsky" in bibliographic references, my text uses Luciano Mecacci's transliteration "Vygotskij" in tribute to the Italian scholar who first translated, free of censorship, *Thinking and Speaking*, (Vygotskij's [1934/1962] unfinished masterpiece.)

CAN MEMORY BE SOCIALLY IMPROVED?
INTRODUCING VYGOTSKIJ'S PERSPECTIVE ON
THE SOCIAL INTERMEDIATION OF MEMORY PROCESSES

From a phenomenological point of view, memory either occurs after a conscious decision, when a person actively seeks for a memory to come to his or her mind, or on the contrary "pops" up abruptly. In this latter case a hint in the external environment (such as a smell, sound, word, or particular view) or in subjective activity (such as a thought, feeling, emotion, or mental image) usually acts as a cue—retracing, through a more or less complicated chain of associative links, the original memory. Memory may therefore be experienced as either a voluntary or an involuntary phenomenon.

Starting from these observations, Vygotskij (1983) proposed dividing memory processes into two main categories:

1. Elemental processes, due to direct associative links between the current situation and past perceptions and experiences, which are responsible mainly for involuntary memories.
2. Superior processes, guided by the active decision and will of the person remembering, in the case of voluntary memories.

Vygotskij (1982) proposed, moreover, that this second, voluntary recalling of memories, which he called "superior processes," distinguishes human from animal memory. For only human (superior) processes are based on the conscious use of intermediations to guide associative chains linking actual cues to past memories; intermediations that, in a famous quotation, he exemplified in activities either meant for private eyes only (such as knotting one's handkerchief) or aimed at chosen communities (such as putting a statue in a public garden to remember a distinguished citizen):

> The very essence of human memory is that human beings actively remember with the help of signs.... As one psychologist [Dewey] has said, the very essence of civilization consists in the fact that we deliberately build monuments so as not to forget. In the knotted handkerchief and the monument we see the most profound, most characteristic and most important feature which distinguishes human from animal memory. (Vygostky, 1931/1983, p. 86)

If we adopt the observation criteria proposed by Vygotskij, the sharp border between individual memory and social environment tends to become fuzzy because the social activities in which we are embedded also influence our memory, facilitating or inhibiting it. This is particularly evident when memories that "pop up" refer to cues present in the individual environment because of some social decision, as in the case of the monuments. In this case, what may seem at an individual level a kind of involuntary recollection (we think of a particular historical episode, for instance, because we happen to pass a street whose name commemorates it) may be seen at a more collective level as a consequence of a voluntary decision, stressing some kinds of historical episode instead of others (choosing to remember a battle won by our own country, say, instead of a defeat, although both are similarly important from an historical point of view). Interestingly, the research group coordinated by Vygotskij chose as its main procedural strategy exactly the interactions between the individual and social cues to explore the development of memory as both a personal and a social ability.

A well-known series of experiments, organized by the research group coordinated by Vygotskij and originally published in 1931 by him (1931/1983) and by his disciple Leontiev (1931/1975), showed how a memory performance may be sometimes ameliorated, sometimes damaged, by the same external cue. In their research design, different groups of participants (younger or older children, adolescents or adults, children disadvantaged by intellectual deficits or children without intellectual deficits) had to remember a list of unrelated words in two different research conditions. In the first condition they had to remember them by heart; in the second, they received external help—looking at a set of pictures linked to the words to be remembered.

Results showed that, when remembering subjects, adolescents or adults were able to organize their own chains of intermediations but they showed a poorer memory performance when, before remembering, the external set of intermediations (the pictures) was shown. In short, these external aids acted as a source of interference, slowing down the autonomous processes of recalling.

Nevertheless, when participants had some memory difficulty (because they were younger and therefore less able to organize their recall autonomously, or because of some intellectual deficit), the presence of an external help, giving them a prearranged possibility of intermediation, clearly ameliorated their memory performance.

Vygotskij's (1983) theoretical explanation for these data was that the use of intermediation was a social practice, first made available by the social environment, which only in a later stage of individual development was gradually internalized by participants. This could explain how the same exterior aid could produce both a poorer performance among adolescents and adults, and a better one among younger or disadvantaged participants.

As is well known, the experimental work of this research group was very soon interrupted by Vygotskij's premature death. Nevertheless, his scientific production, covering only 10 years (1924–1934), shone with such a genial transformation of previous research paradigms that it suggested to Stephen Toulmin, in a famous article (1978), a parallelism between Vygotskij's tragic role in the history of psychology and the one attributed to Mozart in the development of music.

In a certain sense, this short time meant that some features of his research remain partly unexplored or at least too rough to show all the subtleties proposed by his theoretical model. Nevertheless, these data clearly demonstrated that individually impaired memory might be socially improved. This is one facet of the basic Vygotskijan idea of the existence of an area of individual development that may be brought to perfection only through educative tools made available by social structures and therefore changing according to cultural and historical periods.

As is also well known, Vygotskij's theoretical models were furiously attacked by the establishment of his own society, the USSR of the late 1920s and early 1930s, as unduly "idealistic" and "cosmopolitan" (that is, too similar to research trends developing in Europe, because of his frequent quoting of prominent authors such as Freud and Piaget). These were the years when Stalinism entered the life of the scientific community as well as all other kinds of organized social life, cruelly showing its will to persecute intellectual originality of any kind. And this trend was doomed to last long after his death, forcing his former disciples to try to distance themselves from his research, first criticizing his methods and theories as too complex and abstracted from the material conditions of social life, then editing his works and censoring many words or quotations to make them more acceptable to both the Russian and the European political context (Mecacci, 1990).

If we think about this climate of political and personal persecution, it is surprising to discover how, when evaluating his own scientific production, Vygotskij was yet again far in advance of his contemporaries. Commenting in 1932 on his own results and theoretical models, he had the moral and intellectual strength to acknowledge bitterly, "I'm inclined to think that it [his research on memory] represents a colossal oversimplification, even though at first it was often criticized as unduly complex" (quoted in Bakhurst, 1980, p. 392).

It is touching to note how, in the last years of his too short life—while enduring a stupid and violent persecution of his ideas, considered too "immaterial" to fulfil the needs of a Marxist view of psychology (Bakhurst, 1980)—Vygotskij not only had completely overturned many of the limited research paradigms of his time but was also conscious of further changes necessary to understand memory processes better. In particular, he felt that his procedure and tasks, judged by his contemporaries as "unduly complex," were, on the contrary, too simple to capture all the facets of memory processes. A set of words was shown, a set of intermediations was given; then a comparison between the number of words shown and the ones recalled was used as an index of memory performance.

Although recognizing that less complexity was necessary to produce the intelligent simplicity essential to experimental conditions, Vygotskij was aware that, in the case of memory, these research procedures could lead to "colossal oversimplifications." For example, interviewing child participants after one version of his experiments, in which they were allowed to choose which pictures to use as memory aids, he noticed that some children showed a strategy more complicated than a simple chain of associative links. One child chose a picture of a crab near a stone to remember the word "theater," for instance, saying that the animal looking at the stone reminded him of men staring at a stage; another used a camel as a cue for the word "death," imagining a story in which a lost voyager, without food and water, starves in a desert (Bakhurst, 1980). Interviews of participants, therefore, clearly suggested that the efficiency of intermediations was due not only to a more or less complicated chain of associative links but also to a creative way of inserting these cues in a complex strategy that aimed at seizing a relation between the meaning of the stimulus and the meaning of the intermediation used as a cue.

These observations demonstrated how the experimental task, recalling a series of words, captured not only the reproductive aspect of memory but also the reconstruction of the meaning of past stimuli.

Unfortunately, Vygotskij had no time to change these intuitions into new research procedures. Nevertheless, in these same years another researcher, Frederic C. Bartlett, decided to focus his work precisely on this reconstructive aspect of memory, summarizing his results in a book, *Remembering* (1932), destined to become a classic reference in memory research.

MEANING RECONSTRUCTION AND SOCIAL SCHEMATA: BARTLETT'S CONTRIBUTION

Unlike the mainstream memory research of his day, Bartlett (1932) proposed the provocative idea that memories are not copies, more or less accurate, of the past.

It is well known that the development of psychological studies on memory reached an important turning point because of the innovative procedure, created by Hermann Ebbinghaus (1913), of observing the way in which one person (usually the experimenter himself) memorized different lists of nonsense syllables. Using this meaningless material, Ebbinghaus could observe how a different kind of exercise may cause a better or worse performance in recalling the lists studied, in the certainty that no association whatever could link the nonsense syllables to previous knowledge. In other words, by this new procedure, he disentangled the effect of simple rehearsal from that of association between the stimuli used in the experimental tasks and the previous knowledge of the participants. This procedural device not only skilfully isolated amelioration in memory performances due only to exercise, but also—and perhaps most importantly—it proved that basic memory processes could be efficiently investigated by experimental procedures.

The experiments of Ebbinghaus thus expanded beyond the specialized domain of memory studies to become a more general and very effective demonstration of how psychological research procedures had to be distinguished from their traditional philosophical roots if they were to be better placed in the domain of natural sciences—as seen in the positivistic perspective dominating scientific debate at that time (Farr, 1996).

Bartlett obviously recognized the skilfulness of this kind of procedural device, but strongly argued against the ecological validity of these results. To use his own words (1932), the "lifeless copies" of meaningless syllables originated by the Ebbinghaus (1913) procedures could be seen only as "unpleasant fictions," due to the artificial setting of laboratory tasks. In everyday life, one is actually very unlikely to have to study and reproduce meaningless material: On the contrary, memory can be seen as an effort to reconstruct the meaning of past perceptions and experiences, to grasp the gist of the memory itself using all the knowledge presently accessible.

Referring to Bartlett's (1932) definition, then, every act of memory, whenever we recall it, is constructed freshly anew: In other words, it is an act that starts from the present to reconstruct the past, rather than past knowledge influencing the present activity of the mind.

Starting from this theoretical position, Bartlett proposed to study permanent memory by an original methodology, the "repeated reproductions method." After showing participants a meaningful material (a map, drawing, or story, say), he asked them to repeat their reproductions of the original material at different times, and recognized the work of their memory as the gradual shaping—from the first reproduction to the later ones—of a new and more complete meaning, slowly emerging from differences between these repeated reproductions.

This creative transformation of original material was due, from Bartlett's point of view, not to "mistakes," as in theoretical models which view memory as a copy, but to a never-ending effort to understand the gist of original items, slightly changing any reproduction until the memory reaches a stable pattern of meaning. The results of many experiments made using this method showed that any reproduction differed considerably from its predecessor until it took a simplified shape, when it then stayed similar in all subsequent reproductions. In fact, if memories changed quickly and dramatically in the first part of the reproduction sequence, they became stable from the moment this new, simplified organization was reached. Bartlett named these kinds of reorganized and simplified structure as "schemata," which are reactivated whenever a new reconstruction is made, from the moment the schemata themselves were reached—as if, once created, they guided all subsequent reproductions.

If the schemata notion appeared to contemporary critics too loose and somehow elusive an idea, it was later recognized as one of the most useful explanations of not only memory processes but all the cognitive dynamics organizing the never-ending flow of new information which ceaselessly reaches the mind. In a certain sense, this later success of a previously misjudged concept was achieved by emphasizing some aspects of his theory while obscuring others (Mazzara & Leone, 2001). That schemata act as a sort of "freezing" of never-ending changing information fitted very well with the more generalized idea, basic for the cognitive approach, of the mind as a "cognitive miser," always trying to simplify a too complex reality. On the contrary, the stress on the dynamic quality of the mind, able at any time to reconstruct a new memory afresh, and always struggling to reach the meaning of the past, was quite obscured by this emphasis on the mind's tendency to persevere in using preconsolidated schemata. Kintsch (1995, p. xiii), introducing a recent reissue of *Remembering*, noted how "schemata as fixed memory structures that are pulled out for use on demand, as in most applications of modern schema theory," are a new interpretation of Bartlett's original work that, although very widespread, are far from catching the core of his work. Therefore, he comments, Bartlett's original presentation of schema theory "is particularly worth rereading, for what a surprisingly fresh and sophisticated version of schema theory it is!"

To better appreciate the sophistication of Bartlett's original theory, which later disappeared in its more mechanical interpretations, it is perhaps useful to remember that he borrowed the concept of *schemata* from two eminent physiologists of his time, H. Head and G. Holmes, who used it to describe the internal structure that automatically guides the spatial orientation of human bodies. They, in their turn, derived this term from the Greek skhma, indicating "the body's gesture, captured during its movement, not contemplated during its rest; the body of athletes, orators, statues ... a body exerting itself" (Barthes, 1977/1979; my translation of 1979 Italian edition, p. 5), as in the famous statues of Olympic players caught in the effort of athletic performance. In the same way, Bartlett wanted to catch the memory as a mental movement—which was what essentially distinguished his approach from classical experimental paradigms (Leone, 2001). As Kintsch (1995, p. xii) also observed in his introduction to *Remembering*, the originality of Bartlett's research is based on the fact that "he is concerned not so much with the accuracy of recall as with what he terms the 'effort after meaning' ... It is not precise measurements of retention time or interference conditions that matter, but the inference processes the subjects engage in and their attitudes."

To observe this effort to organize acceptable schemata, catching the core meaning of past information, Bartlett's (1932) procedures tried to illuminate both individual and social dynamics. On one hand, individual effort after meaning was made clear through the possibility, for each participant, of reproducing memories several times, having time to add or change previous versions. On the other hand, the social aspects of remembering were made evident in two principal ways.

First, to make the observation of the memory's "effort after meaning" as clear as possible, Bartlett used material coming from other cultures (native Americans and Africans, for instance). Confronted by these unusual contents, the repeated reproduction method showed, from repetition to repetition, a process that Bartlett called *conventionalization:* Material was changed in such a way that any unfamiliar content was forgotten, whereas new and more plausible elements were inserted, producing a final memory very different from the original one, because they were gradually shaped into more familiar (more culturally conventional) schemata.

Second, in some trials Bartlett asked participants to "pass" their memories from one to another: one participant, for instance, having heard a story, had to recount it to a second; the second had to recount what he heard to a third, and so on. Through this different kind of repeated reproduction (*serial reproductions*), Bartlett tried to capture what happens in everyday life whenever we receive secondhand news. By this procedure he replicated in some ways what happens in social phenomena such as the creation of rumors or the spreading of false anecdotes ("urban myths"), but also represented, at a more general level, the deep changes affecting a memory when it is shared with others through a narrative activity.

In short, repeated reproduction and serial reproduction methods suggested that memories frequently rehearsed or recounted to other people are not only made more stable, more accessible, and "alive"; somehow they are spoiled, too. Indeed, reconstructive changes due to the "effort after meaning" made by memory are not only amplified by elaborations during internal rehearsal but changed by the need to put one's memories into words and arrange them into a plausible narrative shape so as to make them comprehensible to listeners. Some participants make this point clear, declaring after the task that they have changed some culturally unusual details to create a more "sympathetic climate" for their listener.

Of course, some of these effects could be a direct consequence of Bartlett's methodological choices. It could be argued that Bartlett's instructions were too loose (asking participants to reproduce the items, he did not stress the need to be as precise as possible) or that his theoretical model emphasized the reconstructive rather than reproductive aspects of memory. Participants could therefore be induced to fantasize (Kintsch, 1995). Nevertheless, his methodological choices, although in some respects too informal, let him discover a set of phenomena extremely important in the everyday uses of memory. By asking participants to repeatedly reproduce the same memory, he highlighted how the rehearsal or sharing activity aims not only to reconstruct the meaning of past experiences but also to "turn around" memories, so as to check and ameliorate them. The spontaneous use of these monitoring activities is another crucial point to consider if we want to grasp the social influences on memory.

TURNING MEMORIES AROUND: THE ROLE OF MONITORING PROCESSES

If we look at memory performances outside laboratory settings, when remembering is used to fulfill the needs characterizing everyday life (see Cohen, 1996; Neisser, 1982), we may see that, although people remember, very often they not only recall a previously perceived content but try to be certain in identifying its essential meaning, what Neisser (1982) called its "gist." Therefore they constantly evaluate and monitor the quality and validity of their memory processes. To better understand this point, let us examine some of these everyday phenomena.

Consider, for example, the situation in which you know perfectly well that some content is present in your memory but you cannot reach to grasp it at the moment (it is "on the tip of your tongue"). This means that, although you cannot temporarily access it from your memory, you are nevertheless somehow able to know that it is stored there.

On the other hand, you may know perfectly well that some content is entirely new to you yet have a strong feeling of having a memory of it (as in the *déjà vu* phenomenon). This completely reverses what happens in the tip-of-the-tongue phenomenon. In this case your awareness tells you that this content is not stored in memory, although you may have a strong illusion of accessing it as a proper memory.

Or consider what happens to a memory that is frequently rehearsed or recounted to other people. Elaborations during internal rehearsal, or the need to put one's memories into words and arrange them into a plausible narrative shape, to make them comprehensible to listeners, are all processes that cause content reformulations to "stick" to original experience, irreparably changing memories forever.

Nevertheless, in spite of being aware that such rearrangements have been made, people spontaneously try to monitor the source of their memories, distinguishing original sensations and perceptions from later imaginations, comments, and thoughts. For instance, through so-called "source monitoring" (Johnson, Hashtroudi, & Lindsay, 1993), the remembering person tries to evaluate, as far as it is possible, differences between really experienced contents and what was later elaborated or imagined, starting from the type of content prevalent in memory itself (more sensations and perceptions, for previous experiences; more considerations and reflections, for further reformulations). In an interesting series of experiences, Mazzoni and Vannucci (1998) clearly demonstrated that, to a certain extent, people manage to distinguish between really experienced and reformulated contents, disentangling what they actually remember from what they know about this same memory.

Strictly linked to these monitoring activities, other phenomena spontaneously shadowing recalling occur, due to the degree of confidence that people show in the accuracy of their own memories. Sometimes people feel that their memories are highly accurate; at other times they seem more doubtful. Interestingly, a large number of researches demonstrate that feelings of confidence are only very loosely related to the actual accuracy of the memories themselves but seem much more linked to the situation in which memory occurs (a testimony during a trial raises more doubts than an informal chat with friends) and to the personality characteristics, the degree of self-assurance, of the person remembering (Ross, 1997).

Nevertheless, although confidence cannot be used as a good way of evaluating accuracy, it is an intriguing phenomenon per se. For instance, we (Ritella, 2004) recently made a series of experiments using a very easy recognition task. Of the four people in each experimental session, three were confederates instructed to make an evident mistake in some critical trials and overtly declare their wrong answer in front of the fourth (the experimental participant), who was obviously unaware that the experimenter had instructed them to give erroneous answers. The results showed that even when socially isolated participants, exposed to the mistaken answers of a unanimous majority of confederates, did not change their correct answer, their degree of confidence in the accuracy of their actually correct memories was significantly diminished. Our results suggest, therefore, that the degree of confidence may be linked more to the social acceptance of memories than to accuracy itself.

Outside psychology labs, in short, people remembering not only recall a more or less high degree of original stimuli but also incessantly check the quality and accuracy of their memories, decide whether to share them with others, and feel deeply touched when they recall particular memories or, more generally, during the sharing activity itself.

All these phenomena cannot be reduced to memory's capacity to copy past reality. Rather, they pertain more to the interpreting function of memory: in other words, to the need to not only replicate reality but also be reasonably certain of having grasped its meaning or core rather than its inessential details.

REMEMBERING WHAT WE ALREADY KNOW: HALBWACHS AND THE MONITORING OF THE AFFECTIVE MEANING OF REPEATEDLY SHARED MEMORIES

Until now we have analyzed how others may influence us in remembering (or failing to remember) particular contents. Yet there is another social use of memory in which sharing information is not the first aim, indeed not an aim at all. Every now and then groups and communities spontaneously engage in an activity that at first glance seems purposeless: remembering what everybody already knows. This same pattern of shared remembering may occur during a dinner in which old friends recall yesterday anecdotes just for old times' sake, or in a serious institutional reunion in which authorities remember a well-known contribution made to the institution by a famous member. In a certain sense we may say that these are, apart from their striking differences in power and consequence, just two examples of the many commemorations, private or public, which we experience throughout our life.

The family is one social context in which this kind of joint remembering of well-known contents is very frequent. Researches based on nonintrusive observation of spontaneous family conversations have estimated that, for every hour observed, from five to seven sequences of communication were based on remembering memories known by everyone (Blum-Kulka & Snow, 1992; Miller, 1994).

This kind of family "social game" was clearly observed and commented on by Maurice Halbwachs. In the famous fifth chapter of his classic essay on the social framings of memory (1925/1992) he described what happens "when a family remembers." He noted that members of the family, when nobody extraneous can hear them, repeatedly share memories of some little episodes of family life, or recall the personality and characteristics of particular members. Moreover, he observed that all family members seem to attribute a fundamental value to these informal moments, sharing these memories as a "private treasure."

Starting from these observations, Halbwachs asked himself what this need was for repeatedly sharing information already mastered by anyone. Moreover, he wondered why some episodes or some family members are more frequently remembered than others. Finally, he tried to understand why, as time passes, family members discover, to their great surprise, that they think more and more about these well-known and apparently trivial anecdotes. The reason for all these, he argued, is that in these memories, family members—and only they—may find "a more or less mysterious symbol of the common ground from which they all originate their distinctive characteristics" (Halbwachs, 1925/1992, chap. 5: my translation of Italian translation in Arcangeli, 1925/1996, p. 35).

These observations seem to illuminate, with great originality, another kind of monitoring activity regarding memory, related not only to the actual accuracy of recollections but also to the emotional meaning of memories themselves. I may recognize that a memory of my past life is accurate and precise but feel that the social context framing this recall has now vanished from my current interests because it refers to a city in which I no longer live, a job from which I resigned a long time ago, a political association to which I no longer give my time, a personal relationship that is not as important to me as it was yesterday, and so on.

This is perhaps why, when we do not remember an anecdote recollected by an old pal, we feel somehow obliged to pretend to remember it because we know that to reveal we no longer remember it may be another way of stating our indifference to this old friendship, because, as Halbwachs (1967) said so simply and well, "the one who loved the most, remembers the most."

In short, this pioneering work by Halbwachs on the social framing of memory shows a new and most important function of shared remembering: the possibility of creating and consolidating the sense of belonging to an affective community that is built up and confirmed through the social sharing of well-chosen memories which are eventually felt to be "our memories."

According to the innovative proposal of Halbwachs, the social sharing of memory may be seen not only as a function of durable groups but also one of the most valuable protective factors a meaningful community may give its members. Repeatedly shared memories create a kind of "affective armor," to use his phrase, constantly reminding participants of the way their groups were able to cope with past difficulties and challenges.

It is not by chance, perhaps, that this trend of study, characteristic of the first part of Halbwachs's career (his essay *Les Cadres sociaux de la mémoire,* containing his first reflection on repeated family recollections, was published in 1925), was renewed when he had to cope with the dramatic climate caused by the Nazi occupation of France. In these days in 1944, the old professor, age 67, freshly created teacher of Social Psychology by the French Academy, the Collège de France, decided to turn back to his first interest in social memory, writing *La Mémoire collective,* the unfinished masterpiece that was to be published and edited by his friends and family members, after his tragic death in Buchenwald in 1945 (Halbwachs, 1950/1967). The dark times in which he lived, and his courage in resisting their terrible historical menace, are silently embedded in the classical serenity of all the unforgettable pages of this little book.

Although never overtly referring to the tragic choices to which, as an eminent French intellectual, he was confronted in those days, he implicitly resumed his moral and political dilemma—between acceptance of the Occupation and active and dangerous resistance—by his theoretical argument in favor of his somewhat provocative thesis on collective memory. In the book, in fact, he (Halbwachs, 1950/1967) proposed that, in spite of the commonsense assumption that sees memory as a faculty of individual minds, any act of memory is socially founded, because of the need of the one remembering to choose the social belonging that frames the reading of reality through which the recollection is made.

Starting from this new point of view, a memory may be not only accurate or inaccurate, accessible or not accessible, but also emotionally meaningful or meaningless—this last kind of monitoring being above all the consequence of currently accepting, or rejecting, that one belongs socially to the old affective community to which the memory refers.

Moreover, in this new theoretical perspective the time voluntarily dedicated to the more or less ritualized social game of remembering well-known episodes together implicitly shows the importance attributed to the social sharing of memories, seen not only as a guarantee for consolidating social groups but also as a way of confirming the affective bonds that substantiate the current social identity of individuals. In Halbwachs's words, repeatedly recollecting particular episodes or persons of one's community is, in fact, another way of saying to all who accept themselves as members of this community, "that is the way we are"; because the "private treasure" of recollections available to community members, and only for them, implicitly shows them the way in which other members of their affective community have coped with everyday difficulties as well as historical challenges. (For a comment on the impact today of the classical theory of Halbwachs, see also Leone, 1998, 2001; for an important reflection on relations between commemorating activities and affective coping, see Frijda, 1997.)

HOW INTERACTION DESIGN MAY FOSTER SOCIAL REMEMBERING ACTIVITIES

In this last section of this chapter I try to speculate about how the classical contributions on the social dimensions of memory, reviewed earlier, may be applied to the new dimensions of interpersonal and collective interaction opened up by interaction design.

Starting from the point made by Vygotskij's theories on memory and the crucial role of voluntary intermediations to guide information recall, it is obvious that new computerized devices for storing and communicating information create a multiplicity of possible associative chains, helping users recall preselected information more easily and quickly.

Certainly, something understressed by Leroy-Gourhan (1964), the tendency to exteriorize information storage, saving it on some material medium (from the simple sheet of paper in a notebook to the virtual space in a hard disk) may be considered a constant feature of human memory.

Nevertheless, the current availability of "virtual intermediations" may be seen as not only a new kind of medium, replacing the older ones, but also an innovative social practice that spreads new opportunities for the socially induced reorganization of self-guided recalling. If people could be taught to use these new kinds of self-organized intermediation, they might develop more competence in crucial areas such as prospective memory for managing formal (work or study, say) or informal (leisure or household) activities.

This use of interaction design focuses on the mastery of new technologies reached by individuals (and it is easy to imagine that this will create new boundaries between well educated and uneducated people, as well as between old and new generations). We may see in these innovative technologies, therefore, another example of the gradual "internalization" of socially induced performances that, according to Vygotskij, is a crucial step in developing potential areas of individual memory. Other important consequences may be envisaged, on the contrary, depending on managing the social and interpersonal potential of interaction design.

We have seen, reviewing the classical ideas proposed by Bartlett, that repeated reproductions "passing" from an individual to another accelerated the process of conventionalizing memory contents. A frequently repeated remembering of events (as during conversations via mailing lists, or within sites devoted to particular problems or topics) may be another important natural setting for noting how in everyday life a memory may dramatically change when repeatedly replicated. Moreover, in interaction aimed at exchanging simple conversations or chats, or informally sharing points of view (as in daily e-mail activity), memory's "effort after meaning" may be amplified and simplified—freed from the limits of distance, unavailability, or the time spent waiting for a reply.

It is challenging to imagine how the new possibilities of frequent and easy exchange of personal memories, due to technological advances, may influence the wide range of monitoring processes discussed earlier. But, referring only to the lesson of Halbwachs, we may suggest the idea that a well-structured interaction design, making possible another way of repeatedly sharing a memory of what we already know, could strengthen the affective bonds of our collective identities, thus accomplishing one of the most crucial tasks that distinguish the psychological sense of community from a merely instrumental belonging to transient associations (Sarason, 1974).

Chapter 15

Memory for Future Actions

Maria A. Brandimonte
Università degli Studi Suor Orsola Benincasa, Naples

MEMORY: MENTAL TIME TRAVEL

> With one singular exception, time's arrow is straight. Unidirectionality of time is one of the most fundamental laws. It has relentlessly governed all happenings in the universe—cosmic, geological, physical, biological, psychological—... galaxies and stars are born and they die, living creatures are young before they grow old, causes always precede effects, there is no return to yesterday.... Time's flow is irreversible.
> —*Tulving (2002, p. 1)*

The singular exception to this is represented by the human ability to travel in mental time. Rememberers can travel back into their pasts and forward into their futures, hence violating the law of the irreversibility of the flow of time (Tulving, 2002).

There are many important psychological consequences of humans' time perspective. One of the most relevant to the organization of human memory is that, according to temporal distance, people form different mental representations depending on whether the information pertains to the near past or to the distant past, to the near future or to the distant future. Individuals form abstract, high-level temporal construals of distant past and distant future events. High-level construals include general, decontextualized features that convey the essence of information about time (Trope & Liberman, 2003, p. 403). On the other hand, people form low-level con-

227

struals for near past and near future events (Trope & Liberman, 2003). Low-level construals include more concrete, contextual, and incidental details. Both high-level and low-level construals are formed for past events as well as for future events.

In terms of memory processes, high-level and low-level construals formed for past events go under the rubric of retrospective memory, whereas high-level and low-level construals formed for future events go under the rubric of prospective memory. The primary focus of this chapter is on prospective memory, that is, on the mechanisms and characteristics of memory for actions that have to be performed in the future.

MEMORY, ACTION, AND TEMPORAL CONSTRUAL

Since the mid 1980s, a new conceptual framework has been developed within which cognitive processes are seen to be deeply rooted in the body's interactions with the world (Koriat & Pearlman-Avnion, 2003, p. 435; for a review, see Wilson, 2002). This view—which has been stimulated by the notions of "embodied cognition" and "situated cognition"—brings action to the forefront of cognitive theory (Koriat & Pearlman-Avnion, 2003, p. 435). By their nature, memory processes are intimately tied to action (Zimmer et al., 2001). Therefore, any memory theory should take into account the question of when and how memory functioning influences the individual's interaction with the world.

An increasingly relevant area of research within which the relation between memory and action has been studied refers to the realization of intentions (Brandimonte, Einstein, & McDaniel, 1996). Everybody knows how important it is that we remember to execute previously formed intentions successfully and at an appropriate time. In our daily lives, we are required to form and initiate several intentions. However, often we cannot execute the action at the moment the intention is formed because we are busily absorbed by another task and are then forced to delay execution until some later time. As a consequence, memory processes are of paramount importance for successfully executing the task. The process of storing and retrieving such intentions is known as prospective memory (Brandimonte et al., 1996) and over the last few years, it has received increasing attention. Everyday examples of prospective memory (from now on abbreviated to PM) include remembering to buy bread on the way home from work, remembering to give friends a message on next encountering them, and remembering to take medication.

It is commonly accepted that there are many different forms of intentions and that each type of intention has its own characteristics and processing requirements (Brandimonte et al., 1996). However, there are also some properties that are common to all PM tasks (Burgess, Quayle, & Frith,

2001; Ellis, 1996). First, any PM task involves a retention interval between the formation of the intention and the time to realize it. This period may last minutes, hours, or days. A second feature is that a PM task involves both an ongoing and a background task. In a typical laboratory PM task, participants are required to perform an ongoing task (memorizing a list of words, for instance, or generating associations among words) while they have to remember to do an action at the appropriate moment (a background task such as pressing a particular key on the computer keyboard when a particular item appears or at a particular time). That is, the paradigm takes the form of a dual task, with a primary, ongoing task that serves as a covering task for the prospective, background task. According to the type of retrieval context—that is, on the appearance of a specific event or at a particular time—researchers refer to these as, respectively, event-based and time-based tasks (Einstein & McDaniel, 1990). Finally, another central feature of PM tasks is that the rememberer must recollect the intended action at the appropriate instance without an agent stimulating retrieval (see, e.g., Craik, 1986; Einstein, Holland, McDaniel, & Guynn, 1992; McDaniel & Einstein, 2000).

Memory for intentional actions to be realized in the future reflects a very special and unique human ability: that of traveling forward in time, mentally anticipating the properties and characteristics an action may have. Typically, the further ahead the action to be performed, the more general and abstract its representation. Therefore, the individual's time perspective will influence the way the intention will be maintained during the retention interval, monitored, and eventually updated as the appropriate time for execution approaches. In general, future time perspective issues have a great relevance for everyday life activities because, for example, many PM failures are among the most common causes of human errors—of omission and of commission.

Consistent with this idea, Gollwitzer and Brandstaetter (1997; see also, Gollwitzer, 1999) showed that forming so-called "implementation intentions" (a concrete plan as to how, when, and where to perform an activity) enhances the likelihood of actually undertaking the activity, as compared to forming more abstract intentions to perform the same actions. There is no doubt that humans perform actions to reach goals—that is, to create or modify some event according to their intentions. Therefore, intentional actions presuppose some kind of conscious or unconscious anticipation of the intended goal event, some knowledge about the goal and how it can be achieved (Hommel, 2003). Some authors introduced the concept of episodic future thinking, which refers to "a projection of the self into the future to pre-experience an event" (Atance & O'Neill, 2001, p. 533). The concept of episodic future thinking is strictly tied to the concept of autonoetic consciousness, which Tulving (2002, p. 2) defined as "a special kind of consciousness that allows us to be aware of subjective time in which events happened" and that extends from the personal past through the present to the personal future. The combination of autonoetic consciousness and episodic memory allows people to travel in their mental time, reexperiencing the past or preexperiencing the future (Atance & O'Neill, 2001). In terms of PM efficacy, episodic future thinking might be relevant to how we develop and implement strategies for remembering to perform an action in the future. Indeed, the more specific the plan for the future action (the when, where, and how of responses leading to goal attainment), the more likely the completion of the task (Gollwitzer, 1999).

Extrapolating from this reasoning, a general prediction in the realm of PM research, susceptible of empirical investigation, is that the better the ability to preexperience the future event (and hence unfold the plan), the more likely the realization of the intention. However, an important constraint on this general hypothesis derives from temporal construal theories. As already mentioned, individuals represent distant future events at a more abstract level than they represent near future events. In addition, when planning future actions, people tend to consider time constraints only when these events are in the near future. This bias toward present and near-future time plays a role in generating inefficient behavior with respect to completing tasks with distant deadlines. As I discuss later in the chapter, this kind of individuals' time perspective may have important practical consequences in everyday life.

PROSPECTIVE MEMORY IN REAL-WORLD SETTINGS

Prospective memory failures are particularly relevant in everyday life. Winograd (1988) noted that if retrospective memory fails, the person's memory is seen as unreliable, but if PM fails, the person is seen as unreliable. In natural settings, people tend to make a variety of errors. Some may be due to PM failures, whereas others can be attributable to different types of failures such as loss of the content of an intention, inappropriate output monitoring, or absent-mindedness.

Any PM task implies a delay between the time the intention is formed and the moment to realize it. In everyday life situations, delay is a critical aspect of PM. The fulfillment of an intention is often delayed because we are absorbed by another task and can execute the action only at some designated moment in the future. Sometimes, we postpone an action because it is inappropriate in the current situation; at other times we do so because the current task is too demanding to respond to immediately (Einstein, McDaniel, Williford, & Dismukes, 2003). On other occasions, we may even forget to do something and forget that we have forgotten. As Reason (1990) argued, PM failures are among the most common causes of human error.

However, with few exceptions, the types of prospective memory error (in contrast to other types of human error; see Reason, 1990) have been mostly studied in situations such as work settings. But individual decision making and medication adherence has been largely neglected. As a consequence, methods and techniques to improve PM skills have not been developed until recently (e.g., Camp, Foss, Stevens, & O'Hanlon, 1996; Chasteen, Park, & Schwarz, 2001; Einstein et al., 2003).

In the next section, therefore, I concentrate on a practical issue that in the near future might represent an important challenge for applications of PM theory; namely, the differential value of external memory aids in the neuropsychological therapy of memory-impaired patients and in economic agents' behavior. To anticipate, whereas external memory aids have proven extremely useful in helping memory-impaired people to compensate for their deficits, welfare-improving memory aids are purchased only by agents who have (or believe themselves to have) poor PM, hence linking the value of external memory aids to the level of PM self-confidence.

THE ROLE OF MEMORY AIDS IN THE THERAPY OF MEMORY-IMPAIRED PATIENTS

In a recent article, Thöne-Otto and Walther (2003) examined the usefulness of external memory aids as tools for improving memory in brain-injured patients and proposed a new electronic memory aid aimed at compensating for both lack of self-initiated retrieval and for problems during the execution of actions or the evaluation of their outcome.

Successful compensation for PM deficits is a relevant predictor of the ability to live independently after brain injury. In the last decade, neuropsychologists have tried to use memory aids for the therapy of memory-impaired patients. A number of tools have been developed, ranging from simple, portable paging systems which can be fastened to the patient's belt like NeuroPage (Hersh & Treadgold, 1994) to voice organizers (Van den Broek, Downes, Johnson, Dayus, & Hilton, 2000), standard mobile phones (Wade & Troy, 2001), palmtops (Kim, Burke, Dowds, Boone, & Park, 2000), and pocket computers (Wright et al., 2001).

Commercially available memory aids have the advantage of being available to everybody, but they can only support mildly impaired patients. In addition, many commercially available electronic aids are usually too difficult to handle. Therefore, a number of modifications have been proposed (reviewed in Thöne-Otto & Walther, 2003). In particular, Thöne-Otto, Schulze, Irmscher, and von Cramon (2001) have recently presented the MEMOS system, currently under construction, which is especially suitable for patients with more severe deficits. This new memory aid seems particularly promising in that it presents many advantages compared to existing tools. Thöne-Otto and Walther (2003) described it as follows:

> It consists of an Internet server that allows the management of several clients, such as patients, therapists, and significant others. In addition there is an application server managing the execution of incoming and outgoing tasks. Appointments can be entered via different computers, which may be organized at a central service interface or in the patient's home. In addition, appointments can be entered via speech input directly into the mobile device. Relevant patient data are stored in a database. From the application server, information is sent to the patient's personal memory assistant, PMA, a mobile memory device similar to a mobile phone. Interactive contact is possible between the application server and the PMA at any time. Thus the patient can be contacted directly in case ... of missing confirmation of relevant intentions (such as important medication). Patients are asked whether the task can be fulfilled or if it needs to be postponed. In the case of postponement, the system automatically looks for other appointments that may conflict with the postponed one. (p. 8)

Apparently, MEMOS presents many advantages: It is easier to encode, patients are interactively guided through the steps of an action and each step has to be confirmed, execution has also to be confirmed, postponement is possible, and the system automatically looks for other appointments that may conflict with the postponed one (Thöne-Otto & Walther, 2003).

To summarize, external memory aids are of great value in compensating for PM deficits shown by brain-injured patients. However, the role of external memory aids in helping normal people to fulfill their intentions remains an open question.

PROSPECTIVE MEMORY OVERCONFIDENCE AND EXTERNAL MEMORY AIDS IN NORMAL POPULATIONS

The need to view ourselves favorably seems to be fundamental in motivating our behavior, so influences many decisions in everyday life. Typically, humans tend to acquire overconfident belief in, for instance, their skills, abilities, or intelligence. (Holman & Zaidi, 2004; Koszegi, 2000). For example, in retrospective memory research it is well-known that eyewitnesses overestimate their memory for the physical details of a criminal suspect (Loftus, 1979). The concept of personal metaknowledge is typically used to refer to the individuals' beliefs about their abilities, personal utility ("ego utility": Koszegi, 2000), and their obstacles to successful performance, and also to refer to the belief system people hold about PM (Dobbs & Reeves, 1996).

Metaknowledge about PM may have fundamental consequences on the quality of life. For example, task importance (Brandimonte, Bianco, Villani, & Ferrante, 2005; Kliegel, Martin, McDaniel, & Einstein, 2001) and personal beliefs about one's own PM abilities may determine if and when this knowledge is put into play (Dobbs & Reeves, 1996). Yet, so far, we know very little about PM metaknowledge. For example, it is clear that there are important individual differences in this kind of metamemory. Children typically overestimate their abilities to remember to perform an action in the future (Beal, 1988) and do not see the need to set plans (Kreutzer, Leonard, & Flavell, 1975). The findings from self-ratings similarly indicate that the elderly tend to overestimate their PM and give themselves higher PM ratings than do young adults (Martin, 1986). Thus, overconfidence seems an important determinant of PM failures. However, until recently, there has been no formalized model of PM that has taken into account PM overconfidence.

Most recently, Holman and Zaidi (2004) developed a baseline model of PM to be applied to decision-making problems. The model focuses on long-term, episodic, step PM tasks. A step PM task has a wide time period for its execution ("Meeting John sometime today"; see Ellis, 1988, 1996), with an externally imposed deadline. The individual will perform the action within the appropriate period if (a) the intention is retrieved, and (b) the expected utility of doing the task in that period is higher than the subjectively-perceived expected utility deriving from procrastinating the action and relying on future memory. When PM overconfidence (defined as either overestimating the base likelihood of recall in the future or underestimating the effect of temporary forgetting on subsequent recall) is introduced into the model, it reduces welfare in that it leads to not only less than optimal rates of task execution but also the prediction that the probability of task execution can vary inversely with the length of deadline (Holman & Zaidi, 2004).

Thus, PM overconfidence explains inefficient behavior with respect to completing tasks with longer deadlines. When agents are overconfident about their PM, they will inefficiently procrastinate PM tasks, overoptimistically relying on their ability to retrieve intentions. After all, people find it difficult to imagine not remembering what they are aware of in the present and therefore they are overoptimistic about their ability to retrieve later a piece of information currently held in their consciousness. One key result in the model is that PM overconfidence increases the likelihood that extending the deadline will be detrimental to the agent.

In an extension of the model, Holman and Zaidi (2004) incorporated memory aids into the model and demonstrated that only people with poor memory will purchase welfare-improving memory aids. Indeed, if memory aids are sufficiently costly, they will only be employed by agents who believe themselves as having a poor PM. These individuals may eventually perform better on PM than on retrospective memory tasks (Wilkins & Baddeley, 1978), but only because they are more likely to purchase and employ PM aids. However, even individuals sophisticated about their PM limitations (that is, who are not overconfident), who correctly realize the value of memory aids, may still not use them. The reason is that most PM aids (calendars, PDAs, and other reminders) involve immediate costs and future benefits, and if individuals are present-biased (in the sense that each day they would rather put tasks off until tomorrow) and naively unaware of this tendency, they may plan every day to use a memory aid but never actually get around to doing so (J. Holman, personal communication, May 24, 2004).

CONCLUDING REMARKS

Taken together, the previously mentioned considerations highlight some important issues that, so far, have played a marginal role in PM theories and their applications. A first issue pertains to the psychological consequences of humans' time perspective. A second, related, issue refers to the practical consequences of forming temporal representations of future actions to implement strategies according to temporal distance. One such practical consequence refers to the use of external memory aids to enhance prospective remembering. The degree of efficacy provided by external memory aids seems to depend on the existence of memory deficits and on the individuals' expectations regarding future recall of intentions.

Chapter 16

Community Memory as a Process: Reflections and Indications for Design

Giorgio De Michelis
Università degli Studi Milano-Bicocca, Milan

THE OLD LADY IN THE VENICE GHETTO

This chapter focuses on memory as a social process, disregarding both the biological and psychological characterization of individual memories and the links between individual memories and social behavior. I begin with two short stories in which memory as a process emerges as a relevant theme.

Campiello was a "connected community" research project funded by the European Commission (EC). Its aim was to design and experiment with an interactive participatory medium, distributed around a geographic area to support the exchange of information and experiences between its visitors and the local community. When we began it in the fall of 1997, the project team visited Venice for two days to become familiar with the city where we were to develop our experiments. Wandering around the city, we arrived at the Ghetto, the neighborhood originally reserved for Jews and from which all other ghettos in the world gain their name.

It is a very peculiar part of the city, its buildings built in accordance with the strict rules that Venice imposed on its Jewish community. In the Campo del Ghetto Nuovo, as we tried to identify the old synagogues among the upper windows of the many tall buildings surrounding it, an elderly lady rose from one of the benches and asked if she could help. She pointed out the Spanish, German, and "Canton" synagogues, adding that they had not been used for everyday worship since the end of the 18th century, when Napoleon canceled religious discrimination and allowed Jews to open a synagogue (which is still in use) at street level. The older ones are used today only for special occasions like funerals and weddings, she said, although the Spanish one has a difficult stairway so is difficult to reach.

The old lady in the Ghetto offered us, as we began a project to revitalize the communities living in "cities of art," a striking example of what we were looking for: helping those communities to share their collective knowledge about their place and its story, inhabitants, customs, and so on, so that they could become its owners again and, as such, offer warm hospitality to visitors. She acted as a member of her community, cordially opening up it and its home territory it to her visitors. The knowledge she exhibited was community memory, continuously renovating itself as it connects past and present, and reshaping its geographic home in accordance with the current practice of its members.

VISITING THE SITE OF AN ACCIDENT

In the late afternoon of April 18th, 2002 a small plane crashed into the Pirelli Tower, the site of Lombardy's regional government and one of Milan's most popular symbols, killing two women and the pilot. The twin towers of New York's World Trade center had collapsed not long before, so strong emotions were stirred and did not decline, although it was soon clear that in Milan this was not a terrorist attack. I had a privileged observation point because I live close to the Pirelli Tower (for a short period it seemed that my family might have to evacuate our apartment) and thousands of people passed under my windows to see the disaster. In fact, the area around the tower was immediately closed to the public, and my street was the closest path to where the effect of the plane's impact was most clearly visible.

The procession of these spectators was most numerous in the evening, and even at night. Generally silent when coming, they became noisier when leaving as they commented on what they had seen. It was not a new phenomenon: when Lady Diana died, and after the World Trade Center was attacked, people exhibited the same curiosity, the same desire to have the most direct and close experience of the event as possible. The fact that I could directly observe the behavior of the people visiting the Pirelli crash site increased my awareness of the phenomenon. Coming to see in person the effects of the accident was a way of not remaining a passive spectator but becoming one of its actors. By saying "I was there yesterday night," people shifted attention from the event of the crash to their visit to the site, embedding the event within their own stories of action and interaction. What we communicate must be something we have lived directly and want to share with others. Our memory, therefore, combines our direct experience with the stories told by other members of our community—sometimes indeed, by means of surprisingly ingenious behavior, we remember ourselves as participants in events we have actually only witnessed through the media.

My Venetian and Milanese stories, although very different, are good examples of how sharing the memory of past events constitutes the texture of our social relationships. On one hand, the old lady welcomes visitors to the Ghetto, sharing with them the memory of the Jewish community living there; on the other, those going to view the Pirelli Tower after the accident are creating personal memories of it that they can share with other people and thus contribute to their common history.

An investigation of memory will deepen our understanding of the social dimension of human lives. So this chapter focuses on the dynamic dimension of a community's memories to learn about the type of support we can design to enhance their vitality and effectiveness. The next section briefly recalls the fundamental features of communities and their memories. The third section hints at how systems might be designed to support communities and their memories. The fourth section focuses on a major aspect of community knowledge, its ontology. The last section summarizes my major points and poses some remaining problems.

THE MEMORY OF A COMMUNITY

"Communities are social entities whose actors share common needs, interests, or practices: they constitute the basic units of social experience" (Huysman, Wenger, & Wulf, 2003, p. xi). From a complementary viewpoint, looking at its phenomenological features, a community is an aggregate of interacting people sharing a place (local communities), a language (as in Italian communities abroad), an experience (communities of practice or interest; see Lave & Wenger, 1991; Wenger, 1998), and a memory (for example, religious communities). Each community is located in this four-dimensional taxonomic space. Its experience, place, language, and memory constitute an unique and inextricable bundle.

Members of a community share their experiences—either living them together, or, more often, conversing about them (as in the Pirelli Tower story). In their conversations, they share a language—or, more precisely, a language game (Wittgenstein, 1953)—which, as well as reflecting their experiences, defines their potential for future actions and interactions. This shared language transforms the portion of space where they live into a place, investing it "with understandings of behavioral appropriateness, cultural expectations, and so forth" (Harrison & Dourish, 1996, p. 69).

In other words, a place is a portion of space decorated with sense. The members of a community share a place, a language, and an experience because they share a memory of the past as well as of the future (see Maria Brandimonte in chap. 15 of this book). As it evolves, in fact, community memory links past experiences to the new experiences within which the latter are recalled and shared, helping to transform the remembrance of a past experience into a new experience. Memory constitutes the cognitive counterpart of the place of a community. Every community has therefore an inner as well as an outer identity. On one hand, the place of a community has boundaries which delimit it, separating members from nonmembers; on the other, its memory becomes the foundation on which members build their individual and collective identities, distinguishing them from other people.

The boundary delimiting a community is generated by its members' greater potential for action and interaction in its place, and, simultaneously, by the physical and institutional filters nonmembers have to pass through to participate in it. It has also to be noted that the boundary is not only delimiting but opens the community to newcomers: Boundaries can be crossed, and memories shared.

In recent years the attention of several disciplines has returned to the concept of community, both to underline the crisis of social experience and to suggest the potential for socialization that human beings can still access. The philosopher Jean-Luc Nancy (1990) has interpreted Heidegger's concept of *Dasein* (being there) as *Mit-Sein* (being with), claiming that our life experience is essentially social and that the place of this experience is the community. Even without adhering to Nancy's radical view that people's personal identity emerges from the *Mit-Sein* within communities, the Californian school of work ethnography has defined the concept of "community of practice" to characterize the communitarian dimension of any work practice (Brown & Duguid, 2000; Lave & Wenger, 1991; Wenger, 1998). This is more than a recognition that work is frequently performed in teams: Communities of practice constitute the social context where we live our working experiences and build our competence in performing them. Communities, in fact, are where people learn to practice, in a process moving from a peripheral to a central participation (Lave & Wenger, 1991).

Even from this very short and schematic resumé of the concept of community, it should appear clear that memory plays a central role in it, linking together its language, space, and experience. On the memory they share, the people of a community ground their common membership, coupling their common identity with their common potential for action and interaction. Community memory is not a static reflection of past experiences, not only because it is both memory of the past and of the future, but also because it is continually changed by the conversations of its members. Within a community, as the Pirelli Tower visitors showed, recalling and sharing an experience is itself an experience. Narrations allow those who did not participate in an experience to share its memory with those who did. Narrations also allow the community's borders to be crossed because, listening to them, its members can open themselves to the experiences of other people and communities—and conversely, through them, share their experiences with others (the old lady in the Venice Ghetto). Community memory is intrinsically narrative.

All this suggests that memory is relevant for communities as a process through which their members share knowledge of past experiences, transform the space where they live in their place, and cocreate the language (game) through which they can interact and give sense to their actions. Memory is not a collection of information about the community's past and future events, but rather the process through which that collection is continuously recreated. Memory is the process through which a community continuously creates its knowledge, or, in other words, through which it capitalizes social value (Nonaka & Takeuchi, 1995).

The crisis of communities in contemporary society is twofold: On one hand there is no longer a shared set of criteria that orders the multiplicity of communities where people live their social experiences, and they seem unable to deal with the diversities this creates; on the other, globalization is reducing the space for communities, weakening their identities and the ties binding their members together (Putnam, 2000). In many parts of the world, moreover, communities react to their decline by aggressively closing themselves and considering nonmembers as enemies. The concern for communities that many observers feel today is, therefore, a fear of what communities can do against social order (as many dramatic cases all around the world testify) and of the impoverishment of social life their decline can provoke (as everyone can see all over the Western world).

From a community memory viewpoint, this means we can point to two different aspects of a community's decline: On one hand, its memory becomes ever more closed to the outer world, so that its knowledge becomes impoverished; on the other, it becomes ever more unable to retain the experiences of its members, so that there is no way to capitalize the knowledge created within them.

SUPPORTING THE MEMORY OF A COMMUNITY

It is evident that the decline of communities is mainly a political question, requiring choices to be taken at the national and supranational levels. This chapter is not the right place and I am not the right person to discuss this issue. But we can observe that, even when these political choices are taken, it may be very difficult to make them effective, because the decline or closure of communities develops as a spontaneous process determined by the concrete conditions of contemporary social life. Mobility (spatial and social), mass communication, international networks, specialized services, and national and supranational institutions, seem all to converge toward voiding the place of communities, weakening their social ties, destroying their roots, and homogenizing their language.

Despite attempts to resist this tendency, the growing complexity of our life makes the survival of communities ever more difficult. Robert Putnam (2000) pointed out at CSCW 2002, the Computer Supported Cooperative Work conference, that information and communication technology can counteract this trend; indeed, he ended his talk with a call to CSCW researchers and practitioners to dedicate their intelligence and efforts to developing systems to help declining communities revitalize themselves. But although we strongly agree with this call, we must underline that technology as such cannot help revive communities; on the contrary, information and communication technology is also a significant means through which globalization induces their decline.

We need, therefore, to understand what requirements a computer-based system must satisfy to be a community support system. This understanding can only emerge from careful observation of human practices, and evaluation of the systems that communities may adopt. We can now list, from the community memory viewpoint, some of the requirements we have discovered in our projects and the systems we have developed to meet them:

1. Supporting community memory requires supporting, on one hand, access and navigation, on the other, renovation and transformation—In most systems devoted to knowledge management or the support of organizational memory, as well as in many systems supporting communities, there is a clear separation between the mechanisms for accessing content and for creating it: The former are designed for generic users and stress user friendliness, the latter are for professional editorial roles. Electronic publishing systems, for instance, can do little to support a community memory, in particular with respect to updating the memory of its members' experiences.

A true community memory-support system is not only knowledge-based, making its content easily accessible, but supports memory as a process: from its enrichment (by creating new knowledge) to its storage (with efficient search mechanisms), and from its presentation (so that users access it immediately) to its diffusion (point-to-point or broadcast). The system must support the whole knowledge circulation process, not only its final stage.

2. Supporting community memory requires supporting users whoever, wherever, and whenever they are—Community and civic networks
emerged in the 1990s, seeking applications of information and communication technology able to enhance socialization (Casapulla, De Cindio,
& Gentile, 1995; Ishida, 1998; Ishida & Isbister, 2000; Schuler, 1996).
But, despite their numerous merits, community networks do not go beyond the narrow focus of offering personal-computer access to information and communication services. So they risk confusing the community
of their users with the community they refer to, and, on the other hand,
creating a gap between those users who like to play with the technology
and those who do not. This problem is more general than the well-known
"digital divide" (Warschauer, 2003). Community memory is pervasively
present in any situation of social life, so any system supporting it must deliver its services to members whoever, wherever, and whenever they are.

This means that systems supporting community memory must be multichannel, each channel providing specific means of interacting with the system, based on its features and the situation of its users. The Campiello
system (Agostini, De Michelis, Divitini, Grasso, & Snowdon, 2002) is accessible through personal computers, large interactive screens in public places
(the "Community Walls" in Fig. 16.1), and coded paper ("Interactive Fliers"). The Milk system, devised in another EC research project, is accessible
through personal computers, large interactive screens located in meeting
rooms and social spaces, and cell phones. Building multichannel systems
such as these requires a rethinking of their architecture, with a clear separation between the knowledge-management system, characterized by multiple services (searching, indexing, profiling, and so on), and the interaction
managers for each channel who use them selectively (the Milk system's
architecture in Fig. 16.2, for example, and in Agostini et al., 2003).

3. Supporting community memory requires presenting records of
past experiences so that the actions and interactions of members become
more effective—The facility to search in the knowledge base for internal
or external knowledge is important in any knowledge management system and therefore in any system supporting community memory. But
when people are engaged in some task for which a piece of data is missing, they are, in general, unwilling to search for it. This is especially so
when people are interacting in groups and there is no time to suspend
what they are doing to search the knowledge base. They really need the
system to present them with all the relevant knowledge, together with
the object on which they are working: on one hand, all the related documents or objects (explicit knowledge), on the other, the links to the people who have competence in it (tacit knowledge). Put another way: on

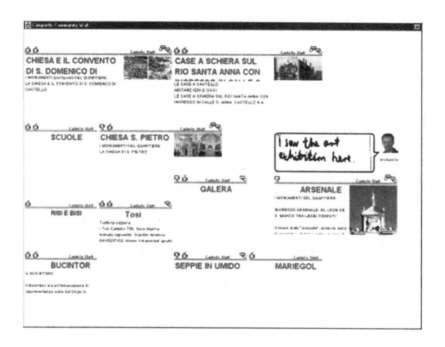

FIG. 16.1. A "community wall." Large interactive screens in public places: part of the Campiello system.

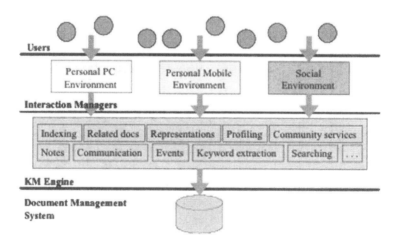

FIG. 16.2. The architecture of the Milk system. In this multichannel system, the knowledge-management engine is clearly separated from the interaction managers for each channel.

one hand, all the documents, messages, and people related to the object (internal knowledge), on the other, all the documents, objects, and people semantically related to it (external knowledge).

In Milk we created for this purpose the "view with context" interaction mechanism, situating any document which users might open in a visual context in which related objects and people are presented and made accessible in an orderly way (Agostini et al., 2003).

4. Supporting community memory should avoid creating distinctions between content creators and consumers—Discussion of requirement number 3 points out an important aspect of the practice of communities. Accessed knowledge makes less sense when people are purely consumers of information than when it is part of an activity where new knowledge is created. Nonaka and Takeuchi (1995) described this process as a "knowledge spiral," a continuously connected internalization and externalization of knowledge.

This means that content-creation and content-access services should be both delivered online, and inextricably linked so that users continuously switch between them. In Atelier, another EC research project, we created a working environment for architecture and interaction design students which exemplifies the variety of functions and services that might be provided for accessing and creating new content. Atelier paid particular attention to integrating physical and virtual objects and spaces—all tools and systems for designing and presenting new artifacts—into an "augmented place" (see Fig. 16.3, last section of chapter, and Binder et al., 2004).

5. Every community has its own ontology. Supporting community memory implies building a knowledge base reflecting that ontology—As was discussed earlier, collecting objects, documents, and links to people is not sufficient for delivering a system supporting community memory: the system must also be able to search, filter, and present knowledge about a given object in accordance with both semantic and pragmatic perspectives. When the community is characterized by professional competence, as in the case of communities of practice or of interest, the semantic component of a "view with context" may be based on the ontology characterizing a specific discipline: In the Atelier project, for instance, we developed the ontology of architecture. This allows external and internal knowledge to be integrated through changes users may make to the system. But the system's pragmatic component must strictly reflect the experiences of the community. It is not easy to define the ontology of a community's practice: on one hand it is semantically closer to the philo-

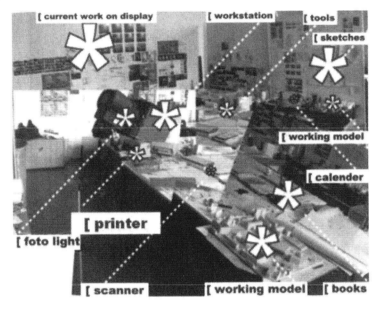

FIG. 16.3.   A student working environment, part of the Atelier research project (see also Figs. 10.2 and 10.3). Physical and virtual objects and spaces are integrated into an "augmented place."

sophical definition of ontology than that typically developed within artificial intelligence and knowledge engineering; on the other, it has to be based on common sense, rationality, and some universal properties of human practice.

The next section discusses the rationale of the pragmatic ontology we designed for Atelier, deriving from a proposal by Leiva-Lobos (1999).

THE ONTOLOGY OF SOCIAL LIFE

When looking at human practice, it may be useful to assume Wittgenstein's viewpoint in his *Tractatus Logico-Philosophicus* (1961), Proposition 1.1, which asserted that "The world is the totality of facts, not of things" (p. 5). In other words, the ontology for which we are looking concerns the characterization of the facts constituting human practice. These facts may be characterized by (a) locating them in time and space, and (b) according to the humans participating in them, or, more precisely, the communities in which they happen. On this assumption, Leiva-Lobos (1999) developed a "3-ontology" as a universal basis for the ontology of social practice, and therefore also of the practice of a community, where every fact has three dimensions: its Place, its Event, and its People (communities).

The three dimensions of a fact are strictly interrelated, because they recall each other to deepen its characterization. Places are spatial entities characterized in time (what happened, happens, or will happen there?) and in people (who lived, lives, or will live there?). Events are temporal entities characterized in space (where did, do, or will they happen?) and in people (who participated, participates, or will participate in it?). And people are social entities characterized in space (where did, do, or will they live?) and time (in which events did, do, or will they participate?).

These three basic categories of 3-ontology are the universal basis of the specific ontology of any community. Within them users can define any subcategory and any link among subcategories. The ontological (in its philosophical sense) nature of the place, event, and people categories is confirmed by their being the basis for three synthetic forms of representation that are today standard all over the world. Spatial entities can be represented in maps, temporal entities in calendars, and social entities in directories. It is not by chance that calendars, maps, and directories are the basic external supports for human memory and that the possibility of sharing them makes them especially suitable for supporting the memory of a community for the benefit of its members, newcomers, or visitors.

CONCLUSION

The Atelier project, whose aim was to design an environment supporting creative learning, developed systems that created augmented spaces in a more radical and explicit form than in previous projects: by combining physical and virtual spaces. It is a line of research opened by the seminal work of Weiser (1993) and gaining interest in different research fields, as is shown in many research programs ("Disappearing Computer" in the EC's 6th Framework Programme, for instance), conferences (like "Ubiquitous Computing"), scientific journals (like *Pervasive and Ubiquitous Computing*), and books (such as Norman, 1998).

Moving within this perspective gives a deeper understanding of the intrinsic connection between the place and the memory of a community. The memory's effectiveness depends not only on the quality of information it contains but also, and this is not a minor feature, on its spatial distribution and representation: to design a system that supports the memory of a community is to design its place. I claimed earlier that the growing complexity of social interactions today makes the survival of communities more difficult: In spatial terms this means that physical space is no more able to host the place of a community. The systems we design for supporting their memory must therefore create new augmented places exhibiting the properties of openness, multiplicity, and continuity which physical spaces cannot have (De Michelis, 1998, 2003). In Atelier, for example, students could configure their working space to support dynamic representations, linking videos or other digitalized information to paper drawings and maps (Fig. 16.4 and Binder et al., 2004).

Another device designed within Atelier has inspired a further conceptualization of the nature of augmented places. The Texture Brush designed by Imagination Computer Services of Vienna for covering a physical model with a virtual texture (Fig. 16.5 and Binder et al., 2004) exemplifies the type of object that should inhabit an augmented place. We have called them prototypical "mixed objects" (Binder et al., 2004) and are currently analyzing their features and qualities as a basis for the future design of systems supporting user communities.

Considering such systems, including those supporting community memory, as mixed objects gives a solid foundation to the multidisciplinary approach with which our projects have experimented from almost 10 years (Agostini, De Michelis, & Susani, 2000). To design systems based on information and communication technology, transform the space where they will be located, and finally, offer new possibilities for the actions and inter-

FIG. 16.4.   Another view of the Atelier workplace. A reconfigurable environment of sensors, artifacts, digital media, and projections supports dynamic linking of paper, three-dimensional, and digital representations.

FIG. 16.5.   The Texture Brush system. A real brush, passed over the surface of real objects, allows different textures to be projected onto them.

actions of their users, requires the collaboration of three different cultures: technology, design, and the social sciences. Multidisciplinary design is not easy because in this context these cultures have little experience in cooperating together. But our experience of the Campiello, Milk, and Atelier projects indicates that it can generate applications of information and communication technology which are entirely new.

ACKNOWLEDGMENTS

This chapter presents ideas which emerged from the author's participation in several research projects, mainly Campiello, Milk, and Atelier (funded by the EC within its Framework Programmes), and Mais (funded by the Italian Ministry for Instruction, Universities and Research within the FIRB Initiative). For everything he learned during these projects, the author thanks the many members of the participating research institutions. He also especially thanks the organizers and participants of Interaction-Ivrea's Foundations of Interaction Design symposium for the opportunity to reflect on the concept of memory in the discussions following his talk.

MARKET

Chapter 17

Bad Design by Design? Economics Meets Other Types of Interactions

Lucio Picci
Università di Bologna

ECONOMICS AND DESIGNING: THEIR COMMON GROUND

A few centuries ago the discipline of economics was not yet a well identified and separate branch of knowledge. Its foundation and its subsequent development required hardships and toil by a group of pioneers coming from disparate fields of knowledge such as moral philosophy, physics, and engineering. Their efforts have produced a rigorous theoretical construction that many consider as a true sanctuary, providing its priests—the economists—with a set of powerful tools to analyze reality. As an economist myself, I share a sense of respect for the intellectual construction, and its foundations, that our predecessors endowed to us.

In the eyes of many detractors of the science of economics, indeed, the project proved to be too successful, because it emboldened economists to the point of arrogance, and they started using their analytical weapons to conquer and dominate bordering fields of knowledge. Other lands of knowledge—such as the political sciences, sociology and geography—have become fair game for economists, who are now seen by their indigenous inhabitants as an invading army with no United Nations approval: Economics is sometimes seen as an imperialistic social science, and it is treated with an equal share of scorn, respect, and envy.

So before any misconceptions arise between the indigenous inhabitants of the honorable design discipline and myself, let me assure you that I come in peace, and that my only purpose is to contribute to a fruitful dialogue between our two seemingly very different fields of knowledge (and maybe to distract you as our troops get ready).

As I argue, this difference is mostly one of looks, and may just follow previous habits, because designing and economics increasingly share a common ground. In particular, both have to do with describing interaction systems, an endeavor where economists have learned lessons that are often subtle and surprising. For example, sometimes the optimal design of a product is not the one that provides the interested parties with the best possible interaction experience, but is an intentionally "bad" design. Such an outcome may seem illogical to all those designers whose mission is to do their very best in the service of the final user of a product. Contrary to that belief, we sometimes observe examples of "bad design by design."

Another issue that I raise here is the organization of the design, prototyping, and production process, seen in the light of open-source software production, an emergent and interesting designing and production method. I draw some parallels between open-source software production and the design and prototyping process familiar within the interaction design community, to discuss their similarities and their differences.

As a prerequisite for both discussions we need to understand in what sense economists increasingly consider themselves as designers and engineers.

"MECHANISM DESIGN": THE ECONOMIST AS AN ENGINEER AND AS A DESIGNER

The objects that economists design are called "mechanisms," sets of rules to solve a problem of allocation (of goods, other resources, or people) in a situation where there are conflicting goals. A mechanism is an interaction system in the sense that it gives form to, and orientates, a set of interactions between different economic agents. An example clarifies the point.

Assume that Interaction Design Institute Ivrea organizes an internship program for its students. When the numbers are not too big, this type of task can be effectively carried out informally, trying to juggle all the requests coming from the different actors—the students, and the organizations who participate in the program. However, many conflicting goals are present, and the allocation problem of an internship program is conceptually challenging. Some students are better than others, and so are some of the organizations. The students would like to do their internship within a good organization, but firms also would like to have the best students, particularly when they look at them as potential future employees. Moreover, students differ among themselves in other dimensions: They live in different places and may prefer an internship not too far from home. Some, moreover, have spouses and possibly children, so their degrees of freedom are limited by these facts of life.

Guaranteeing the best match between students and organizations is not easy, and "matching theory"—a branch of economics—is devoted to understanding these types of problems. When this theory is used not just to understand but to solve a practical matching problem, by indicating a set of rules to be followed when managing it, we have an instance of "mechanism design." More details on the "internship problem," incidentally, are in Varian (2002), with reference to Roth (2002).

PRICE DISCRIMINATION AS MECHANISM DESIGN

The next example of mechanism design has to do with what economists call "price discrimination," and it leads us to the possibility of rationally desiring a "bad" design for a product.

Producers of goods (or services) would rather not sell at a uniform price, but prefer instead to impose a higher price on those willing to pay more for the product—because they are rich, or because they very much like it. As long as they recover (marginal) costs, producers are also more than happy to sell at a lower price to people who would not otherwise buy the product. "Price discrimination," as this activity is called, is very common. But it is not straightforward to achieve, because potential customers will not easily let the producer know their willingness to pay: People who are asked how much they would like to pay for a given product, knowing that what they declare will reflect itself in the price tag, probably understate the truth.

However, it turns out that there *are* ways to know, at least approximately, how much people are willing to pay for a product, or, in economists' jargon, to obtain a "truthful revelation of preferences." This knowledge is often obtained by creating options where, by choosing according to their self-interest, people implicitly (and usually without being aware) declare their willingness to pay. It is another instance of "mechanism design": When the mechanism is designed properly, customers end up self-selecting themselves into the appropriate category, where they are made to pay according to their true willingness to pay.

A few examples clarify this point. In the market for books we can roughly assume that there are two types of readers. Some are avid readers of a particular author's books and will pay much for them; sometimes, indeed, the publication day of a new book by a well-known author amounts to a memorable day in the life of his or her fans (ask my nephew about Harry Potter). Other readers, by contrast, are less interested and will only buy the book at a lower price.

The publishing house wants to keep these two types of customers separate. This is achieved by distinguishing between the hardcover edition and the paperback. The fact that the former has a better binding, and is often printed with a bigger and more readable font, is in fact irrelevant: The difference in cost between the two editions is negligible and in no way justifies the difference in price. What is relevant is that the cheaper paperback edition is typically published after an intentional delay. The self-selection of customers is obtained through the difference in their impatience: The eager reader wants the book as soon as possible, whereas the less interested reader can wait.

Mail-in rebates are another instance of price discrimination. A rebate is offered to the customers of a product who take the time to fill in a form and send it to the producer. Those willing to pay more, who tend to be richer, value their time more than others, and will not bother to mail the rebate: They pay the full price. On the other hand, the people who do apply for the rebate, by taking the time to do so, implicitly declare their low willingness to pay for the goods, and end up paying the lower price.

One last well-known example of price discrimination by self-selection is provided by the air travel industry. The price of the same seat on an airplane can vary significantly, depending on the type of ticket, on how far in advance of the flight it is bought and, more generally, on the restrictions it carries. A typical restriction involves stopping over at the destination on a Saturday night. There is no technical reason for this—airplanes are not required to rest away from home on those nights. The rationale has to do with price discrimination: Tourists—the people with a low willingness to pay— do not mind, and often actually prefer, staying out on Saturdays; business people, on the other hand, who are willing to pay more for their tickets, want to be back from their trips for the weekend.

What about "bad design?" All these examples of price discrimination, and many others, involve the introduction of some type of bad designing. Consider mail-in rebates. If the purpose of the seller is to give a rebate of, say, $2 on a $10 purchase, there is an obvious way to better design the interaction with the customer: Write "$8" on the price tag, and avoid the hassle of mailing in (and processing) the rebate. However, by choosing this hypothetical "better" design, the discount would go to everybody, including the people who would buy the product even at the higher price: there would not be price discrimination.

Paperback books are also an example of bad design: They could be improved at very little cost, and could be published promptly and without making people wait. In other words, they could be their hardcover version. Similar considerations apply to cheap, but highly restricted, airplane tickets. Without the restrictions, the interaction between customer and product would obviously be better. In all these cases, however, the inconvenience of the cheap version of the goods is for a reason: to obtain price discrimination.

Let us generalize these examples. With mail-in rebates the bad designing is about the way a piece of goods is priced. With books and restricted airplane tickets, the bad designing also results in some inconvenience for the customer. The designing involves not a single product but a whole suite of different versions of a product, so that they induce self-selection among different types of customers—as in first- and second-class transport in trains, for instance. This is obtained by artificially creating some flaw in the lower price version(s) of the product, so that it is still appealing enough to the low-willingness-to-pay customers, who pay less for it, but is bad enough for the high-willingness-to-pay customers, who prefer to spend more for the better version.

In some cases, such as in the hardcover and paperback versions of books, the difference in cost to the producer between the different versions is minimal. Sometimes, indeed, the more expensive version is cheaper to produce, like the IBM Series E laser printer at the beginning of the 1990s, whose slower and lower price version was obtained by introducing a "slowing" integrated circuit into the faster and more expensive version (cited in Deneckere & McAfee, 1996). So, to price discriminate, producers often actually have to design *bad* products. The job of the designer is to design the suite of versions jointly, one in relation with the other, to obtain for the worst version(s) what we could call an "optimally bad design."

HOW JUSTIFIABLE, AND HOW RELEVANT, IS "BAD DESIGNING?"

There is a common perception, on whose cultural and historical origin I will not speculate, that the price of something has to be "fair." If there is such a thing as a fair price, it must be unique, and selling the same product (or two very similar products) at different prices to different people seems unjust, unethical. So, is price discrimination "unethical?" Is this state of affairs good only for the producer, who reaps a higher profit at the expense of the customers? Would it make sense to forbid price discrimination so that, for example, we could all afford the luxury of hardcover books and unrestricted cheap flights? Not quite. Let us consider this issue from the point of view of the welfare of customers.

It turns out that price discrimination often serves the interest of not just the producer but customers too, because it allows the market to serve the people with a low willingness to pay. For example, if airlines were forced not to price discriminate, chances are that the cheap fares at the base of most international tourism, and of some international academic projects, would not exist.

And it is not only some people who would suffer. Many industries would not exist without the possibility of price discrimination, because competition on a unique price would drive the price below the level that allows producers to recoup their fixed costs. This is particularly true for industries characterized by a cost structure in which the fixed cost for setting up the business and building the "first copy," or prototype, of a product is high, but where additional copies of that product are inexpensive. Writing the manuscript of a book, for instance, or writing, executing, and recording a piece of music, requires much effort by highly skilled workers; but printing a book or copying a CD has a very low unit cost. When the cost structure for production has such pronounced "increasing returns to scale," as economists would say, price discrimination is indeed a crucial issue not only for the producers, for them to stay in the market, but also for the industry to survive and for customers to be served.

Most information goods, and many information technology products, share such a cost structure. Because these goods are increasingly important, it follows that price discrimination is a practically very relevant issue, and the idea of having a "bad design by design" is much more than just an anomaly.

SERVICE DESIGN VERSUS MECHANISM DESIGN

Before I move on to a different topic, I would like to clarify the relation between the activity of "mechanism design" and design proper. A mechanism is not a physical good but an interaction system. Such a definition resonates with what designers call "services." So can we translate "mechanism design" with "service design?" Not really, because the two concepts are not identical, and their difference is of interest to us here.

Designing services has mostly to do with taking care of the interaction experience once its rules have been laid out. Mechanism design, on the other hand, has to do with designing the interaction rules. Mechanism design can be seen as the "wired" part of the interaction machine. Service design is the way the machine looks. Mechanism and service design represent two different levels of what we could call an "interaction machine," by which I mean the practical implementation of a designed interaction system.

In fact, thinking about two separate levels within the interaction machine only represents a first attempt to relate service design with mechanism design. I believe that the wired components and their "look and feel" can't be decoupled. For example, restraints in the latter may endanger the implementation of a given mechanism design, and could require its rewriting. On the other hand, the presence of possibilities at the service level, not foreseen by the "economist-engineer" working on the design of the mechanism, could open new possibilities in designing the "wired" part of the interaction machine.

If these two levels are so interdependent, we could develop a unified vision of interaction machines, of both their wired and their look-and-feel parts. Such a unified approach could be advantageous to both disciplines: mechanism design and service design. Or, should these two disciplines really be thought of as separate?

THE ORGANIZATION OF DESIGNING *AND* MANUFACTURING: OPEN SOURCE EVERYWHERE?

I have not mentioned so far that the study of organizations is one of the fields of knowledge that economists managed to colonize. For this purpose, economists developed the so-called "transaction-cost theory of the firm" (and of other types of organizations). According to this theory, in deciding whether to do something by interacting in a market, or within the boundaries of an organization, of paramount importance are the costs incurred in carrying out the transactions; if these costs are relatively high in an "arms-length" market relationship, there is room for setting up an organization where the same transactions are carried out by its members within the boundaries of the organization itself, and without using a market (see Williamson, 1975).

This knowledge, besides exposing my credentials as an economist in speaking about organizations, allows me to consider what interaction design could learn from a mode of organizing production that is receiving much academic attention: the production of open-source software—that is, of software distributed under a license that permits users to freely share or modify it.

It is a type of production with very little (apparent) structure, very horizontal in its organizational relations, and allowing much room for collaboration and to experimentation. The tight relation between production and experimentation is shown by one of the open-source community slogans: "deliver early, deliver often." There is no clear distinction between the planning and the production phase of a product.

Also, open-source developers usually do not pay much attention to the codified tenets of software engineering: The discipline that establishes how software projects should be conducted and determines, among other things, that the requirements of the software to be developed should be analyzed formally and at length. Open-source software development usually starts with the purpose of solving a problem that the developers themselves face, but does not include a proper analysis of requirements.

There are noteworthy analogies between open-source software production and some ideas familiar within the designing community. Consider the issue of governance. At least in principle, within an open-source software project the way decisions are taken does not preclude anybody's contribution. In this sense, open-source software production is an example of what has been defined as "participatory design," a theme of some relevance within the designing community (see Chapter 10 by Pelle Ehn in this book). Another trait in common between open-source software production and interaction design is the common emphasis on prototyping (see Chapter 19 by Bill Moggridge).

These are just analogies, but suggestive ones that raise issues for consideration. One issue has to do with incentive systems: What makes people willing to play the open-source game? We know that many open-source programmers work for free, and economists' analyses are based on the idea that people are self-interested, so that observing highly qualified professionals not receiving a pecuniary retribution for their services amounts to a puzzle. An explanation of this apparent conundrum is based on the observation that programmers' productivity can vary enormously depending on their skills, and that these skills are hard to communicate to potential employers. By participating in an open-source project, high quality programmers are able to reveal themselves as such, to not just their fellow programmers but the world at large. While working for free, they acquire a reputation as a good programmer, which has a significant market value (see Lerner & Tirole, 2002).

The first message that the open-source experience delivers to interaction design, then, has to do with incentives. If designers want to experiment with more participatory forms of design, they should think hard about the incentives for participation, possibly taking suggestions from the highly successful open-source community. There, persons may even work for free, as long as their good work contributes to building a good reputation that can be utilized in the future, for example in the job market. A crucial aspect of the open-source community is that individual good work does not "get stolen," to the point that one of the greatest "crimes" is to deny someone's contribution. The incentive system of open-source software production is based on acknowledging personal contributions to the project.

Another issue worth analyzing is the relations between designing an item and its actual industrial production. This distinction is simply not present in open-source software production: The first version of a program is often made to meet the needs expressed by the programmers themselves, and is used right away, at least by the programmers, who can put up with its early idiosyncrasies and, almost always, lack of good documentation. Designing, experimenting, producing, and using are all meshed together, and in this respect the open-source software production method itself represents an impressive interaction system at work.

Within more traditional contexts, on the other hand, the participatory part of the design process and its experimental emphasis, when present, only occur during the prototyping phase. Once the product has been designed, it changes hands and goes to the factory, which to the designer is to some extent a black box with impregnable walls. When designers have finished their job, they move on to a new project.

Open-source software production suggests new questions and a change in perspective. Could its mode of production, with such a tight integration between the designing and the production of goods, be extended to goods other than software? Could we have an open-source car, bottle-opener, or chair? Is open-source production an interesting curiosity only good enough for software, or can we have "Open Source Everywhere?"

Although it is not easy to answer these questions, I find such a possibility very intriguing and worth exploring. Researchers from the interaction design world who are today thinking about how designing activity is carried out, will breath some fresh air by considering the evidence on open-source software production. A few analogies do not make a relevant case, and more research work is needed. However, we should welcome the adoption of an expanded vision of the designing *and* of the manufacturing problem, where the distinction between the two, designing and production, is not so obvious, nor so inevitable.

CONCLUSIONS

Economics, I have argued, is increasingly about the design of interaction systems, and its analytical tools allow new insights to the field of interaction design. I have provided a couple of examples to make my point. First, I have shown a case of "mechanism design"—price discrimination—where it makes sense to produce goods that are intentionally and prima facie suboptimal. Observing "bad design by design" is not an intellectual curiosity without practical relevance. On the contrary, we observe it frequently in a world where "increasing returns to scale" are ubiquitous.

"Bad design" allows for price discrimination, which should not be considered unethical, because it often serves societal needs. It follows that implementing price discrimination is in accordance with the objective of contributing, through good design, to human happiness: to make a good design, sometimes what is needed is a "bad" design. Foul is fair, then? Not quite: Simply, things have become a little bit more complicated than they used to be, and this is healthy, because it means that we have enlarged our perspective.

I have also argued that a contribution to interaction design could come from economics as a discipline that studies organized behavior. The example of open-source software production casts light on the process of designing and, in particular, on the distinction between designing and manufacturing. True, it sounds futuristic at best to think that the open-source mode of production could be extended to other realms; that we can apply such a playful way of dealing with what we want to design, experiment, produce, and use; and that we do all this at the same time, without a clear distinction of phases. However, because the discipline of interaction design is thinking about its future, I believe that some science fiction is not out of place here.

The interaction design of tomorrow, then, should be able to consider not just the interaction between the goods it produces and the people who use them, but the whole set of actors of the system: users, producers, *and* designers. There are many complicated interactions involved, so a theory is needed to provide a framework and make the problem's complexity manageable. Economics can give a hand in this effort.

An appropriate analytical framework would allow us to see more clearly through the different issues I have presented. For such an effort, I am convinced, we would receive a double reward: Not only would we understand better, we would be able to do better. We would be more prepared to fulfill what I see as the highly ethical goal of interaction design: to make the whole interaction experience with the goods, material and immaterial, that enrich our lives, as enjoyable and beautiful as possible.

ACKNOWLEDGMENTS

The author of this chapter thanks Sebastiano Bagnara, Tom Erickson, and Raimondello Orsini for their helpful suggestions.

Chapter 18

From Interaction Costs
to Interaction Design

Maurizio Franzini
Università degli Studi di Roma "La Sapienza"

THE ECONOMICS OF INTERACTION

Interaction as mutual or reciprocal action or influence has been studied extensively in both natural and social sciences. It encompasses a broad array of situations and widely varying subjects: individuals, objects, animals, plants, and organs. Almost all of these can interact between and among themselves. Interaction can take place in very different contexts, and can be conflictual or cooperative. Humans in particular can interact directly or through more or less sophisticated objects. Sometimes humans interact by means of other humans who act as intermediaries. Given the choice, humans may not want to interact. On the other hand, they might want to interact, but not be able to.

More important, the results of interaction can be extremely varied, ranging from very good to very bad. In economics, the most studied interaction is carried on in the market by a seller and a buyer. Typically this is a conflictual interaction between selfish agents driven by reciprocal advantage, the lure of an ex post facto situation better for both parties. Interaction of selfish individuals can make everybody better off. This is the miracle epitomized by Adam Smith's invisible hand (1776/1904). Self-interest is harnessed to a greater good, "led by an invisible hand," to promote an end which was no part of anybody's intention. This is probably one of the most

261

striking cases of virtuous interaction. However, it can be much more problematical and less favorable to social welfare.

Interaction, therefore, is a very complex phenomenon and it would be futile to search for a general theory. This has clear implications for the subject at hand: interaction design. It is essential that we draw a perimeter and choose a well-defined perspective.

As an economist, the approach I suggest in this chapter is founded on the notion of interaction costs and on the assumption that interaction design should be geared to making those costs as low as possible. Reducing those costs will, in general, enhance social welfare. This is the reason why interaction design may be considered as a socially relevant task. In what follows, I introduce the notion of interaction costs, stressing the link of such costs to risk. I ask whether these costs have been increasing over the past few years as a consequence of important developments related to technical progress and globalization. Then, I consider the origins of those costs and the possibilities of reducing them. I stress that there are cases in which the reduction of interaction costs is of high social value, although by no means easy to achieve. Interaction design, I suggest, is particularly necessary.

Starting from interaction costs it is possible to define quite precisely the object of interaction design and to clarify the contribution it can make to the progress of society.

INTERACTION COSTS

Interaction normally brings benefits to the interacting agents but it may also give rise to costs. Those costs are one of the main reasons why interaction may be only partially effective or may even be lacking.

Over the past few decades economists have been much concerned with transaction costs (Rao, 2002; Williamson, 1979; Williamson & Masten, 1999). These are costs that arise not out of the process of production but in relation to the organization of transactions between individuals. There are various types of these costs and they are not only monetary in nature. They may include the time and effort spent by the transaction parties, typically—but not exclusively—a buyer and a seller transacting in a market. Most of these costs are borne to overcome all sorts of information problems.

Interaction costs can be considered a particular type of transaction costs. As with transaction costs in general, some of these arise before interaction takes place: It may be costly to locate potential partners (both as partners in an exchange of goods or services or information) but also time consuming. Sometimes what is required is a comparison of alternative partners and this may not be an easy task at all. Therefore there are costs also in making decisions. Interaction costs are incurred also in the process of learning how to extract the highest utility from an object that is only partially visible to the consumer.

A good example is what economists call *experience goods*: that is, goods the true quality of which becomes visible to the consumer only after a process of learning by using them (Nelson, 1970). Time devoted to learning how to best interact with the object is a cost, as is the period that must sometimes elapse before the consumer can get full satisfaction from the object. This is a vivid example of the well-known dictum that time is money. The invisible properties and qualities of the object become visible after a while. An obvious alternative would be a design that made the objects more fully "visible" from the beginning.

All these costs can be seen as obstacles to effective and good interaction. In particular, they can hamper any type of interaction and make the net benefits of interaction much smaller. In the consumption sphere, an example of the former is the decision not to buy too "invisible" an object, whereas the failure to exploit fully the interactive potentialities of objects is an example of the latter.

Interaction costs have some interesting connections with risk and uncertainty. Indeed, a large part of those costs arise from the fact that information and knowledge are far from perfect. Acting under a veil of ignorance about the effects of our conduct means that we are trapped in a risky situation, where risk means that different outcomes—some positive, others negative—may result from our actions and we do not know for sure, in advance, which one will materialize. If you do not know exactly what will happen when you push a button—because the link between that action and its effect is not known or, to use a more suggestive example, is invisible—then you face a risky situation. Usually individuals do not like bearing risk: they are risk adverse in a precise technical sense.

To reduce risk, either you refrain from using the object or you bear the costs necessary to get more refined information. In other words, you may not interact at all or you derive a smaller amount of net benefits than possible from interaction. Risk reduction and trust are closely related, although it is more appropriate to mention trust when interaction takes place between individuals. Therefore, making interaction more effective implies that trust is stronger. In a network it is easy to connect but not necessarily to interact, precisely because trust is a problem (Basili, Duranti, & Franzini, 2004). People may relate to each other but not interact: They do not exchange goods or reciprocally crucial information, they do not share their knowledge, and so on.

Interaction can give rise to costs of other types. In particular, we can have situations in which the interaction between two individuals—or between an individual and an object—may be extremely positive for them but generates costs to others. You can take great satisfaction from your interaction with your drums but your neighbors' life may worsen quite a lot. These costs are usually called "external" to indicate that they fall on people outside the considered interaction (Papandreou, 1994). Despite their importance, in the following pages I am concerned almost exclusively with "internal" costs. As already mentioned, these costs may cause losses of individual welfare either because they absorb resources or because they prevent the full exploitation of the potentialities of interaction. In the most extreme cases they can block interaction altogether. For example, an object that is very costly to interact with will not be bought and an object that is not bought will no longer be produced. In such a case, to stay in the market, a new design capable of making interaction costs markedly lower is necessary.

In more general terms, interaction costs should be a primary concern for interaction design. Effective and workable interaction design should aim at making them as low as possible. This may also include the creation of institutions that act as a connecting element and allow interaction to start. Cost reduction, however, may not be so easy to achieve. Sometimes it may also be advisable not to try to achieve it. These issues are taken up shortly.

ARE INTERACTION COSTS INCREASING?

An interesting question is whether interaction costs and the effectiveness of interaction have changed, and in which direction, over the past few years. If one could argue that interaction has become more costly or less effective, there would be good reasons for investigating why interaction design attracts more interest today. Unfortunately, this is an interesting but extremely difficult question. Our theoretical knowledge is inadequate and we have no good tools for exactly measuring those costs. Moreover, the relevant phenomena are far from uniform and there may be situations where costs have increased and situations where the opposite has occurred. However, let us consider some broad developments and their likely effects on the problem with which we are concerned.

Globalization has made some types of interactions easier whereas it has greatly complicated others, when it has not led to their disappearance. The latter category includes the traditional type of social interaction, typical of small communities where the sharing of customs and culture made interaction easier and not risky. Social norms are a very useful lubricant for interaction. They make the world of our relations more visible and other people's behavior much less unpredictable (Elster, 1989).

By altering these types of relations, globalization has probably increased the costs of interaction. This is not to deny that, coupled with technological advances, globalization has eased other forms of interaction and led to the diffusion of networks. But, as mentioned, to interact in an effective way is not simply to relate.

Other important developments are, on the one hand, the tendency toward the specialization and fragmentation of knowledge and, on the other, the fact that we allow our happiness to depend more and more on rather complex objects. Complexity raises interaction costs also for educated consumers, although the competence of some consumers may be so high that they become the best advisors of the producer. Fragmentation of knowledge, for its part, may obstruct that combination of different competences and skills that are often necessary for reducing interaction costs at other levels. Indeed, high interaction costs in the production stage may be one of the causes of high interaction costs in the consumption stage.

Taking all these into account, the guess is that, despite growing networking, the fragmentation of knowledge and social relations and the uneven complexity of objects may have resulted in the substitution of relatively cheap and effective interactions with costlier and relatively ineffective interactions. This may hold for many people, not all, because some may have gained from these developments. But this could be enough to strengthen the demand for better interaction design.

All this is purely tentative and more research is required to put these opinions on a sounder ground. However, the hypothesis that interaction costs have increased for many people is coherent with the idea of some authoritative social scientists that, in our societies, individual risks are much higher because interaction costs and risk are strictly connected (Beck, 1992).

There are good reasons for arguing that interaction design cannot be divorced from interaction costs. Indeed its primary purpose should be to make interaction costs as low as possible. But here are many difficulties and sometimes it may be better not to try to reduce these costs. Let us see why.

INTERACTION COSTS AND THE TASK OF INTERACTION DESIGN

Suppose that an agent bears interaction costs that amount to X and gets a given satisfaction from the process. Economists would argue that if agents are rational they will be ready to pay up to X for any solution that would allow them to save the interaction costs while ensuring the same final satisfaction. As an example, let us consider the learning costs referred to earlier. If an additional service of any type could allow the consumer to achieve the same results without the loss of time and effort implied by the learning process, he or she will be ready to pay up to the amount of the saved costs for the additional service. This willingness to pay is proof that a better design has a value, but not an unlimited one. Interaction costs create profit opportunities for firms and, more generally, opportunities for greater social welfare. But these opportunities have an upper limit: Consumers will not be willing to pay more than the amount of the interaction costs they are incurring in the original situation.

As Nobel prize winner Milton Friedman (1975) once said, there's no such thing as a free lunch. Reducing interaction costs is not without costs itself and, if these costs are high enough, the willingness to pay for a "better design" will not elicit a matching supply.

Briefly, the point is that the costs to the supplier of a better design should not exceed the interaction costs that the buyer can save.

Having this in mind, we can draw a very important distinction: defective or costly interaction cannot be eliminated, either because it would cost more than the benefits it determines or because some obstacles do not allow it despite the fact that benefits are greater than costs. The two cases are very different. Under the surface of a defective interaction may lie several reasons with different implications for social welfare. In the former case, to ask for better interaction is tantamount to asking for something that costs more than what anybody is ready to pay. Under standard hypotheses of rationality, this would imply that the joint welfare of the two parties is not increased. The latter case, on the contrary, is a cause for strong concern precisely because the lack of better interaction determines a loss of social welfare. Let us go a bit further into these differences.

INTERACTION COSTS AND SOCIAL WELFARE

Consumers of the same objects differ in many respects. They may face very different interaction costs and they may pursue very different goals in their interaction with the same object. Think, for instance, about the impact of different education and literacy segments. Multiplicity of consumer types raises serious problems.

In general it may be rational on the part of the supplier to choose the solution that fits best the largest number of user needs. Some needs, however, will go unmatched or will be satisfied only if the consumer concerned is willing to undergo a lot of learning in the sense already specified. Here we have a latent demand for better design. Can it be matched in the market? In case this demand remains unsatisfied, would this be a serious failure?

To increase the variety of uses of a given object or to make the relevant features flexible enough to support a wider array of user-specific tasks is a costly activity and there is no guarantee that these costs will not exceed the price the market as a whole is willing to pay. The key point here is the willingness of the market as a whole to pay. In fact, because it is impossible to discriminate prices according to use, even those who are satisfied with the current degree of variety should pay more. However, if these consumers resist price increases, the market as a whole can give the thumbs down to the project despite the presence of users ready to pay more. When users have different preferences and different demands for the same object, it is almost inevitable that the market leaves some consumers unsatisfied.

Indeed, it is not simply a matter of designers becoming more aware that there is an unsatisfied demand for flexibility and variety. Although some consumers would benefit from designing products that support more user goals, the costs of the solution are an obstacle. The solution would be to make price discrimination viable. But it is really hard to implement such a solution. Alternatively, if the neglected user goals are of social relevance in and of themselves, a public authority should take care of the problem, interfering with the market. In conclusion, in this case we do not have a really serious failure of interaction design. Let us now consider different instances.

As to the latter case, we economists know that there are many circumstances under which a transaction that seems to be reciprocally beneficial does not take place. In this case, we are in presence of a failure and if the event is being played out in the marketplace, the expression we use is *market failure*. There may also be interactions that do not take place (or interactions less satisfactory than possible), although they might deliver higher benefits than the costs they command. Opportunities for improving social welfare are lost and we would consider this a serious failure.

Mutually beneficial actions are not pursued for various reasons. I shall focus on three of them: opportunism, bad distribution of costs and benefits of better interaction, and locked-in consumers.

The distinguishing feature of opportunistic behavior is the attempt to take an unfair benefit from an advantage that in most cases originates from better or fuller information. Interaction can cause opportunistic behavior because the supplier and the buyer do not share the same information on the interactive properties of the transacted object. Let us see why. Suppose the buyer attaches to a better designed object a higher value than the cost the producer has to sustain to supply it. This means that there are the conditions for a reciprocally beneficial transaction. Suppose, moreover, that the buyer is not able to ascertain immediately the interaction property of the object, as in the case of experience goods. The producer could be tempted by opportunism, and could try to sell as highly interactive an object which is not. If the buyer trusts the producer, the producer will get higher profits, to the extent that a less interactive object has lower production costs. Knowing this, the buyer will be suspicious and will not trust the supplier when the latter advertises a highly interactive object. The final outcome may be that only badly designed objects are produced and used. Opportunism and lack of trust are the reasons for this failure (Akerlof, 1970).

To overcome this difficulty and win consumers' trust, suppliers should build up a reputation as producers of highly interactive goods. This means that they have to abstain from opportunistic temptations in the short run and rely on the future demand of loyal consumers who, through experience, have come to know that the object really has the advertised features. But in many cases reputation-building remedies may not work and failure will not be averted.

A related case is that of switching costs and lock-ins, which are quite common especially in computer markets (Shapiro & Varian, 1999). Once consumers have spent a lot of effort and time to learn how to use an object, they will not be willing to switch to another object, although there are good reasons for expecting some benefits from the switch. An excellent example is provided by the effort and time needed to master an operating system. When a large number of consumers is locked in to the old object, the producers of more interactive goods may reach too limited a market to compensate for the full costs. Lock-ins and switching costs are obstacles to the effective implementation of better design—especially if it is not radically innovative—although its value to the consumers may well exceed its costs, except for switching costs. It is worth noticing that the higher the effort and time spent in the past (that is, the less interactive were the objects), the stronger this effect may be. Therefore there may be a sort of path dependence in bad design.

The last general reason for the persistence of less-than-optimal interaction design is the distribution of costs and benefits stemming from the improvement. In this case, too, we can assume that the total benefits of a better design exceed its total costs. However, their distribution can be ill-determined in the sense that those who have to bear high costs (or the largest share) will get little benefits. This may be enough to prevent a social-welfare-enhancing change in design. Situations like this are quite frequent when cooperation among several agents is involved. Suppose that agents with different skills and competences have to cooperate to improve the interaction features of an object. The value of this improvement as established by the market is, by assumption, higher than its cost to the full set of cooperating agents. However, the share of benefits for one of them is much higher than his or her cost, whereas the opposite holds for the other. The person who gets less than his or her cost will not participate in this activity and if that person is in some sense indispensable, failure will be the sure outcome.

CONCLUSIONS: WHICH FOUNDATIONS FOR INTERACTION DESIGN?

It has been repeatedly asserted that a lot of interaction design has been made without a theory that helped to understand what was being done and why. This is correct, but I would rephrase this sentence in a slightly different way: Several activities called interaction design have been carried on lately. What do they have in common that justifies their grouping under the same heading? Theoretical investigation can help us to answer that question, as it points to problems and identifies interpretative categories and tools to tackle those problems. As an economist, this is what I tried to do in this chapter.

According to the framework I sketched earlier, to understand interaction design, one must start from interaction costs. Interaction design is the set of methods and devices that allow interaction to proceed as smoothly and as cheaply as possible. By so doing, interaction design contributes in a very concrete way to social welfare. Investing in ergonomic chairs, to take an example from a closer field, has been welfare-enhancing because the costs of ergonomic objects were much lower than the benefits they delivered, also in terms of saved health expenditures. The same should apply to interaction design in general.

This framework is very broad but not broad enough to encompass all the different meanings that interaction design takes on. Indeed my perspective is somewhat limited, although I tried to make it as coherent as possible. I believe that there is much to gain from clearly setting the boundaries of interaction design. In this respect, my conclusions are based on the preceding analysis.

First of all, "better" design should not be necessarily considered as interaction design. Nicer objects increase welfare but not through better interaction. I would not include this in interaction design. As a consequence, I would suggest not linking interaction design too closely to marketing, only because its purpose is not to design objects that have or may have a market. In particular, interaction design should not try to persuade or deceive consumers. On the contrary, it must make people's lives happier by giving better solutions to real problems.

Second, interaction design is particularly valuable from a social point of view when it delivers benefits that exceed costs and does that by overcoming the obstacles that prevent this result. It is of social value also when it allows the situation of particularly deserving people to improve, although total benefits may be less than total costs. In this case interaction design can promote social equity.

Third, incentives created by the market may be too weak to carry on all the interaction design that may be valuable for society as a whole. In many cases the most important cause of defective interaction is the weakness of these incentives rather than the simple lack of awareness or the narcissism of the designers.

Fourth, it is sometimes unclear whether interaction design should refer to objects or artifacts rather than to service. The approach in terms of interaction costs highlights that it is not a matter of object or services. On the contrary, the unifying element is the reduction of interaction costs, which in some cases may be the result of a new or improved object and in others of a better organized service. Therefore, those who are engaged in interaction design may be experts in very different disciplines but what makes them part of the same group is concern for interaction costs. Although I have not stressed this aspect, I believe that organizational issues, as the optimal degree of centralization and decentralization, should be a vital concern for interaction design. From this I would venture to suggest that interaction design experts should respect their separate competencies even if they agree on the general principles that should guide interaction design.

Finally, as I noted, left to its own devices the market may not guarantee the needed type and amount of interaction design. Moreover, in many instances those who suffer from poor interaction are people who deserve social help and consideration. This poses a crucial and thorny problem: the role of public bodies in supporting interaction design. But I leave this problem for the reader to ponder.

Chapter 19

Interaction Design: Six True Stories

Bill Moggridge
IDEO, Palo Alto, California

The decades ahead will be a period of comprehending biotech, mastering nature, and realizing extraterrestrial travel, with DNA computers, microrobots, and nanotechnologies the main characters on the technological stage. Computers as we know them today will a) be boring, and b) disappear into things that are first and foremost something else: smart nails, self-cleaning shirts, driverless cars, therapeutic Barbie dolls, intelligent doorknobs that let the Federal Express man in and Fido out, but not 10 other dogs back in. Computers will be a sweeping yet invisible part of our everyday lives: We'll live in them, wear them, even eat them.... Yes, we are now in a digital age, to whatever degree our culture, infrastructure, and economy (in that order) allow us.
—Nicholas Negroponte ("Beyond Digital," 1998)

STORY 1: THE GRID COMPASS LAPTOP—
THE NEED FOR A NEW DISCIPLINE

This chapter tells six stories about interaction design. First, one of my own.

I glimpsed the implications, for designers, of Negroponte's prediction when, as an industrial designer in 1980, I had the opportunity to design the first laptop, the GRiD Compass Computer. Back then there were a few luggable computers but they were more the size of a sewing machine than a laptop, and the IBM personal computer and Apple Macintosh were still unborn. In the mid-1970s, the Xerox Palo Alto Research Center (PARC) researcher Alan Kay had imagined the Dynabook, a portable notebook computer to be used for education, but Xerox had been unwilling to fund its development as a real product because it believed its market strength was in office equipment, not educational tools. Still, John Ellenby, who had also worked at PARC, seeing how computer components were steadily shrinking, had the vision to realize the huge potential market of people who need to move around for their work, and would like to carry the information with them in their computers. So he founded GRiD Systems to build a truly notebook-size mobile machine.

He first asked me to visualize a design, to be shown to venture capitalists and potential employees, that would help them understand what he imagined. We felt that a three dimensional model would communicate much more powerfully than drawings, so I designed a concept and made a completely real-looking model (Fig. 19.1) out of very unreal materials. A shape like a fat dictionary opened like a clamshell to reveal a flat display on the top half and a keyboard on the bottom. The model did its job—the money was raised and a team assembled.

My contribution was the physical design of the enclosure, and how the screen was hinged to fold down over the keyboard for carrying. This geometry was only one of the 43 innovation items in the utility patent we were awarded. Most of these items, today taken for granted, were new then: the flat electroluminescent graphic display, for example, low profile keyboard, bubble memory, and die-cast magnesium enclosure. The metal housing combined strength with lightness, creating an amazingly tough machine that, as well as being used by executives, was also dropped from military helicopters and sent up in the space shuttle (see Fig. 19.2).

When the first 25 working prototypes were ready, I took one home and started using it. I soon realized that the physical design I had spent so much time on was insignificant compared to the design of the operating system and interactive software. I only noticed the physical design for a few minutes in each session, but for hours at a time I was buried in the delights and frustrations of the interactive behaviors. That made me decide to learn how to design interactions.

Thinking about the experience of using this laptop as a whole, both

FIG. 19.1. The first sketch, 1980, of the final design of the GRiD Compass computer. Designed by Bill Moggridge, this was the world's first laptop computer.

FIG. 19.2. The GRiD Compass computer, 1981.

the physical machine and the software, I realized that the human and functional values of design applied equally to both the physical and digital aspects of the problem. But although I knew much about the techniques and tools for designing physical objects, I had few equivalents for designing in the digital realm. The need for a new discipline, interaction design, was clear.

ALL THINGS DIGITAL

What does "interaction design" mean? Broadly, the design of everything digital: digital objects, digital services, and digital experiences.

Digital objects are those that include or are enabled by electronic technology. We usually think of this technology as something expressed in the design of the personal computer, with its keyboard, mouse, and screen, but digital objects are much more pervasive. Think of interactive toys, greeting cards with chip-generated verbal messages, or toasters equipped with fuzzy logic. Or cars, of whose cost electronic technology represented one-third by the 1990s, a proportion that continues to grow: computer systems control instrumentation, fuel economy, emissions, and emergency devices like airbags and antilock brakes. Few people are aware how much this technology has invisibly colonized the vehicle and thus subtly altered their driving experience.

Figures 19.3 and 19.4, showing two "future concept" projects by IDEO imagining products for the year 2010, indicate the territory.

By digital services, I mean the aspects of services that are enabled or enhanced by electronics, including everything that makes use of the Internet, as well as significant parts of simple everyday experience. The wisecracking waiter in your local restaurant may well compile your order on an electronic notepad and transmit it wirelessly to the kitchen: You still enjoy the hu-

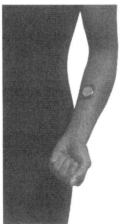

FIG. 19.3. A workstation for the year 2010. In this "future concepts" project, prepared by IDEO for *BusinessWeek*, a large high-resolution screen is operated by stylus or touch,combining the traditional precision of pen on paper with the versatility of the digital desktop and windows.

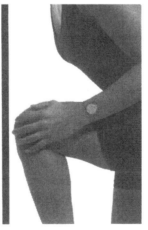

FIG. 19.4. An exercise monitor for the year 2010. In this "future concepts" project, prepared by IDEO for *BusinessWeek*, a monitor, worn on the surface of the skin, gives real-time feedback about time, pulse rate, and other physiological functions.

man-to-human interaction but thanks to the service's digital element, your food comes sooner.

When we design services at IDEO we try to think about designing the whole experience, and it is surprising how often the digital aspect occurs. Designing the Amtrak Acela high-speed train service between Boston and Washington, for instance, the team identified 10 steps on the journey—shown in Table 19.1, which summarizes the overlapping of the physical and digital aspects of designing the service. Notice that the traditional idea of designing a train, both interior and exterior, only applies to the Boarding and Riding steps in the journey, but the opportunity to enhance the experi-

Table 19.1

Designing a Train Service

Steps	Physical Aspects	Digital Aspects
1. Learning	Advertising, Travel Agent, Word of Mouth	On-line, Auto Phone
2. Planning	Station Staff, Travel Agent, Brochure, Phone	On-line, Auto Phone
3. Starting	Other forms of transportation	—
4. Entering	Station Architecture	Signage
5. Ticketing	Ticket Office, Travel Agent	On-line, Auto Phone
6. Waiting	Waiting Room, Station Facilities	Signage
7. Boarding	Doors and Luggage Storage	Auto Doors
8. Riding	Seats, Meal Services	Info., Media, Comms
9. Arriving	Station Architecture	Signage
10. Continuing	Other forms of transportation	—

ence with digital technology applies to all of them, although in an unknown form for the Starting and Continuing steps.

Digital experiences, finally, is a catch-all phrase for the design contexts that do not fall neatly into the object or service categories. Think of everything that is more than an object or a service, where bits might help. In the computer realm it includes software for operating systems and applications, applying, for example, to the following:

- Computer and video games, and other forms of media and entertainment where digital technology is present.
- Environments mediated in some way by digital technology; think of museum spaces, exhibits, hospitals, libraries, trade shows, and art installations.
- Web sites and navigational structures on the Internet that are focused on information and communication, rather than services.
- Personal computers, laptops, palmtops, and increasingly, as Internet access and message services gain in popularity, cell phones—which play a central role in several of my stories later in this chapter.

STORY 2: THE THREE STAGES OF TECHNOLOGY ADOPTION

David Liddle, the project manager for the pioneering Xerox Star at PARC, later founded Metaphor Computers and led Interval Research. He has described (in Moggridge, in press) how new technologies are typically adopted first by enthusiasts, then by professionals, and finally by consumers:

> The normal progression is first to "enthusiast" users, who actually love and appreciate the technology in an aesthetic way, who enjoy exploiting it.... The enthusiast phase is really important because the enthusiasts take the technology far beyond what the inventors and designers imagined could be done with it.... If you're an enthusiast you're somewhat proud of your ability to manage all of the complexities and difficulties. Early automobiles broke down every four or five miles, and you had to stop and pump up the tires, or re-crank the starter or something, but that was a good part of the fun. It was after all just a Sunday afternoon thing that you did. The 35mm cameras used by the astronauts in the 'fifties nearly required a PhD in optics to operate them.

> [But] once enough enthusiasts have their hands on a technology, sooner or later one of them will say "I can use this in my work!" They get a clever idea about how they're going to do something really practical with it.... As this begins to happen there is a great change in the priorities of the developers of the technology. For one thing they become more focused on costs and prices, not because it's going to become inexpensive, but because it will now be judged to some extent by how practical or useful it's going to be.... In the case of the 35mm camera, when it went from very expensive exotica, to being broadly used by professional and serious amateur photographers, it suddenly stabilized. The viewfinder was in one place, you exposed by pushing with your right index finger, you wound the film with a lever with your left hand, and you focused in a particular way. ...

> After a product has built up big enough volumes through this "professional" phase, ... one begins to reach a price point where it's practical for consumers to buy it.... In the "consumer" stage the priorities for the product have dramatically changed, and one thing that we always see is that most of the important controls become automatic ... today when you buy a 35mm camera, if it even has film in it, it will read the film speed automatically and set the exposure automatically and set the flash automatically, and actually a chimpanzee can take pretty good photographs with today's highly automatic 35mm camera.... Computing is the same way. In this third stage we see prices that allow easy consumer decisions, the automation of the most subtle and important of the controls, and a great emphasis on the compatibility of the lifestyle of the purchaser with the image of the product.

A clear understanding of these three stages in the adoption of digital technologies is valuable for entrepreneurs and business people in today's economy. To some extent this phased sequence allows the product or service to evolve incrementally, building on successful small initiatives and learning from (hopefully) small mistakes. Because economies of scale do not operate fully until the consumer stage, it is only then that opportunities for successful businesses suddenly expand. But because digital technology is still so young, however, most industries have not got past the enthusiast and professional stages, so experts in the consumer stage are increasingly sought after.

STORY 3: I-MODE—A FOCUS ON CONSUMERS

I now offer two examples of how cell phones, given new force by interaction design that creatively empathizes with consumers' needs and desires, can appreciably enhance everyday life and experience.

In interaction design, which Fig. 19.5 locates in relation to its disciplinary relatives, the designer operates in the technological domain of hardware and software rather than three-dimensional objects and spaces. It resembles industrial design in that it too focuses on human qualities. But, precisely because it deals with behaviors, with interactivity, successful interaction design practice benefits from unusual organizational setups and an intense and imaginative study of what people do and might want.

A remarkable example of these was the birth and flourishing of NTT DoCoMo's i-mode service. Its creation is described by Mari Matsunaga in her volume (2000), from which I draw much of the following story.

In the case of i-mode a fast-developing technology meant that commercial success depended on skipping the enthusiast and professional adoption stages and diving straight into the consumer stage. In 1997 NTT's president Koji Oboshi gave Keiichi Enoki the task of starting a new mobile phone service. He asked him, indeed, to start a completely new venture, appointing him general manager of the corporate sales department of a business that did not yet exist, and to find and hire his own team from both

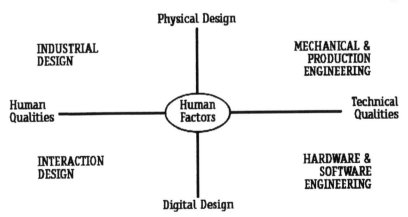

FIG. 19.5.   Interaction design and its disciplinary neighbors. Interaction design resembles industrial design but its context is digital rather than physical, and the designer operates in the technological domain of hardware and software rather than three-dimensional objects and spaces.

within NTT and outside. Such a "spin-out" (where the mother company keeps the ownership by providing the venture capital, but gives the appointed leader enough freedom to escape the weight of the corporate structure and culture of the parent) allows the kind of start-up behavior so fertile for innovation. A small team of dedicated people, highly motivated to succeed, are supported by the parent's money and technology but unencumbered by its size and complex history.

In those days the market for cell phone services in Japan was assumed to be only business people, but Enoki had observed the behavior of his children and their friends. His son Ryo, in junior high school, never needed to refer to the instruction manual when he got a new video game or program for his computer. And his daughter Kyoko, a high school student, was "always exchanging e-mail with her friends. Even when we're having a meal, once she knows she has e-mail, she's on tenterhooks until she can check it. Why does e-mail interest her so much? What makes her so excited? You can get some interesting business ideas pondering over why she acts the way she does." Enoki wanted to create a service so valuable and accessible to everyone, not just businessmen, that "even children will use it!"

He realized that this could be achieved if the normal phone service was enhanced by messaging and Internet access, and designed so as to appeal to both impatient youngsters and adult technophobes. He therefore invited onto his team Mari Matsunaga who, as former chief editor of "employment opportunities" magazines, knew how to condense messages into the few words of a classified ad. This design skill, later crucially needed to fit effective communication into the confined space of a 50-character liquid-crystal cell phone screen, was probably a major contributor to the service's consumer appeal, and thus its extraordinary and sudden business success (Figs. 19.6 and 19.7). After the first three years of the service, 33 million people, about a quarter of Japan's population, were subscribers.

STORY 4: FLUID~TIME—A DIGITAL SERVICE FOR MANAGING TIME

Here is my second example of experience-enriching interaction design that greatly expands the potential of the cell phone. Under its director Gillian Crampton Smith, Interaction Design Institute Ivrea, known as Interaction-Ivrea, has particularly studied the future of digital service design from the consumer's viewpoint (Crampton Smith, Mattioda, & Tabor, 2003; Mattioda, Norlen, & Tabor, 2005). A pioneering example, exploiting existing cell phone technology, is its Fluid~Time project. Its design team, led by Michael Kieslinger, aimed to design time services and tools to help busy people manage their time more flexibly. People currently have limited access to timely information about public services or even private appointments, and

FIG. 19.6. The first generation of i-mode mobile phones, 1999.

FIG. 19.7. The third generation of i-mode mobile phones.

are left wondering when their bus will arrive or whether their doctor's appointment list is still on schedule. In their private and working lives they increasingly exhibit new time-related habits, notably using the cell phone to quickly schedule or change appointments. However, aside from the phone, they have few tools or services to support this new way of life, especially when they interact with public or private services. Fluid~Time supports flexible planning by providing them with personalized, accurate, time-based information directly from the real-time databases of the services they seek.

The team started by creating two services and their interface prototypes. The first was for the 20,000 people who on average use Turin's public transport each day. Not knowing when the bus will arrive can make the experience frustrating. So the team took advantage of the internal bus-tracking system implemented by Turin's transport authorities, and developed a service to make this data visible to travelers in real time. The information fits on a small portable screen, such as a cell phone or connected PDA, and can also be seen on a computer or mechanical display unit. So wherever you are, you can see the number of minutes you can expect to wait, and because the buses are also shown in a perspective view, you can judge at a glance how far away they are (see Fig. 19.8).

The other service was a personalized and flexible scheduling system to help Interaction-Ivrea's 50 researchers and students organize their use of the communal washing machine. It is irritating and inefficient to have to remember to book a time slot on a noticeboard, bring in your laundry, keep the appointment, and check the machine to see if it has finished. So the Fluid~Time service allows you to book on your cell phone and automatically reminds you in the morning to bring in your laundry; it then tells you when your slot is due, the progress of the wash cycle, and when it is finished. Because the system knows the users' profiles and how busy the day is, it can adjust its reminding behaviors from forgiving to strict (see Fig. 19.9).

THE NEED FOR INTERDISCIPLINARITY: CELL PHONE DESIGN

Interaction-Ivrea is a multidisciplinary institute, and the cell phone demonstrates why: The context in which interaction design is practiced is often so complex that a single person cannot fully grasp it, let alone respond with a good solution.

FIG. 19.8. Fluid~Time bus display. Constantly updated information about when the next bus is expected appears on your cell phone or connected PDA (courtesy, Interaction Design Institute Ivrea and Michael Kieslinger).

FIG. 19.9. Fluid~Time washing machine display. Through their cell phones, students of Interaction Design Institute Ivrea can book slots on their communal washing machine, are reminded when to bring in their laundry, and monitor the washing cycle remotely (courtesy, Interaction Design Institute Ivrea and Michael Kieslinger).

Imagine designing a cell phone and the service to which it is attached. The phone itself is a little technological miracle in the palm of your hand, with high resolution color display, an array of buttons, a pointing device, electronic circuitry and storage, main and backup batteries, transmitter and receiver. So the design team must include experts in many aspects of engineering and technology, with specialists in each of the components, plus manufacturing engineering, integrated circuit, hardware and software engineering know-how. On the human side it needs a contribution from HCI, industrial and interaction design, and the business disciplines of marketing and management.

Just for the handset, in short, an interdisciplinary team is essential. But beyond the handset lies a confusing hierarchy of overlapping systems: Each call is supported by the infrastructure of the cell network and the overall phone system of lines, exchanges, optical networks, microwaves, and satellites. All this should ideally be transparent to the user, who just wants to make a call, but in practice it often becomes annoyingly visible. Why? Because the service provider is designing an offering in a competitive environment, often leading to an escalating range of features that soon become impossibly complex; because the service provider is separate from the handset vendor; and because the handsets take longer to develop than the services, and are more intimately linked to the electronic behaviors of the input and output devices and the chips that drive them. And once you connect to Web-based services, each site or service accessible through the phone has interactions designed by separate teams, each trying to provide a version of their offering to fit the scale of the interactions that work well on the phone, but often knowing little in detail about the interactive capabilities of the handset itself.

No wonder the modern cell phone is difficult to use. To design a service that truly serves the consumer, a different interdisciplinary team is needed—with a similar mix of technical, human, and business disciplines but a different set of priorities. Focus on the consumer's viewpoint is the first key concept of interaction design.

STORY 5: THE GOSSAMER CONDOR—ITERATIVE PROTOTYPING

I finish with the second key concept of interaction design: iterative prototyping. Nobody gets it right the first time, so the "Try, try, and try again" strategy is essential for successful development. The fastest and easiest prototyping method is often the best, particularly if the final design solution must be reached fast.

In the field of (admittedly not mass-produced) product design, a dramatic story of successful iterative prototyping methods is that of human-powered flight. In 1959 the British industrialist Henry Kremer offered

£5,000 for the first human-powered aircraft that could demonstrate the same degree of aerodynamic control as the early Wright fliers by tracing a figure of eight around two markers four fifths of a kilometer apart. In 1977, after several failed attempts by enthusiasts, the Gossamer Condor succeeded, pedaled by the cyclist Brian L. Allen.

The Condor's design team head, the technologist Paul B. MacCready, Jr., had used iterative prototyping to develop it. The plane was built of a delicate skeleton of thin tubes and tensioning wires, covered by transparent plastic sheeting taped in place—a flexible kit of materials that allowed the design to be changed and a new prototype built as often as once a day. This rapidly repeated prototyping was so effective that the Condor easily won in the competitive tests, and a later generation of its design was flown by pedal power for great distances, including a crossing between England and France.

STORY 6: THE APPLE PULL-DOWN MENU—ITERATIVE PROTOTYPING IN INTERACTION DESIGN

An example of iterative prototyping in interaction design, finally, was the formative step in the design of the user interface for computers: Bill Atkinson and Larry Tesler's design of the pull-down menu structure for the Apple Lisa in 1981.

The pair developed a round-the-clock working relationship. Atkinson worked nights and Tesler worked days. During the nights, Atkinson made prototypes of user interface concepts, written in code robust enough to support some form of testing. Then Tesler ran user tests during the day. It was easy to find participants, as many of the new Apple employees had never used a computer before. Tesler gave Atkinson a report at the end of each day, saying what he had learned from the tests. Then they would brainstorm for ideas and decide what to try next, so that Atkinson could spend the night programming. In the morning he would bring in a new version, then go home to bed.

They used this iterative method over several intense weeks, during which they designed the arrangement of pull-down menus across the top of the screen so familiar today. Tesler (cited in Moggridge, in press) described the breakthrough:

> One day, I said to Bill, "Why don't you try moving the labels to the top of the screen. Then, you'll always have the full height and width of the screen to display the menus. The screen is small enough that people may be able to associate the menus with the window they affect."

> Atkinson went home to try it. In one night, he developed the entire pull-down menu system! Everything! He hadn't just moved the labels to the top of the

screen. He had the idea that, as you scanned your mouse across the top, each menu would pop down. They would ruffle like cards as you went back and forth. You could scan them all with one sweep of the mouse. When Atkinson came in the next day, he went directly to Steve Job's office to give him a demo, and then came to show us. I went, "Oh right! This is what we want. We want everything to be apparent!"

He had come up with highlighting commands as you moved your mouse down inside them, command shortcuts, and a way of showing which window was active and associated with the menu bar. He'd thought up and implemented the whole thing in one night! I can't imagine what happened there in his home that night.

I can. What happened that night was that iterative prototyping had combined with an imaginative focus on the user's viewpoint and experience, giving birth to a radical innovation that still benefits us all.

Part
VIII

IN SEARCH OF A FRAMEWORK

Chapter 20

Interaction as an Ecology: Building a Framework

Sharon Helmer Poggenpohl
Hong Kong Polytechnic University

SEARCHING FOR DEFINITIONS

Although there has been considerable interest and discussion about the concept of interaction, it remains a slippery term begging for some structured understanding that can serve at least as a straw man or, even better, as a fundamental position from which to build a detailed understanding. Many disciplines engage with interaction, but from different perspectives and different levels of abstraction or specificity, for example, HCI, human factors, psychology, sociology, and virtually all design disciplines. What follows is a practical approach to thinking about such a framework from the perspective of communication design.

Often a discussion begins with a request for someone to define terms. The response is usually a "here we go again" groan or a pedantic smile. In this classic situation, everyone takes a step back as the context, meaning, and use of a term is presented—usually with a telling example. This is the situation with the term *interaction*. There is no agreed definition and the term is even a subject of controversy.

A couple of years ago, in the process of trying to establish a meaningful vocabulary with which to tag design research, some doctoral students and I tried some experiments to form a social and (dare I say it?) interactive method for defining key terms. One of the experiments engaged five well-known international design professors to collectively (but anony-mously) and iteratively define the term "interaction." This was not easy and the consensus we hoped would organically form had to be forced. The syn-thetic definition that formed after several rounds of discussion and amend-ment is as follows:

> Interaction is a process of mutual or reciprocal influence among the vari-ables or parts of a system. Interactions are a succession of actions, each re-sponding to prior actions and each being responded to by succeeding action. By identifying and studying interaction patterns in this succession, we can design interventions that provide material support for desirable in-teraction patterns to emerge. The essential concept of interaction is recipro-cal action, influence, or effect. (Poggenpohl, Chayutsahkij, & Jeamsinkul, 2004, p. 603)

The problem with one definition is that interaction is complex and may be more like a framework in which its various elements are foregrounded or backgrounded or completely absent, depending on the context of use. But this framework does not exist although aspects of it pop up in various at-tempts at definition. In the light of this speculation and the existence of many definitions, I begin with a sampling of definitions from which tenta-tive framework elements can be extracted.

But first—here in a nutshell is the controversy mentioned earlier. In the process of questioning what is new about new media, Espen Arseth (2003, p. 418) answered that "interactivity, hypertext, and virtuality offer partial, inconsistent, ideological answers to the question of newness." He went on to give a brief history of the use of the word *interactive*, first in terms of the change from "batch" modes of operating to "interactive" modes, and then in terms of interactivity being "better." He suggested that interactive could be replaced with *digital* in most texts and the mean-ing would not change; and he went on to argue that "interactivity" has no descriptive power. However, based on his analysis of various definitions, he placed the definitions of interactivity into three categories: (a) a phe-nomenon involving the exchange of information between two equal part-ners, typically human; (b) a situation involving a feedback loop and response; and (c) composite definitions that talk of either degrees or com-ponents of interactivity (p. 425). (Of the three terms he equated with new-ness, only *virtuality* survived his analysis.)

Separating the words *interaction, interactive,* and *interactivity,* is the goal Dag Svanaes (n.d., p. 5) sets for himself. He arrived at the following distinc-tions: "An interaction involves at least two participants. In the context of hu-man-computer interaction, the human is interacting with the computer. I define an artifact to be interactive if it allows for interaction. I further use the term interactivity to denote the interactive aspects of an artifact." Although his conception is clearly stated, it has little explanatory power. His discussion went on to focus on issues of representation in which the "feel" of system use is an aspect of interactivity. Of particular interest is his discussion regarding shifts of human attention due to disturbances and breakdowns. This he likened to switching from the task-at-hand (writing this paper digitally, for example) to searching for an interface operation (how to get an æ ligature, for example). At a more sophisticated level, he used the term ready-to-hand (in use) as opposed to present-at-hand (dealing with the tool) after Heidegger's distinction (Svanaes, n.d., pp. 45–47). Identification of this sequentially fractured aspect of human attention strikes me as important and it may be inherent in all human use of tools. This is a high level concept.

Still within the range of HCI, but moving down to a detailed level, the reciprocity between objects and people is also a major issue. Disambiguating aspects of interaction between people and computers is the goal of Chujit Jeamsinkul's careful analysis (2002). Her diagram (Fig. 20.1) analyzes what happens in less than the blink of an eye in terms of the shift from human to computer and back again.

This analysis, at a fairly fine level of conceptual granularity, was constructed to support work on the relation between motion affordance and its ability to complement software function in relation to common human understanding of the meaning of various kinds of motion. Here the model of interactivity is based on a particular intent and within action theory's separation of communicative action initiated by people from instrumental action generated by objects or machines (Jeamsinkul & Poggenpohl, 2002).

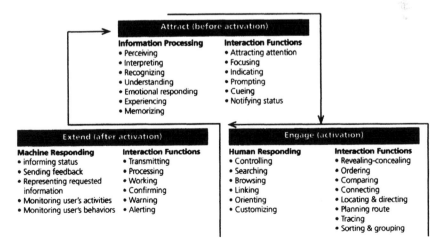

FIG. 20.1. Analysis of human–computer interaction (adapted from Jeamsinkul, 2002).

Games are another specific domain in which interaction plays an essential role. Gunnar Liestøl's interest is *gameplay*, but although he is suspicious of the term, he finds he cannot avoid interaction (2003, p. 402). In the same article, he references Brenda Laurel, who describes interactivity in computer games based on three variables: "frequency (how often can one interact), range (how many choices are available), and signification (how much the choices really affect matters)" (p. 402).

Taking games further, a recent book (Salen & Zimmerman, 2004) quotes communication theorist Stephen Littlejohn, who defined how interactivity emerges from a system: "Part and parcel of a system is the notion of 'relationship'.... Interactional systems then, shall be two or more communicants in the process of, or at the level of defining the nature of their relationship" (p. 58). Salen and Zimmerman went on to say that "something is interactive when there is a reciprocal relationship of some kind between two elements in a system. Conversations, databases, games, and social relationships are all interactive in this sense. Furthermore, relationships between elements in a system are defined through interaction" (p. 58). From their game-based analysis of interaction, they defined four modes of interactivity (pp. 59–60):

1. Cognitive interactivity or interpretive participation (psychological, emotional, intellectual).
2. Functional interactivity or utilitarian participation (functional, structural interactions with material components).
3. Explicit interactivity or participation (design choices and procedures).
4. Beyond-the-object-interactivity or participation within the culture of the object (co-construct communal realities).

Obviously, the definitions cited here are not exhaustive, but they demonstrate the broad range of interest in interaction and some attempts to come to grips with its meaning. These definitions range from very general to specific, locate themselves in specific contexts of use, and focus on different elements or themes within interaction. This reinforces the idea that a framework supports investigation more thoroughly than a definition and that a definition at this time may be premature because variation in context of interaction alters focus and requires a more dynamic and adaptable definition. Nevertheless, attempts at definition from various contexts and disciplines contribute to a richer and more complete understanding of interactivity.

Returning to the idea of interaction as a framework, what can be extracted from the previously mentioned information?

• From the first experimental attempt at definition comes reciprocal influence, parts of a system, and successive actions.

- From Arseth—exchange of information, feedback loop, and response.
- From Svanaes—the idea of shifting human attention (task versus operation).
- From Jeamsinkul—a listing of sub-components of interaction (see Fig. 20.1).
- From Laurel—characteristics of interaction in terms of frequency (time), range (scope), and signification (effect-affect).
- From Salen and Zimmerman—modes of interactivity: cognitive-interpretive, functional-utilitarian, choice-procedure, and coconstructive-cultural.

What is suggested here is a method that uses the best ideas regarding interaction to construct a framework within which various researchers or practitioners might find their niche and the conceptual elements important to them, and identify other people and their contributions working in the same arena. The construction of such a framework is beyond the scope of time and this chapter, but it could contribute to establishing and understanding a discipline of interaction. Furthermore, it would establish a context for identifying not only patterns of interaction but its various contexts.

SHIFTING FROM TECHNOLOGY AND SYSTEMS
TO PEOPLE AND ECOLOGIES

One of the early and significant contributors to thinking about everyday people in the context of design was the sociologist Abraham Moles. His conception of micropsychology (1976) develops an accounting for human use based on microanxieties, micropleasures, microstructures, microevents, and microdecisions. Surely interaction relates to this accounting from the standpoint of how well it supports or diminishes these qualitative aspects of use.

Moles (1986) described the use of secondhand information:

> In a world which is the product of artifice, design more and more explicitly seeks to render the image of that world equivalent with the use project the individual may apply to it: It is in this equivalency that it finds the measure of its success. Wanting a legible world, design seeks to transform visibility into legibility, that is, into that operation of the mind that arranges things in the form of signs into an intelligible whole in order to prepare a strategy for action. (p. 48)

Everyday people are seen by Moles to wander through an information environment created by design, and now increasingly by technology with or without design. His concern is the generalized cost of wandering and the microanxieties this entails. Generalized cost concerns elapsed time, mental effort, stress of uncertainty, and the success or failure of a microdecision. These surely are subtle measures of interactive quality.

Empirical evidence of people's search strategies when looking for information on the Web and making decisions with regard to taking action (such as buying a toy, booking a vacation, or finding a remote weather forecast) was the subject of a recent doctoral dissertation (Sawasdichai, 2004). This qualitative work recorded people's microanxieties—the time spent, the changes in search strategy, their methods of comparison based on web information and design—ending with their ability to make a decision and get satisfaction from the site. Such empirical work is essential to developing a better understanding of how people actually use computer tools (Sawasdichai & Poggenpohl, 2003).

Following in Moles's footsteps, Nardi and O'Day are also people-oriented in their critical volume (1999). Societal and technological critics such as Neil Postman, Jacques Ellul, and Bruno Latour appear here, so their book is not hyperbole in the service of technology as a value-free and undeniably good enterprise. Although much thought has been given to technological development, only recently has the human side of this system been given much attention. Nardi and O'Day defined "information ecologies" as "a system of people, practices, values, and technologies in a particular local environment. In information ecologies, the spotlight is not on technology, but on human activities that are served by technology" (p. 49). Describing the connection between interaction and information ecologies, they noted "examples of responsible, informed, engaged interactions among people and advanced information technologies. We think of the settings where we have seen these interactions as flourishing information ecologies" (p. 24).

This provides an interesting context in which to develop an interaction framework because it sets the stage for a deeper look into the human side of interaction. Shifting from *systems* (a term that calls up rather dry, complex, technical, and contextually stable design) to *ecologies* (implying adaptation and co-evolution) seems like a better conception in the dynamic and changing situation in which we try to understand and map interaction.

THEORIZING INTERACTION FROM TWO PERSPECTIVES: EMOTION AND COMMUNICATION

After centuries of separation between mind and body in Western philosophy, neuroscience is empirically discovering and theorizing the mind–body relationship as a unity.

Unfortunately, the empirical discoveries are the result of studying people with brain injury, using various imaging technologies to reveal brain differences between normal and injured individuals. Enormous strides have been made in mapping brain structures and understanding its functional importance in terms of human emotion and behavior. One of the leading theorists in this area is Antonio Damasio (2003), who postulated that emotion is aligned with the body and physiological changes, whereas feeling is aligned with the mind. He further stated that emotion precedes feeling. He defends this position: "It is legitimate to ask ... why emotions precede feelings. My answer is simple: We have emotions first and feeling after because evolution came up with emotions first and feelings later. Emotions are built from simple reactions that easily promote the survival of an organism and thus could easily prevail in evolution" (p. 30).

We tend to forget that we are animals who have passed through eons of successful evolution. Few of us fully appreciate the complexity of our physical reactions to various stimuli, much less understand the underlying physiological and neurological systems. Damasio created a hierarchy that puts physiological and psychological constructs in order. From simple to increasingly complex, they are based on a nesting principle in which simple components at the bottom are folded into more elaborate ones higher up (see Fig. 20.2).

Feelings, according to Damasio, are "thoughts with themes consonant with the emotion; and a mode of thinking, a style of mental processing, in which increases in the speed of image generation [neural mapping] make images more abundant" (p. 84). This is with reference to good feeling; negative feeling has decreased image generation. One thinks here of Csikszentmihalyi's concept of "flow" (1990). Both would agree that people are drawn to harmonious action and seek pleasure rather than pain. Neural "[m]aps of a certain configuration are the basis for the mental state we

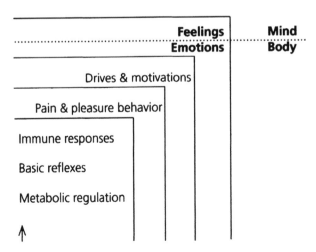

FIG. 20.2. Damasio's nested hierarchy of physiological and psychological constructs (adapted with permission, from Damasio, 2003).

call joy and its variants, something like a score composed in the key of pleasure" (Damasio, p. 137). "Feelings are mental manifestations of balance and harmony, disharmony and discord. They do not refer to the harmony or discord of objects or events out in the world, necessarily, but rather to the harmony or discord deep within the flesh" (p. 139). (This relates to Svanaes's shifting of attention between task and tool.) Damasio offered a provisional definition: "a feeling is the perception of a certain state of the body along with the perception of a certain mode of thinking and of thought with certain themes" (p. 86). And later he amplified feelings as revelations about the state of a person and the mental events in a conscious mind that "helps solve nonstandard problems involving creativity, judgment, and decision-making that require the display and manipulation of vast amounts of knowledge" (p. 177).

Emotions and feelings color decision making. Based on how we categorize the situations we experience—how we structure various scenarios and the importance they have in our life story—we get different options for action associated with different emotions and feelings. Of importance to my chapter here is Damasio's conception of the "emotionally competent object" as one that can initiate an emotion-feeling cycle. Such objects can be actually experienced or recalled from memory. They remain at a high level of abstraction in his writing, but for designers the challenge is to more deeply understand what characterizes them.

J. J. Gibson's (1979) theory of "affordance," too, is a reminder of our reading of objects and environments for indications of their usefulness and perhaps even pleasure. Gibson understood "affordance" as "a radical hypothesis, for it implies that the 'values' and 'meanings' of things in the environment can be directly perceived" (p. 127). Affordance refers to the complementarity of person and environment. Although he developed his ecology of visual perception and considered the nature, understanding, and integration of the physical environment with human action, the concept can be extended to the computer environment. Here appears another environment, one full of artificial signs and signals that afford various possibilities of information access and action. Although Gibson attended to physical bodily extensions and what qualities an environment presents, this can be transferred to the artificial screen environment as an extension of human mental life, in which signs, by their design, display certain affordance and promote certain kinds of interaction.

Currently we are increasingly aware of cultural difference, but at a fundamental level we are all alike as human animals. It is important that we do not forget this, and Damasio's work is a keen reminder. It would be a mistake to focus solely on human cultural difference. A study mentioned briefly in the first part of this chapter (Jeamsinkul, 2002) began from an embodied perspective to see if people interpreted basic motions on screen in a similar way, and further studied whether they attached similar emotional characteristics to the motions. The results were impressive for some motions. This is the kind of empirical work that needs to guide interaction design; work in neuroscience will suggest such practical investigations. Such work will provide a basis for the development of cognitively sympathetic interaction systems.

Like neuroscience, communication theory is another area with a short history (Rogers, 1994). Because we live in communication, it was difficult to grasp and theorize; it developed after World War II with the work of Claude Shannon and Norbert Wiener, who created, respectively, the first diagram of message transmission (see Fig. 20.3) and the concept of feedback and cybernetics (see Fig. 20.4). Both were concerned with issues of technical transmission and control rather than the human dimension of communication.

Thus the early years of communication theory focused on technology and fidelity in broadcast media. Somewhat later, models emerged that focused on more human dimensions such as discourse or the problem of establishing rapport; others attended to gratification or sustaining interest through expectation, motivation, and emotional experience; yet others examined change (or innovation) through consideration of interpersonal influence, social norms, or persuasion; and finally some examined context

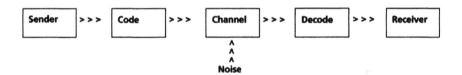

FIG. 20.3. Claude Shannon's model of communication. The first diagram of message transmission.

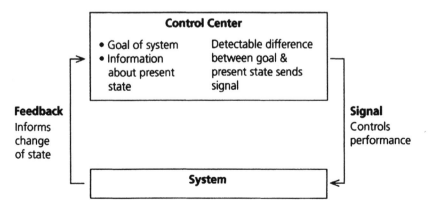

FIG. 20.4. Norbert Wiener's model of feedback.

closely to reveal media characteristics, everyday routines, feedback opportunities, assessment, and so on. (For a detailed description of these models, see McQuail & Windahl, 1993.) These models describe different contexts for communication much as I suspect interaction will reveal itself through multiple models.

Like interdisciplinary interest in interaction, communication models also span many disciplines, from late 19th-century linguistic and philosophical perspectives (Fernand de Saussure and Charles Saunders Peirce), through perceptual psychological models in the 20th century (Wilhelm Wundt, Wolfgang Kohler, and J. J. Gibson, for example), to the already mentioned engineering and technological perspectives (Claude Shannon and Norbert Wiener). Behavioral perspectives came to the fore in the last half of the 20th century (Charles Osgood, Wilbur Schramm, and a host of others). And now it is interdisciplinary perspectives from philosophy, linguistics, cognitive science, computer science, HCI, and design that seek communication's newer arena for theoretical development—interaction.

Much existing theory relates to mass communication in the context of broadcast or one-way media transmission. Although instruments that support interaction have been around for a long time (the telephone, for example), little if any theory focused on interaction itself. What is obvious is that turn-taking is a phenomenon of interaction, as is feedback and interpretation. Whether it occurs person-to-person or person-to-computer, the situations are similar, the difference being the level of variability in the human context in contrast to the limited and programmed interaction in the person-to-computer context.

Three theorists offer concepts of use to thinking about interaction. The first of these is Donald MacKay (1969), a contemporary of Herbert Simon. He envisaged information as a kind of tool that operates on the recipient's "state of conditional readiness for goal-directed behavior." He identified three kinds of meaning: (a) effective meaning to a recipient, (b) conventional meaning to a standard recipient, and (c) meaning that is inseparable from use. It is the last meaning that is of interest. (I return to this shortly.)

The second theorist is Charles Saunders Peirce (Parmentier, 1994) and the semiotic distinction between sign types: icon, index, and symbol. Icons are prevalent in computer application as they have an isomorphic relation to what they represent. Symbols are abstractions that require learning both the form and the reference, thus they require more time and attention to process. Of these three, indices deliver actual connection between the representation and the object or action of reference and its dynamic or changeable state. To show the progress of information being loaded, for example, the movement of a horizontal band, filling in from left to right, is superior to the circle that rotates, because the circle gives no span of time and marks no progress. The horizontal band is also abstract and conventional, but it delivers more than just a sign of existence: it marks progress through time—it is dynamic feedback. This connects with MacKay's (1969) conception of meaning that is inseparable from use. I suspect that dynamic icons and indices that reveal their state of being with reference to function or performance will become more prevalent; further, that they will more easily reveal meaning through use.

The third theorist is W. Barnett Pearce (1989), who examined communication through the lens of coordination, coherence, and mystery. Although his concern is solely human communication, in particular forms of social, cultural, and political exchange, I believe this lens is useful to considerations of interaction. The conception of coordination leads to adjusting to the communication partner through attending to feedback: repairing misunderstood information or changing vocabulary, for example, and tuning into the emotional or feeling state of one's partner and coming to some social exchange, understanding, or agreement that results in constructing a shared reality. Coherence leads to telling a consistent "story"—often a story embedded in one's culture—or telling a story with clear or at least competent logic. Mystery leads to the infinite number of stories that can be expressed through changing perspective or interpretation; it is about the open-endedness of communication and the fallibility of the process of constructing reality.

Investigation of interaction has primarily focused on coherence—creating logics, often from a computer science or design point of view rather than from trying to accommodate the user's perspective. Coordination is the dimension that acknowledges a partner who may have other interests or ideas; it is active and adaptive. Mystery relates to unfolding technological developments and experimentation that leads to new understanding with regard to how people use and understand these changes. Coordination, coherence, and mystery work together, but these elements can be emphasized or downplayed in various ways. So far, computer-mediated communication has attended primarily to technical and formal coherence rather than to the more open perspective of coordination.

These three theorists bring us different ideas in relation to interaction. The ideas themselves are at different levels of generality or specificity:

- MacKay delivered the idea of meaning in terms of demonstrable use as opposed to a focus on conventional meaning to a standard recipient.
- Peirce's indexical sign opens new territory for exploration in the area of dynamic display, challenging the dominance of static icons of limited information utility.
- Pearce's coordination as opposed to a focus on coherence alone, points to greater recognition and adaptation for the partnership aspect of interaction.

These theoretical ideas are not new but they are revealing in the context of interaction. They serve to position a current attention on people, and flag unexamined opportunities for investigation. From a communication perspective, we might ask why the creation of an interaction framework is important now. One answer is that communication and information permeate our lives and is a primary tool for learning.

CHALLENGING THE TEXTBOOK

One example will have to suffice: learning with a textbook versus learning with an interactive program. First, a critical look at textbooks: They are a one-size-fits-all tool with little regard for different styles of learning, different levels of achievement, or different kinds of interests. Lev Vygotsky (1980) delivered a profound insight regarding the learning activity when he wrote about the "zone of proximal development." This zone is just beyond where the individual is knowledgeable, comfortable, and able to act; it provides a reasonable stretch or context for growth. Within the zone, frustration is minimized and harmonious feelings of success in mastering new knowledge are possible. The only adaptation to individual difference a textbook can provide is variable speed—reading quickly or slowly. The textbook is always a reading exercise with minor opportunities to support action beyond reading and writing. It is fundamentally about acquiring and storing knowledge in order to reproduce it; it is not often about the use of knowledge except in a secondhand way.

Looking past the medium of the textbook to the institutional system based on it, we find textbook adoption standardizes learning in subject areas across entire states in the United States with the largest (California and Texas) influencing, if not dictating, textbook adoption in entire regions of the country. This is institutional power. Another system outcome is the test—the regurgitation of information previously assimilated, but not necessarily put into a context of use as the measure of achievement. Escaping the lock that textbooks have on learning is a formidable challenge.

If we look at textbooks from Pearce's (1989) three dimensions of communication, we find very strong coherence in the storytelling, limited coordination on an individual level, but strong coordination on an educational system level, and next to no mystery. Attempts to break the tradition of the textbook meet high resistance, in part because those controlling education have been educated in the textbook tradition. It is difficult to step aside from the often hidden visual and pedagogical forms that the textbook presents. Bolter and Grusin's discussion of "remediation" (1999) reminds us of the hold previous structures and visualizations have on us. Today one only needs to look at CNN or other news channels to see the "windows" concept migrating from computer to television screen. In a similar way, computer-based learning is often a remediation of textbook characteristics.

From a speculative standpoint, interactive computer-based learning opportunities should be able to redress the limitations of the textbook by going beyond static text and picture—traditions of show and tell—to more discovery-based learning constructed on interactive choice, experimentation, or trial and (yes) error. It should be able to support the social construction of learning, which of course requires interaction. Such learning may present alternative entry points to understanding ideas. For example, to understand the concept of co-evolution one might explore the relation between the dodo and the Calvaria tree, or the dynamics of mice population and acorn production. One's interests, level of knowledge, and learning style can be accommodated, thus opening learning to a more tailored process that respects individual zones of proximal development. This glimpse of possible change in the learning environment is a revolution equal to the mechanical production of books and the pedagogy that developed from them. Understanding interaction is an important component of this revolution.

ENDING AND BEGINNING

This chapter has suggested the need for an interaction framework rather than a general definition; it has demonstrated that telling elements and theories can be identified which might help construct such a framework. Beyond identification is a process of synthesis (Owen, 2001) in which elements are clustered based on similarity or its absence. For example, issues relating to interactive quality can cluster. Thus we find that Svanaes's attention shifts between task and tool in relation to Damasio's conception of emotion, Csikszentmihalyi's conception of flow, Moles's notion of microanxiety, and Gibson's idea of affordance. The synergy among these ideas may stimulate better ideas with which to ensure an improved quality of interaction.

Development of an interaction framework will necessarily cross disciplines, experiences, and contexts, and require the thinking and collaboration of many individuals. Such a framework is expected to identify variations in context, support the development of useful theory, reveal a larger conception of interactive methods, and integrate relevant knowledge. An interaction framework is possible, but it will not be conclusive. It needs to be responsive to new knowledge and revisited periodically to provide an up-to-date reference for those interested in newer, more relevant interaction-based approaches to communication, knowledge acquisition and dissemination, learning, and human work.

Chapter 21

Five Lenses: Toward a Toolkit for Interaction Design

Thomas Erickson
IBM T. J. Watson Research Center, Yorktown Heights, New York

THE ROVING TRIBES OF INTERACTION DESIGN

This volume is concerned with establishing foundations for interaction design. "Foundations" strikes me as an ambitious metaphor, suggesting, as it does, a solid base on which a single, unified edifice will be erected. And, following the metaphor a step further, it assumes the existence of a stable, well organized community with a shared set of values that is ready to embark on such a construction project.

I don't believe these assumptions hold up. To me, the state of interaction design feels more primitive. Rather than being an organized community, interaction design feels closer to being composed of a number of roving tribes who occasionally encounter one another, warily engage, and, finding the engagements stimulating, remain open to other encounters.

If this is the case, how do we make progress? I suggest that rather than trying to construct a unified, coherent account of interaction design, we would do better to take a more syncretic approach, gathering appropriate concepts and exploring their interplay without, however, insisting on resolving their tensions and contradictions.

In this chapter, I explore these issues. I begin with a definition, and illustrate my approach to partitioning the terrain of interaction design using five conceptual "lenses." In so doing, I cover most of what I see as the theoretical roots of interaction design. I then turn to the role of theory in interaction design, and suggest that a good way to begin is to assemble a toolkit of concepts for interaction design that consists of appropriately sized theoretical constructs.

INTERACTION DESIGN

I define interaction design quite broadly:

> Interaction design has to do with the design of any artifact, be it an object, system, or environment, whose primary aim is to support either an interaction of a person *with* the artifact, or an interaction among people that is *mediated by* the artifact.

Although some see interaction design as particularly concerned with digital systems—either computer systems, or artifacts with embedded computational capabilities—I see no reason to exclude humbler artifacts. The forces that shape our interactions, from perceptual and motor processes such as seeing and touching, to social and cultural phenomena such as imitation and fashion, are agnostic with respect to whether an artifact contains digital components. Indeed, much of what we understand about the design of nondigital artifacts—whether it be how to make a switch with a satisfying "click," or how clothing functions as a means of expressing identity—are applicable, as well, to digital systems. Finally, as computer systems become increasingly embedded in our artifacts and environments, and even the most mundane objects are tagged and tracked by digital systems, our ability to discriminate between the digital and the nondigital will fade, even should we wish to maintain it.

THE TERRAIN OF INTERACTION DESIGN

Figure 21.1 shows a series of chess games in Washington Square, New York City. In the foreground we see a chessboard, the players rapt in concentration. To one side of the board a few captured black pieces are gathered together; to the other is a pair of chess clocks that meter out the players' allotted minutes. Farther back we see other chess games, each with its circle of spectators. Still farther back we see passersby, most of whom are oblivious to what is going on, but a few of whom may be drawn into the circle of spectators, and then, perhaps, into playing a game or two themselves. And in the far background we discern trees and buildings, and see that the games are taking place outdoors in a city square.

To me, this picture represents, in miniature, the terrain of interaction design. As such, I'll use it to describe how I go about making sense of interaction. As a designer, I'm continually confronted with new sites and situa-

FIG. 21.1. Chess games in Washington Square, New York: the terrain of interaction design (photo © 2004 Project for Public Spaces, Inc., http://www.pps.org).

tions, and for each site I need to come up with a way to see it, to analyze it, to design for it, and to understand the consequences of what I have designed. I find that I work best when I orient to the site or situation in which the interaction takes place; for me the site comes first, and the conceptual framework and methods and tools come later. As a designer, my principal challenge is to make sure that I don't get too fixated on a single aspect of the situation, that I don't get trapped in a particular perspective or approach. Rather than find a single conceptual framework that fits the situation, my aim instead is to stay grounded in the concrete reality of the site, and to bring a range of conceptual lenses to bear on it.

FIVE LENSES

So let us return to the picture in Fig. 21.1. Let us walk through the image, taking a look through each of the set of lenses that I bring to bear on the sites with which I engage as a designer.

Mind

I begin, perhaps as a consequence of my early training, with the mind, envisioning the game in purely cognitive terms. Playing chess, viewed through this lens, involves a cycle of perception, cognition, and action. This is the domain of cognitive psychologists, such as Donald Norman (1986), and is concerned with issues such as how people might go about learning chess, what sorts of errors they might make while doing so, how players develop strategies, why people find games of this sort engaging, and so on. This is the lens most often deployed by interaction designers versed in HCI, and is of critical import in the design of screen-based applications.

Proxemics

Moving on, we deploy a new lens, shifting our focus from minds to bodies and the ways in which we use our bodies to interact with one another. In the picture we see a number of bodies: the player in the left foreground, his face rapt in concentration as he gazes at the board; the spectator in the right foreground, gazing at the game, his posture suggesting that he has settled down to watch for a while. In the next game back, a player is reaching to move a piece, after which he will quickly slap the chess clock to stop his time and start his opponent's; that game, too, has spectators, although they seem less intent on the game and more interested in talking with one another. This is the domain of ethnomethodologists such as Adam Kendon (1990), sociologists such as Erving Goffman (1963), and anthropologists such as Edward Hall (1983), who focus on the role of expression, posture, gaze, gesture, and timing in interactions within small groups. This lens is important for those concerned with designing material artifacts—especially large artifacts such as control panels, rooms, and buildings—as well as those designing digital systems that support mediated (i.e., disembodied) interaction.

Artifacts

Next we shift our view to the artifacts in the picture. We see a chessboard arrayed with white and black pieces; off to one side we see a cluster of captured black pieces, and off to the other a pair of chess clocks. These artifacts play a variety of roles, interacting with the views from other lenses. One role of artifacts, which Norman explored in *Things that Make Us Smart* (1993), is to ease the cognitive load: The board and the pattern of pieces on it serve to preserve the state of the game, enabling players to focus on planning their next moves. Another role of artifacts is their status as objects that are manipulated by the participants. Although the manipulation of chess pieces is a relatively simple matter, ethnomethodologists like David Sudnow demonstrate that the ways in which people physically interact with objects is incredibly subtle. In his book, *Ways of the Hand*, Sudnow (2001) gave an exquisitely detailed account of the process of learning to improvise jazz on the piano, and the ways in which his hands (not his mind) learned to traverse the keys. A third role of artifacts is depicted by Ed Hutchins in *Cognition in the Wild* (1995), in which he explored the view that cognition is not just a property of minds, but can be seen as a global property of systems of people and artifacts. A fourth role of artifacts is a social one, in that the pair of clocks substitute for a human time keeper. This view is explored by Bruno Latour (1992), who eloquently made the case for a sociology of artifacts, suggesting that it is artifacts that stabilize and extend human interaction patterns. This lens—with the glimpses it gives of artifacts and their varied roles—is important for those who design material artifacts, as well as for those who aim to replace material objects with digital "equivalents."

The Social

Now we move to a level of analysis that is not grounded in anything that can be explicitly seen in our picture. The social lens examines relationships, both among people and between people and objects, and tries to take notice of the norms and rules that underlie them. Thus, in our picture, we see not just people, but people who stand in relationship to one another—players, spectators, passersby—and who are obeying rules as a consequence. Of course, the game of chess has a set of rules associated with it, but of more interest are the unwritten rules being adhered to. Thus, one chess player does not shout at the other as he ponders his move (something which is permissible in games like baseball), and he does not, after capturing a piece, toss it into the dirt beneath the table. There is an unarticulated notion of "proper" behavior in play, and one that, furthermore, extends beyond the game. Thus, the onlookers watch quietly and refrain from offering advice (again, unlike some other games), and one, standing nearby, appears to be waiting his turn to take on the winner, thus participating in an unarticulated but mutually understood notion of turn-taking. This is the realm of social psychology, sociology (Goffman, 1963), ethnomethodology (Heath & Luff, 2000), and anthropology (Whyte, 1988). This lens is essential to any interaction designer wishing to reflect on ways in which a newly designed artifact may disrupt situations in which it is introduced, or the ways in which—as with a Web-based chess game—the digital equivalent of a face-to-face interaction may have very different social effects.

The Ecological

The last lens I discuss gives, by far, the broadest view. It is the view of the interaction as it is situated in its larger context. Here we look not just at the chess game and its audience, but at its temporal and spatial location. Temporally, these chess games are a fixture, recurring nearly every day in the same location—outdoors in a public square. By virtue of its location, passersby, on their ways to other places, become aware of the game, and, over time, notice that it is a recurring event. Perhaps, another day, when on less urgent business, one passerby may pause to watch and even to play, thus helping the game, as an ongoing event, to sustain and extend itself. Even if the game fails to interest most passersby, it still contributes to the liveliness and interest of the urban space. This lens, looking at the ways small interactions like the chess game flourish (or not) in the context of other interactions, is exemplified by the work of urbanists like Jane Jacobs (1961), urban designers like Kevin Lynch (Banerjee & Southworth, 1990), architects like Christopher Alexander (Alexander et al., 1977), and anthropologists like William Whyte (1988). This lens is crucial for the interaction designer who creates artifacts for use in public places, and who desires to create self-sustaining interactive systems.

About the Lenses

I do not wish to argue that these are the five and only five lenses of use to interaction designers; others may wish to suggest additional lenses, or to partition things up differently. The main point is that there are multiple perspectives from which interaction designers can analyze the sites or situations with which they are confronted, and that designers will fare best when they are able to pick up one lens, then another, and then a third. It is the ability to fluidly shift perspective that is, in my opinion, of most value to interaction designers.

THE ROLE OF THEORY

Now I'd like to turn to the question of the role of theory in interaction design. As I've said, I think it's too soon to try to create a unified theory or framework for interaction design; instead, I suggest that a more productive way to proceed is to syncretically assemble a toolkit of theoretical constructs and methods, such that for any of my five lenses (or other lenses to be suggested), there are a number of theoretical constructs and methods that might be brought into play.

Choosing Theories

In my opinion, the key question is how to select theories and so on that are likely to be useful. I believe the problem is one of scale. It is not clear what is the proper scale of theoretical construct, and often we err by seizing on apparently useful concepts without sufficiently understanding their contexts. As an example, consider the notion of "affordance." Affordance, a concept developed by ecological psychologist J. J. Gibson (1979), is now commonly misused in interaction design. As initially defined, it was a relational concept, denoting the possibility of an interaction between an organism with particular characteristics and an artifact with particular characteristics. Gibson developed a sophisticated argument—drawing on a number of concepts ranging from "affordance" to "agent" to "ecology"—that organisms perceive their environment in terms of affordances. "Affordance," as Gibson used it, has little to do with its popular use in interaction design as a visible indication that something can be done (visibility has nothing to do with affordances), and it does not make any sense to talk about an artifact affording something without also specifying the sort of entity to which the affordance applies. The problem is that "affordance" has been plucked out of the theoretical framework that gave it its power and nuance, and used in isolation has become a bit of jargon with little value.

At the same time, we need to be cautious about adopting full-fledged theories from other disciplines. The reason is that theories play multiple roles. At its most basic level, a theory is a useful simplification, a mechanism for imposing a framework on the blooming buzzing confusion that is reality. To the extent that its basic components are understandable and memorable, theories serve as common frameworks, lingua francas that allow insiders and outsiders to speak to one another using a common language and shared concepts. Thus biological concepts such as "disease," "bacteria," "virus," "germ," "infection," "antiseptic," and "antibiotic" provide both specialists and layfolk with a common ground on which they can understand and discuss basic medical issues. However, theories play other roles within a discipline. In particular, a theory can serve as a framework for debate within a discipline, and, as a consequence, over time the theory is articulated and refined in response to the debate, resulting in a more complex theory, or possibly multiple versions of the theory.

These two roles of theory stand in tension to one another: The utility of a theory for promoting debate and further articulation of itself within a field may actually interfere with its utility in communicating beyond the field. The requirements for promoting articulation within a field involve supporting the creation of distinctions and nuances that can serve as the ground on which positions can be established, whereas the requirements for communicating beyond a field require the ability to depict the conceptual framework in a few bold and broad strokes of the brush. Although the ability of a framework to support the finely detailed nuance is not necessarily at odds with the ability to also serve as a simplifying framework, it often is.

What this boils down to is that we need to think carefully about the theoretical constructs we choose to use in interaction design. We need constructs that are neither so large that they bring along all the analytical baggage developed in response to internal disciplinary debate, but not so small that they lose the ability to provide a useful framework for dealing with complexity that makes them useful in the first place. In short, we need a conceptual middle ground, a repertoire of theoretical constructs that are larger than "affordance" or "breakdown" or "flow," and that are smaller than "activity theory" or "distributed cognition" or "ethnomethodology."

Towards a Conceptual Toolkit. What sort of theories and methods belong in a "toolkit" for interaction designers? What is the right size or scale of a theory or method? How do we go about finding them?

One possibility is that we need to take theories developed by other disciplines and simplify them for our purposes, pruning away the complexity generated for internal disciplinary purposes. This is something along the lines that Don Norman has suggested in his proposal for an applied discipline of cognitive engineering (1986). Perhaps, just as cognitive engineering could serve as tool when applying the "mind" lens, other theories might be simplified for use with other lenses. Another candidate—an area of economics known as mechanism design that examines the ways in which systems of incentives are designed to shape large-scale group behavior—is discussed by Lucio Picci in chapter 17 of this volume.

Another possibility is that interaction designers might, by drawing on the work of multiple disciplines, develop design-oriented theories that are targeted at particular areas of interaction design. Such design theories would span several lenses, but by virtue of being targeted at a particular design domain, would retain some simplicity. For example, over the last several years, my colleagues and I at IBM's T. J. Watson Research Center have been developing the construct of social translucence, which is a design approach to designing of systems that support human–human collaboration (Erickson & Kellogg, 2003). Similarly, Katie Salen and Eric Zimmerman (2004) have made an impressive attempt to develop a theory of game design, drawing from a wide range of disciplines.

A third possibility is that a more radical form of simplification is needed: Elsewhere I've proposed that adapting the notion of pattern languages from architecture (Alexander et al., 1977) might provide a way of creating a lingua franca for interaction design (Erickson, 2000a, 2000b) that would foster communication among the diverse constituencies that make it up.

CONCLUDING REMARKS

I began this chapter by objecting to the synthetic program of trying to create a unified and coherent foundation for interaction design. Rather than an organized field with the shared values necessary for such a project, interaction design feels much closer to a confederation of nomadic tribes who occasionally come together. Instead of joining together to construct foundations, we would be better advised to proceed syncretically by sharing our tools—theories, concepts, and techniques—and trying to apply them in our own territories. When we encounter one another again, by virtue of our attempts to use some of the same tools for different ends, we'll have a bit more common ground, and a new set of experiences to share.

Chapter 22

A Trickster Approach
to Interaction Design

Isao Hosoe
Isao Hosoe Design, Milan

Indeed, since the book was published, a whole academic field has grown up around the idea of "design methods"—and I have been hailed as one of the leading exponents of these so called design methods. I am very sorry that this has happened, and want to state, publicly, that I reject the whole idea of design methods as a subject of study, since I think it is absurd to separate the study of designing from the practice of design. In fact, people who study design methods without also practicing design are almost always frustrated designers who have no sap in them, who have lost, or never had, the urge to shape things. Such a person will never be able to say anything sensible about "how" to shape things either.
—Christopher Alexander, *Notes on the Synthesis of Form* (1971a, pp. 3–4)

THE TRICKSTER

We live in an era of late positivism. Our historical phase is dominated by the supremacy of science. According to Comte (1830/1936), each branch of our knowledge passes successively through three different stages: theological (religious), metaphysical (philosophical), and positive (scientific). The first stage is the necessary point of departure of intelligence, the second only represents the transition, whereas the third stage is the stable, definitive one. In this rigorously linear and mono-directional hypothesis, our industrial society finds it very difficult to determine new goals. In a circular hypothesis that is much closer to Eastern philosophy, the point of arrival could also be the point of departure. Having gone through science, we could direct ourselves anew toward a theological stage and research the intimate nature of beings. Our culture, especially design culture, would be enriched by this unforeseen, dynamic hypothesis. The issue thus involves determining who shall guide this change of path. Designers are particularly appropriate because they speak two languages: the technological language of central power, and the dialect of the periphery. The only language that science speaks is mathematics: the linear, digital language. But the human world, where design is located, is made of unforeseeable things and speaks an impure language. Not everyone knows how to move between these two worlds and understand their two languages; design can do it if it assumes the role of the "Trickster."

Masao Yamaguchi (1975) elaborated a hypothesis on the cultural role of the Trickster as a dynamic catalyst within the binominal center and periphery. Norms—scientific power—require anomaly, madness, and discontinuity because these elements introduce a flexibility which allows the norm to adapt and evolve. Therefore Tricksters, instead of presenting a threat to the center, are a survival requirement: Their messages contain a communication vital for ensuring the balance of the system. In any kind of society, absolute homogeneity can endanger social integration instead of guaranteeing it. What has to come into play is the knowledge that anomalies are the productive element of our current society. How will designer-Tricksters act in interaction design? We cannot know this. Tricksters are very quick and unpredictable, and move in a way that is surprising and does not always follow the program. They can be invisible and transform themselves, but in any event they enjoy themselves through their fleeting appearances. They can only be seen at a glance and always in different garbs.

DESIGN COMMON SENSE

In 1987 I had occasion to interview two eminent professors of ergonomics: Etienne Grandjean and Sebastiano Bagnara. The editor of *Ufficiostile* had chosen me as interviewer because a year previously I had declared on the pages of that magazine that "Ergonomics is not a science but a design common sense." That declaration provoked both consensus and dissent, and the editor wanted the interviews to reopen the case. I now invite the reader to follow edited extracts from the interview:

> *Isao Hosoe:* What kind of research methods do you use in ergonomics? Do you think that laboratory science, used mainly in the natural sciences by virtue of its reversibility and repeatability, can be adapted for ergonomics? Don't you think that ergonomics needs new tools to study living human beings?

Etienne Grandjean: Actually in our institute [ETH: Swiss Federal Institute of Technology, Zurich], scientific tools like questionnaires are used to obtain information which includes the subjective sensations and emotions of the users. I don't think ergonomics is a laboratory science, even if it has been so in the past, because now more than half our research work can be defined as field science. Laboratory science has the advantage that constant parameters can to be repeated, whereas in the field sciences the circumstances of research are not reliably constant. But field science gives us the opportunity to discover important phenomena which could not have been foreseen.

IH: I think that one of the principal problems regarding field science, over and above questions about the adequacy of the measurement tools, remains that the observer may influence the behavior of the observed and the real conditions of the environment examined.

EG: This kind of criticism is very frequent from those who prefer the laboratory method. We adopt several expedients which may guarantee the validity of the research. First, we work with a large number of people so as to compensate for the different reactions of large populations to the "interferences" provoked by a single observer. It is also possible to distribute the questionnaire without interference by the observer. Personally I prefer the individual meeting.

IH: I understand that you are well known as the father of ergonomics and in contact every day with young ergonomists around the world. What is your opinion of their scientific level? Are you satisfied with the global scientific talent of the researchers of this field?

EG: It depends on the university they come from. The situation is so pluralistic that it is very difficult to give any judgment on this. In some countries, where ergonomics is equivalent to psychology, I believe they should study physiology more deeply. At ETH Zurich my successor is a "computer person," so in my opinion the institution probably needs more input from psychology. For me ergonomics is an interdisciplinary science and it is necessary to try to discover and investigate new fields. When in one country a new frontier opens, other countries follow its example. There will be always be a new ergonomics, which I'm sure will not come from the laboratories. (Sias, 1987, pp. 14–19)

During the interview, Grandjean talked about the new area of ergonomics—the ergonomics of software—whose study is particularly problematic for the psychological order. He remembered Sebastiano Bagnara as one of the most eminent researchers in this field:

Isao Hosoe: Today's culture is characterized by the attention moving from hardware to software. Ergonomics as it has been conceived until today has became old. The path of your research is very interesting because it denotes the changing demands on ergonomics.

Sebastiano Bagnara: I worked successively in three fields. The first concerned hemispheric specialization and the performance of each cerebral hemisphere in predominantly visual tasks. Then I worked on attention. Now I'm studying error experimentally.

IH: Does optimization as the basic criterion of ergonomic intervention derive from economy in the quantitative sense? And in consequence are the instruments used for this research quantitative rather than qualitative?

SB: No, not exclusively. I can give an example. To study a simple mental process I can use mental chronometry methods to measure the time of execution and the number of errors made. But if I include the problem of the modality of control, which the subjects have over their actions, and how they improve their controlled performance, I have to use a qualitative method to analyze the verbal protocol. Behavioral descriptions are not necessarily obtained by quantitative methods.

IH: A basic condition of laboratory science is the reversibility and repeatability of the object. But the actions of humans are in many respects irreversible and unrepeatable. How can you resolve this basic contradiction in order to ensure that your laboratory research is valid?

SB: You have to know that human behavior is very often more repeatable than you might imagine, so statistical instruments can identify constants in respect to variability. In the laboratory the principal objective is not to describe the aspects which comprise the peculiarity of a single individual but to know the constants characterizing general behavior.

IH: Is ergonomics a interdisciplinary science, or design common sense?

SB: Being interdisciplinary sounds so static; it suggests the application to one object of different competences. The logic of design is something very different: the conquest and the interaction of different viewpoints, the discovery of new things. (Sias, 1987, pp. 20–23)

ABDUCTION

In one of his Harvard Lectures, delivered in April 1903, Charles Sanders Peirce (1998, p. 205) distinguished between three kinds of inference:

> These three kinds of reasoning are Abduction, Induction, and Deduction. Deduction is the only necessary reasoning. It is the reasoning of mathematics. It starts from the hypothesis, the truth or falsity of which has nothing to do with the reasoning; and of course its conclusions are equally ideal. The ordinary use of the doctrine of chances is necessary reasoning, although it is reasoning concerning probabilities.
>
> Induction is the experimental testing of a theory. The justification of it is that, although the conclusion at any stage of the investigation may be more

or less erroneous, yet the further application of the same method must correct the error. The only thing that induction accomplishes is to determine the value of a quantity. It sets out with a theory and it measures the degree of concordance of that theory with fact. It never can originate any idea whatever. No more can deduction. All the idea of science comes to it by the way of abduction. Abduction consists in studying facts and devising a theory to explain them. Its only justification is that if we are ever to understand things at all, it must be in that way.

In the early 1980s, Massimo Bonfantini and Gianpaolo Proni (Bonfantini, 1987, 1990; Bonfantini & Proni, 1980) tried to abduct Peircian Abduction by dividing it into three types of abduction: (1) The applicable mediations law for inferring a case from a result is mandatory and automatic or semiautomatic, (2) the applicable mediation law for inferring a case from a result is sought by selection from the available encyclopedia, and (3) the applicable mediation law for inferring a case from a result is constituted *ex novo*, invented. I now show my abduction of Bonfantini and Proni's abduction of Peirce's abduction.

My first abduction method is *play*. Play needs a certain rule to begin and keep playing. Your play performance becomes faster as your familiarity with the play increases. Very often you make a mistake and that mistake gives you an idea you could not otherwise have imagined. In his Introduction to the Japanese translation of Radin, Jung, and Kerènyi's *The Trickster* (1956), Yamaguchi (1975) pointed out the possible connection between the concept of play described in Johan Huizinga's *Homo Ludens* (1950) and the Trickster who never stops playing. Movement sustains play, and movement needs vacancy. This was affirmed by Laotsu, who has been considered the philosopher of fluidity in the same way that Confucius may be defined as the philosopher of solidity or power. Henri Bergson, considered the philosopher of human beings as living entities, said the following in the Introduction to his *Creative Evolution* (1941, p. v): "We shall see that the human intellect feels at home among inanimate objects, more especially among solids, where our action finds its fulcrum and our industry finds its tools; that our concepts have been formed on the model of solids; that our logic is pre-eminently the logic of solids; that, consequently, our intellect triumphs in geometry."

My second abduction method is *bricolage*. This use of the term comes from *La Pensée sauvage* by Claude Lévi-Strauss (1962). The bricoleur, with a strong curiosity and an excellent body, enters the forest where he can find all kinds of existing things. He does not invent but gathers objects which stimulate his curiosity. In the evening he comes back to his house with a bag full of things gathered in the forest, empties the bag onto a big table, and recognizes that among them are a few highly interesting combinations he could never have hoped to see.

The third abduction method is *folly*. Folly here means what is not normal, average, or standard, not the dogma accepted by the majority and sustained by normality. According to Erasmus (1974), writing in 1509, Folly said: "Whatever is generally said of me by mortal men, and I'm quite well aware that Folly is in poor repute even amongst the greatest fools, still, I am the one and indeed, the only one—whose divine powers can gladden the hearts of gods and men."

It is interesting that Erasmus said "gods and men" not "God and men." It is clear that in his mind there were many generous Greek gods rather than an exclusive, monotheistic god. His intention in writing this book seems to be to criticize monotheistic culture, which puts heavy limits on intellectual activity. Half a millennium after him, Michel Foucault in his *L'Histoire de la folie* (1961, pp. 29–31) revealed that Erasmus was right: Monotheistic culture and modernity tried to exclude diversity, the mother of human creativity. Abduction differs from induction and deduction because, although the latter two are based on the logic of continuity and reversibility, abduction may be neither logically continuous nor reversible. Abduction is human inference, and based on nonlinear and irreversible logical structure. It is like the dragon considered as the creative divinity of the Indus river, who suddenly jumps up to the sky and comes back when he likes, and thus plays the role of communicator between two different worlds, the ground and the sky. Abduction is like a fisherman who knows where the fishes live even if the water is turbid, or a shaman hunter of the Siberian tundra who, unarmed, lets the animals gather in tranquility to be captured by the hunter.

THE EIGHT SENSES

In a 1,000-page handbook on senses and perception, 700 pages are devoted to vision, 120 to hearing, 60 to touch, 40 each to taste and smell, and 20 each to the senses of equilibrium and of time. The fact that 70% of this handbook is occupied by the visual sense is worthy of deeper thought.

Writing in 350 BC, Aristotle started his monumental *Metaphysics* (2005) with the following:

> All men by nature desire to know. An indication of this is the delight we take in our senses; for even apart from their usefulness they are loved for themselves; and above all others the sense of sight. For not only with a view to action, but even when we are not going to do anything, we prefer seeing (one might say) to everything else. The reason is that this, most of all the senses, makes us know and brings to light many differences between things. (Book 1, Pt. 1)

"Brings to light many differences between things" leads directly to the scientific paradigm on which the Western modernity has been constructed.

Aristotle taught us to concentrate attention on the visual sense. We may say that the Western culture is fundamentally based on vision.

In the early 9th century, the Japanese monk Kukai (774–835) absorbed from the Chinese master Keika the tradition of Esoteric Buddhism and began the theological and philosophical construction of Shingon (mantra) teaching. A significant part of one of his monumental works regarding the structure of the Shingon universe (Kukai, 1983) describes the quality of water in Shingon Paradise. Shingon Buddhism knew that Aristotle's postulate on the sense of sight was limited and through the Esoteric teaching tried to "see" the part that was invisible. In his description of the universe, Kukai touched exhaustively on every detail. Speaking about the qualities of water he said that eight characteristics guaranteed its excellence: It must be sweet, cool, soft, light, pure, odorless, and pleasant to the throat, and must not hurt the stomach. The sense of sight clearly plays a secondary role. Aristotle (2005, Book 1, Pt. 2), on the other hand, quoted Thales in a hasty way, referring to the importance of water inasmuch as it is an essential quality for life, deducing this only from the fact that seeds are moist.

In the Introduction to his *The Senses Considered as Perceptual Systems* (1966), James J. Gibson asked the following:

> What are the senses? It has always been assumed that the senses were channels of sensation. To consider them as systems for perception, as this [book] proposes to do, may sound strange. But the fact is that there are two different meanings of the verb to *sense*, first, to detect something, and second, to have a sensation. When the senses are considered as perceptual systems the first meaning of the term is being used. In the second meaning of the term there is the vast difference between sensations and perceptions. (p. 1)

Now let me talk about *Yuishiki*, definable as Buddhist psychology, or better, the "depth psychology of Buddhism" (Okano, 1990). Whereas to vision, hearing, smell, taste, and touch, Freudian psychology adds the unconsciousness as a sixth sense, and Jungian psychology adds the collective unconsciousness as a seventh, Yuishiki is composed of eight senses: eye-sense, ear-sense, nose-sense, tongue-sense, body-sense, consciousness, *mana*-sense, and *alaya*-sense (Fig. 22.1).

Consciousness for Yuishiki is considered the sixth sense because it filters and controls most of the information entering through the first five senses. These six senses belong to the area that we pretend to understand.

Yuishiki says that beyond the sixth sense is a hidden heart strictly attached to the ego. This heart is the seventh sense, *mana*, which in Sanskrit means to measure or doubt. In the depth of your heart you are always thinking about putting yourself, your ego, in the midst of your thought. When Descartes declared "cogito ergo sum," his mental background was the sense of mana whose center was his ego. If we go deeper into our sensorial world, we en-

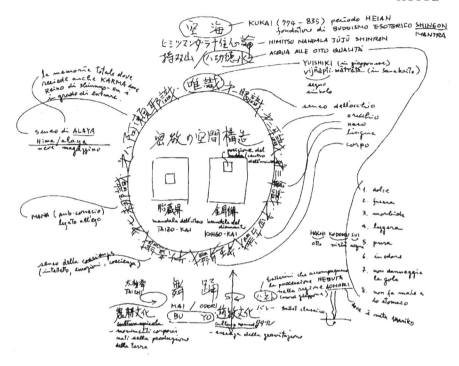

FIG. 22.1.   The cosmology of Buddhism's depth psychology. The key to enter in the tacit dimension.

counter the final sense, *alaya,* the "repository" of sense (*Himalaya* in Sanskrit means the repository of snow). The sense of alaya includes all the memories you have gathered during your life experiences through the other seven senses: what you have seen, heard, smelled, tasted, touched, thought, and felt in your mind. Furthermore, your sense of alaya includes not only your own memories but all the memories of your ancestors, which may cover hundreds, thousands, or a million years. Most Buddhist practices such as meditation, *zazen* ("sitting Zen"), ablution, offering, service, or talking about teachings, are considered as means of communicating with all eight senses. Most of the knowledge we believe we have can be reached exclusively through our practical experiences with the help of our eight senses.

BEHAVIORAL ENERGY

Behavioral energy is a term I coined some years ago when preparing my workshops for several companies and university institutes. The energy used by me in this context cannot be considered in terms of kilowatts or horsepower but may instead be understood, with a certain degree of ambiguity, as the original meaning of *activity*, derived from both the Latin *energia* and Greek *energeia*. I suppose that activity in Latin and Greek has been considered as human activity, but let me extend the term to all those activities not necessarily human but which may still reach human perception through the eight senses I described earlier.

Gibson (1979, p. 7) introduced his theory of affordance by defining the field of affordance as follows: "I have described the environment as the surfaces that separate substances from the medium in which the animals live. But I have also described what the environment affords animals, mentioning the terrain, shelters, water, fire, objects, tools, other animals, and human displays."

Schematically, behavioral energy is counterposed to material energy, but both forms of energy come from the same environment where we live. To be clear in our context, the environment we speak of here concerns interaction design. Thinking of the civilization in which we are speaking of interaction design is to define, even if hypothetically, the idea of our past, present, and future. Roughly speaking, our ancient ancestors lived as nomad hunters for a quarter million years, as farmers for a thousand years, and in our industrial age for 200 years. It seems that each time span diminishes so drastically that it is hard to imagine the age that will succeed ours, and how long it will last. However, we may take for granted that the advent of information technology has generated a different epoch even inside the industrial age. Let me call this epoch *neo-nomadic*, as if the evolution of our civilization is not linear but cyclic, or better, spiral (see Fig.22.2).

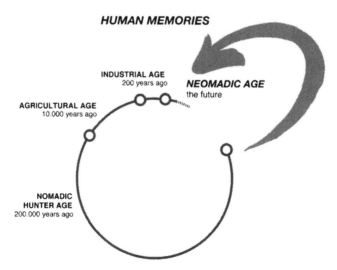

FIG. 22.2. The cycle or spiral of civilization from nomadism to neomadism.

Consider the huge time span of nomad-hunter society compared to its agricultural and industrial successors, and imagine how much memory is conserved in our eight senses, especially alaya, the last sense. I understand that over the course of successive cultures many memories are destroyed or dispersed. But many memories that had to hide themselves for their future survival, or transform themselves to adapt to their time, wait patiently because one day they will in some way be appreciated.

The desire to possess is today considered one of the most important human desires. But it is very easy to understand that possession had no great importance in nomad-hunter society because there was little to possess. *Consumption* is a word created for industrial society. It is true that the word was used before the industrial age, but the context of its use was circumscribed. One consumes foods, one consummates a marriage: in both cases nothing remains after the consumption. *Dedication* is the right word for agriculture, not only for farmers but also the lifestyle of the people and society sustained by agriculture. I believe that even industrial society could start to grow into maturity thanks to the dedication of the workers who came from agriculture. Agricultural dedication in industrial society has been well sustained by Protestant dedication in Europe and Confucian dedication in Asia. Competition replaced dedication when the capitalist system was aware of the market, whose core function is competition. Work as dedication-competition in the agricultural-industrial scenario may be transformed in the neomadic-society scenario into the work as performance. As for the human senses, whereas touch is for agriculture and sight is for industry, the eight senses are for neomadic society. Regarding human memory, past memory is for agriculture, present memory is for industry, and future memory is for neomadic society (see Fig. 22.3).

Now let me talk about project as synonym of design. "Pro-ject" means to throw forward, to act for the future, to perform on future memory. The last word I would like to mention in this comparison of cultures is time. Time for agriculture has been cyclic, industry preferred linear time, but fluid time is for neomadic society. In his *Dance of Life*, speaking of the quality of time, Edward T. Hall (1982) referred to North European monochronic time and South European or Indian polychronic time:

> Particularly distressing to Americans is the way in which appointments are handled by polychronic people. Being on time simply doesn't mean the same thing as it does in the United States. Matters in a polychronic culture seem in a constant state of flux. Nothing is solid or firm, particularly plans for the future; even important plans may be changed right up to the minute of execution. (p. 127)

Let us come back to the discussion on the behavioral energy and the material energy. Energy meant here activity, human activity or the result of the

CULTURAL COMPARISON

	AGRICULTURAL CULTURE	INDUSTRIAL CULTURE	NEOMADIC CULTURE
RESOURCES	• cultivate • possess • attach	• construct • consume • exploit	• gather • share • respect
ECONOMY	• soil as goods • work as dedication	• everything as goods • work as competition	• absence of goods • work as performance
SENSES	• sense of touch	• sense of sight	• eight senses
LIFE	• feminine • care • cyclic time	• masculine • control • linear time	• fusion of the sexes • symbiosis • fluid time

FIG. 22.3.   Agricultural, industrial, and neomadic cultures compared.

human activity, whereas behavior means something that is able to communicate with human senses, and material means something that is not able to communicate with human senses (see Fig. 22.4).

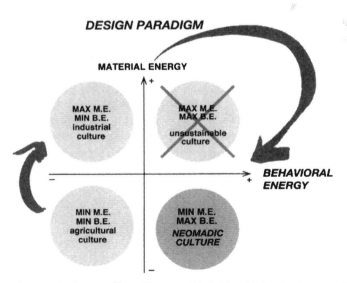

DESIGN PARADIGM

FIG. 22.4.   A design paradigm. The material (ME) and behavioral energy (BE) of agricultural, industrial, unsustainable, and neomadic cultures.

If the total amount of energy spent on something to be designed is constant, let us make the effort to maximize behavioral energy and minimize material energy. In another words, let us put more quality and intelligence in interaction design, involving all the cultural issues and our eight senses, and also encouraging the tacit dimensions, above all our design common sense (see Fig. 22.5).

All this can be done so that transversal information transfer becomes a common design tool, and all who are involved in interaction design become the Trickster—who, by nature, moves everywhere, seeing things that common people do not.

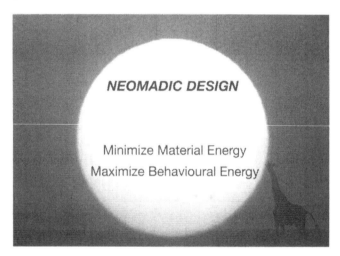

FIG. 22.5.   A design strategy for neomadism.

References

Abbott, K. R., & Sarin, S. K. (1994). Experiences with workflow management: Issues for the next generation. *Proceedings of the AMC CHI'94 Conference on Human Factors in Computer Systems,* 113–120.

Agostini, A., Albolino, S., Boselli, R., De Michelis, G., De Paoli, F., & Dondi, R. (2003). Stimulating knowledge discovery and sharing. In M. Pendergast, K. Schmidt, C. Simone, & M. Tremaine (Eds.), *Proceedings of the ACM Conference GROUP 2003* (pp. 248–257). New York: ACM Press.

Agostini, A., De Michelis, G., Divitini, M., Grasso, M. A., & Snowdon, D. (2002). Design and deployment of community systems: Reflections on the Campiello experience. *Interacting with Computers, 14,* 689–712.

Agostini, A., De Michelis, G., & Susani, M. (2000). From user participation to user seduction in the design of innovative user-centered systems. In R. Dieng, A. Giboin, L. Karsenty, & G. De Michelis (Eds.), *Designing cooperative systems* (pp. 225–240). Amsterdam: IOS Press.

Akerlof, G. A. (1970). The market for "lemons": Quality uncertainty and the market mechanism. *Quarterly Journal of Economics, 84,* 488–500.

Alexander, C. (1964). *Notes on the synthesis of form.* Cambridge, MA: Harvard University Press.

Alexander, C. (1971). The state of the art in design methods. *Design Methods Group Newsletter, 5,* 1–7.

Alexander, C., Ishikawa, S., Silverstein, M., Jacobson, M., Fiksdahl-King, I., & Angel, S. A. (1977). *A pattern language: Towns, buildings, construction.* New York: Oxford University Press.

Allbeck, J., & Badler, N. (2002). Toward representing agent behaviours modified by personality and emotion. In *Proceedings of the Workshop Embodied Conversational Agents—Let's specifiy and evaluate them! at AAMAS 2002, Bologna, Italy.* ACM Press.

Arcangeli, B. (Ed.). (1996). *Memorie di famiglia* [Family memories]. Rome: Armando. (Original work published 1925)

Aristotle. (1985). *Nicomechean ethics.* Indianapolis, IN: Hackett.

Aristotle. (2005). *Metaphysics* (W. D. Ross, Trans.). Retrieved February 23, 2005, from http://classics.mit.edu/Aristotle/metaphysics.1.i.html

Arseth, E. (2003). We all want to change the world: The ideology of innovation in digital media. In G. Liestøl, A. Morrison, & T. Rasmussen (Eds.), *Digital media revisited: Theoretical and conceptual innovations in digital domains* (pp. 418–425). Cambridge, MA: MIT Press.

Atance, C. M., & O'Neill, K. D. (2001). Episodic future thinking. *Trends in Cognitive Science, 12*, 533–539.

Aviles, W. A. (1990). Telerobotic remote presence: Achievements and challenges. *Proceedings of Engineering Foundation and NASA Conference on Human–Machine Interfaces for Teleoperators and Virtual Environments, Santa Barbara, CA, 38*, 38–69.

Baeker, R. M., & Buxton, W. A. S. (1990). An historical and intellectual perspective. In J. Preece & L. Keller (Eds.), *Human–computer interaction: Selected readings* (pp. 3–26). Hemel Hempstead, England: Prentice Hall.

Bakhurst, D. (1980). *Social memory in Soviet thought.* In D. Middleton & D. Edwards (Eds.), *Collective remembering* (pp. 203–226). London: Sage.

Banerjee, T., & Southworth, M. (Eds.). (1990). *City sense and city design: Writings and projects of Kevin Lynch.* Cambridge, MA: MIT Press.

Baron-Cohen, S. (1995). *Mindblindness: An essay on blindness and theory of mind.* Cambridge, MA: MIT Press.

Barthes, R. (1979). *Frammenti di un discorso amoroso* [Fragments of an amorous conversation]. Turin: Einaudi. (Original work published 1977)

Bartlett, F. C. (1932). *Remembering: A study in experimental and social psychology.* London: Cambridge University Press.

Basili, M., Duranti, C., & Franzini, M. (2004) Networks, trust and institutional complementarities. *Rivista di Politica Economica, 91*, 159–180.

Beal, C. R. (1988). The development of prospective memory skills. In M. M. Gruneberg, P. E. Morris, & R. N. Sykes (Eds.), *Practical aspects of memory: Current research and issues* (Vol. 1, pp. 367–370). Chichester, England: Wiley.

Beck, U. (1992). *Risk society: Towards a new modernity* (M. Ritter, Trans.). London: Sage. (Originally work published 1986)

Beckers, R., Holland, O. E., & Deneubourg, J.-L. (1996). From local actions to global tasks: Stigmergy and collective robotics. *Artificial Life IV: Proceedings of the International Workshop on the Synthesis and Simulation of Living Systems,* 181–189.

Bellotti, V., Ducheneaut, N., Howard, M. A., & Smith, I. (2003). Taking email to task: The design and evaluation of a task management centered email tool. *Proceedings of the ACM CSCW 2003 Conference on Computer Supported Cooperative Work* (pp. 345–352). New York: ACM Press.

Berdichevsky, D., & Neuenschwander, E. (1999). Toward an ethics of persuasive technology. *Communications of the ACM, 42*, 51–58.

Bergamasco, M., Allotta, B., Bosio, L., Ferretti, G., Parrini, G. M., Prisco, F., et al. (1994, May). An arm exoskeleton system for teleoperation and virtual environment applications. *Proceedings of the IEEE International Conference on Robotics and Automation, 2*, 1449–1454.

Bergamasco, M., De Micheli, D. M., Parrini, G., Salsedo, F., & Scattareggia Marchese, S. (1991). Design considerations for glove-like advanced interfaces. *Proceedings of 91 ICAR: 5th International Conference on Advanced Robotics, Pisa, Italy, 1*, 162–167.

Bergson, H. (1941). *L'évolution créatrice.* Paris: Presses Universitaires de France.

Bernstein, J. M. (1992). *The fate of art: Aesthetic alienation from Kant to Derrida and Adorno.* University Park: Penn State University Press.

Beyer, H., & Holtzblatt, K. (1998). *Contextual design: Defining customer-centered systems.* San Francisco: Kaufmann.

Binder, T., De Michelis, G., Gervautz, M., Iacucci, G., Matkovic, K., Psik, T., et al. (2004). Supporting configurability in a tangibly augmented environment for design students [special issue]. *Personal and Ubiquitous Computing, 8*(5).

Binder T., Ehn, P., Eriksen, M. A., Iacucci, C., Kuutti, K., Linde, P., et al. (in press). Opening the digital box for design work: Supporting performative interactions, using inspirational materials and configuring of place.

Blum-Kulka, S., & Snow, C. E. (1992). Developing autonomy for tellers, tales, and telling in family narrative events. *Journal of Narrative and Life History, 2*(3), 187–217.

Boland, R. (Ed.). (2004). *Managing as designing.* Palo Alto, CA: Stanford University Press.

Bolter, J., & Grusin, R. (1999). *Remediation: Understanding new media.* Cambridge, MA: MIT Press.

Bonfantini, M. A. (1987). *La semiosi e l'abduzione* [Semiosis and abduction]. Milan: Bompiani.

Bonfantini, M. A. (1990). Semiosis of projectual invention. *VS, 55/56,* 133–142.

Bonfantini, M. A., & Proni, G. (1980). To guess or not to guess. In U. Eco & T. A. Sebeok (Eds.), *The sign of three: Dupin, Holmes, Peirce* (pp. 119–134). Bloomington: Indiana University Press.

Bradley, M. (2000). *Emotion and motivation.* In J. T. Cacioppo, L. G. Tassinary, & G. G. Berntson (Eds.), *Handbook of psychophysiology* (pp. 602–642). Cambridge, England: Cambridge University Press.

Brandimonte, M. A., Bianco, C., Villani, M. G., & Ferrante, D. (2005). *Pro-social prospective memory: The effects of subjectively perceived task importance on the realization of intentions.* Manuscript in preparation.

Brandimonte, M. A., Einstein, G. O., & McDaniel, M. A. (1996). *Prospective memory: Theory and applications.* Mahwah, NJ: Lawrence Erlbaum Associates.

Bratus, B. S., & Lishin, O. V. (1983). Laws of the development of activity and problems in the psychological and pedagogical shaping of the personality. *Soviet Psychology, 21,* 38–50.

Breazeal, C. (1998). A motivational system for regulating human–robot interaction. Proceedings of AAAI98, Madison, WI.

Breazeal, C. (2003). Toward sociable robots. *Robotics and Autonomous Systems, 42,* 167–175.

Breazeal, C., & Fitzpatrick, P. (2000). That certain look: Social amplification of animate vision. *Proceedings of the AAAI Fall Symposium on Socially Intelligent Agents: The Human in the Loop* (pp. 18–22). Amsterdam: Elsevier.

Brown, A. L. (1992). Design experiments: Theoretical and methodological challenges in creating complex interventions in classroom settings. *Journal of the Learning Sciences, 2,* 141–178.

Brown, J. S., & Duguid, P. (1991). Organizational learning and communities of practice: Towards a unified view of working, learning and innovation. *Organization Science, 2,* 40–57.

Brown, J. S., & Duguid, P. (1994). Borderline issues: Social and material aspects of design. *Human–Computer Interaction 9,* 3–36.

Brown, J. S., & Duguid, P. (2000). *The social life of information.* Cambridge, MA: Harvard Business School Press.

Burgess, P. W., Quayle, A., & Frith, C. D. (2001). Brain regions involved in prospective memory as determined by positron emission tomography. *Neuropsychologia, 39,* 545–555.

Burrell, G., & Morgan, G. (1979). *Sociological paradigms and organizational analysis.* London: Heinemann.

Calabi, C. (1996). *Passioni e ragioni* [Passions and reasons]. Milan: Guerini.

Camp, C. J., Foss, J. W., Stevens, A. B., & O'Hanlon, A. M. (1996). Improving prospective memory task performance in persons with Alzheimer's disease. In M. A. Brandimonte, G. O. Einstein, & M. A. McDaniel (Eds.), *Prospective memory: Theory and applications* (pp. 351–368). Mahwah, NJ: Lawrence Erlbaum Associates.

Carofiglio, V., & de Rosis, F. (2003). Combining logical with emotional reasoning in natural argumentation. In C. Lisetti, C. Conati, & E. Hudlicka (Eds.), *Proceedings of the UM'03 Workshop on Affect*. Berlin: Springer.

Carroll, J. M. (1995) *Scenario-based design: Envisioning work and technology in system development*. New York: Wiley.

Carroll, J. M. (Ed.). (2003). *HCI models, theories, and frameworks: Towards a multidisciplinary science*. San Francisco: Kaufmann.

Casapulla, G., De Cindio, F., & Gentile, O. (1995). The Milan Civic Network experience and its roots in the town. In *Proceedings of the 2nd International Workshop on Community Networking* (pp. 283–289). New York: IEEE Press.

Castelfranchi, C. (1992). No more cooperation, please! In search of the social structure of verbal interaction. In A. Ortony, J. Slack, & O. Stock (Eds.), *Communication from an artificial intelligence perspective: Theoretical and applied issues* (pp. 206–227). Berlin: Springer.

Castelfranchi, C. (1996). Reasons: Belief support and goal dynamics. *Mathware & Soft Computing, 3,* 233–247.

Castelfranchi, C. (1998). Modeling social action for AI agents. *Artificial Intelligence, 103,* 157–182.

Castelfranchi, C. (2003, April). *Implicit behavioural communication in imitation.* Poster session presented at the second International Symposium on Imitation in Animals and Artifacts, AISB'03 Convention, University of Wales, Aberystwyth, UK.

Castelfranchi, C., & Giardini, F. (2003a). Will humans and robots ever coordinate each other via behavioral communication? *Proceedings of the 1st Symposium on Evolvability and Interaction: Evolutionary Substrates of Communication, Signaling, and Perception in the Dynamics of Social Complexity, Queen Mary University, London,* 25–30.

Castelfranchi, C., & Giardini, F. (2003b). Silent agents: Behavioral implicit communication for multi-agents coordination and HMI. *Second annual Symposium on Autonomous Intelligent Networks and Systems, Menlo Park, CA.*

Castelfranchi, C., Giardini, F., Lorini, E., & Tummolini, L. (2003). The prescriptive destiny of predictive attitudes: From expectations to norms via conventions. *Proceedings of CogSci 2003: 25th Annual Meeting of the Cognitive Science Society, Boston.*

Castelfranchi, C., & Parisi, D. (1980). *Linguaggio, conoscenze e scopi* [Language, knowledge and intentions]. Bologna: Il Mulino.

Chasteen, A. L., Park, D. C., & Schwarz, N. (2001). Implementation intentions and facilitation of prospective memory. *Psychological Science, 12,* 457–461.

Chekhov, A. P. (1997). *The Plays of Anton Chekhov* (P. Schmidt, Trans.). New York: HarperCollins.

Cialdini, R. B. (1993). *Influence: The psychology of persuasion.* New York: Morrow.

Ciborra, C. (1996). The platform organization: Recombining strategies, structures, and surprises. *Organization Science, 7,* 103–118.

Ciborra, C. (2002). *The labyrinths of information: Challenging the wisdom of systems.* Oxford, England: Oxford University Press.

Clancey, W. (1977). *Situated cognition: On human knowledge and computer representation.* Cambridge, England: Cambridge University Press.

Clark, H. H. (1996). *Using language.* Cambridge, MA: Cambridge University Press.

Clark, H. H., & Schaefer, E. F. (1989). Contributing to discourse. *Cognitive Science, 13*, 259–294.

Cohen, G. (1996). *Memory in the real world.* Hove, England: Psychology Press.

Comte, A. (1936). *Cours de philosophie positive, Vol. 1.* Paris: Larousse. (First published 1830. Translated in G. Lenzler, Ed., (1975), *Auguste Comte: The Essential Writings.* New York: Harper.

Conte, R., & Castelfranchi, C. (1995). *Cognitive and social action.* London: UCL Press.

Contu, A., & Willmott, H. (2003). Re-embedding situatedness: The importance of power relations in learning theory. *Organization Science, 14*, 283–296.

Cooley, M. (1988). From Brunelleschi to CAD-CAM. In J. Thackara (Ed.), *Design after modernism* (p. 197). New York: Thames & Hudson.

Crabtree, A., Hemmings, T., & Rodden, T. (2002). Pattern-based support for interactive design in domestic settings. *Proceedings of DIS2002: Conference on Designing Interactive Systems,* 265–276. New York: ACM Press.

Craik, F. I. M. (1986). A functional account of age differences in memory. In F. Klix & H. Hangendorf (Eds.), *Human memory and cognitive capabilities: Mechanisms and performances* (pp. 409–422). Amsterdam: Elsevier.

Crampton Smith, G., Mattioda, M., & Tabor, P. (Eds.). (2003). *Interaction design almanacco 2003.* Ivrea, Italy: Interaction Design Institute Ivrea.

Crang, M., & Thrift, N. (Eds.). (2000). *Thinking space.* London: Routledge.

Cross, N. (Ed.). (1984). *Developments in design methodology.* Bath, England: Wiley.

Cruz-Neira, C., Sandin, D. J., & DeFanti, T. (1993). Surround-screen projection-based virtual reality: The design and implementation of the CAVE. *ACM Computer Graphics, 27*, 135–142.

Csikszentmihalyi, M. (1990). *Flow: The psychology of optimal experience.* New York: Harper & Row.

Damasio, A. R. (1994). *Descartes' error: Emotion, reason, and the human brain.* New York: Putnam.

Damasio, A. R. (2003). *Looking for Spinoza: Joy, sorrow, and the feeling brain.* New York: Harcourt.

Dautenhahn, K. (2002). Design spaces and niche spaces of believable social robots. In E. Prassler (Ed.), *Proceedings of IEEE ROMAN 2002: International Workshop on Robots and Human Interactive Communication* (pp. 192–197). Hoboken, NJ: Wiley-IEEE Press.

De Carolis, B., Carofiglio, V., Bilvi, M., & Pelachaud, C. (2002). APML: A mark-up language for believable behavior generation. *Proceedings of the AAMAS 2002 Workshop on "Embodied Conversational Agents: Let's Specify and Compare Them,"* Bologna, 758–765.

Deetz, S. (1996). Describing differences in approaches in organization science: Rethinking Burrell and Morgan and their legacy. *Organization Science, 7*, 191–207.

De Michelis, G. (1998). *Aperto, molteplice, continuo: Gli artefatti alla fine del novecento* [Openness, multiplicity, continuity: Artifacts at the end of the twentieth century]. Milan: Dunod.

De Michelis, G. (2003). The Swiss Pattada: Designing the ultimate tool. *Interactions, 10*, 44–53.

De Monticelli, R. (2000a). *L'Avenir de la phenomenologie: Méditations sur la connaissance personnelle* [The future of phenomenology: Thoughts on personal knowledge]. Paris: Aubier-Flammarion.

De Monticelli, R. (2000b). *La persona, apparenza e realtà: Testi fenomenologici 1911–1933* [The person, appearance and reality: Writings on phenomenology 1911–1935]. Milan: Cortina.

De Monticelli, R. (2003). *L'ordine del cuore: Etica e teoria del sentire* [The order of the heart: The ethics of theory and feeling]. Milan: Garzanti.

Deneckere, R. J., & McAfee, R. P. (with McAfee, P.) (1996). Damaged goods. *Journal of Economics and Management Science, 5,* 149–174.

Desmond, W. (1995). *Perplexity and ultimacy: Metaphysical thoughts from the middle.* New York: State University of New York Press.

De Sousa, R. (1987). *The rationality of emotion.* Cambridge, MA: MIT Press.

De Sousa, R. (1995). The body is in. *Semiotic Review of Books, 6.*

Dobbs, J., & Reeves, H. (1996). Prospective memory: More than memory. In M. A. Brandimonte, G. O. Einstein, & M. McDaniel (Eds.), *Prospective memory: Theory and applications* (pp. 199–225). Mahwah, NJ: Lawrence Erlbaum Associates.

Dourish, P. (2001). *Where the action is: The foundations of embodied interaction.* Cambridge, MA: MIT Press.

Dourish, P., Holmes, J., MacLean, A., Marqvardsen, P., & Zbyslaw, A. (1996). Freeflow: Mediating between representation and actions in workflow systems. *Proceedings of the CSCW'96 Conference on Computer Supported Cooperative Work* (pp. 190–198). New York: ACM Press.

Dreyfus, H. L. (1991). *Being-in-the-world: A commentary on Heidegger's Being and Time, Division 1.* Cambridge, MA: MIT Press.

Dreyfus, H. L. (1993). *What computers still can't do: A critique of artificial reason.* Cambridge, MA: MIT Press.

Ducheneaut, N., & Bellotti, V. (2001). Email as habitat: An exploration of embedded personal information management. *Interactions, 8,* 30–38.

Dunne, A. (1999). *Hertzian tales: Electronic products, aesthetic experience and critical design.* London: Royal College of Art Publications.

Dunne, A., & Raby, F. (2001). *Design noir: The secret life of electronic objects.* London: August/Birkhäuser.

Durling, D., & Friedman, K. (Eds.). (2000). *Doctoral education in design: Foundations for the future.* Stoke on Trent, UK: Staffordshire University Press.

Ebbinghaus, H. (1913). *Memory: A contribution to experimental psychology.* New York: Teacher's College Columbia University (Reprinted Bristol: Thoemmo Press, 1999).

Eco, U. (1976). *A theory of semiotics.* London: Indiana University Press.

Ehn, P. (1988). *Work-oriented design of computer artifacts.* Hillsdale, NJ: Lawrence Erlbaum Associates.

Ehn, P. (1995). Informatics: Design for usability. In B. Dahlbom, *The infological equation: Essays in honor of Börje Langefors* (The Studies Report No. 6, pp. 159–174). Gothenburg, Sweden: University of Gothenburg.

Ehn, P. (1998). Manifesto for a digital Bauhaus. *Digital Creativity, 9,* 207–216.

Ehn, P., & Badham, R. (2002). Participatory design and the collective designer. In T. Binder, J. Gregory, & I. Wagner (Eds.), *Proceedings of PDC 02: Participatory Design Conference* (pp. 1–10). Palo Alto, CA: California Parks and Recreation Society.

Ehn, P., & Löwgren, J. (1997). Design for quality-in-use: Human–computer interaction meets systems development. In M. Helander, T. K. Landauer, & P. V. Prabhu (Eds.), *Handbook of human–computer interaction* (pp. 299–313). Amsterdam: Elsevier.

Ehn, P., & Löwgren, J. (Eds.). (2003). *Searching voices: Towards a canon for interaction design.* Malmö, Sweden: Malmö University Studies in Arts and Communication, 1.

Ehrenzweig, A. (1971) *The hidden order of art: A study in the psychology of artistic imagination.* Berkeley: University of California Press. (Original work published 1967)

Einstein, G. O., Holland, L. J., McDaniel, M. A., & Guynn, M. J. (1992). Age-related deficits in prospective memory: The influence of task complexity. *Psychology and Aging, 7,* 471–478.

Einstein, G. O., & McDaniel, M. A. (1990). Normal aging and prospective memory. *Journal of Experimental Psychology: Learning, Memory, and Cognition, 16,* 717–726.

Einstein, G. O., McDaniel, M. A., Williford, C. L., & Dismukes, R. L. (2003). Forgetting of intentions in demanding situations is rapid. *Journal of Experimental Psychology: Applied, 3,* 147–162.

Ellis, J. A. (1988). Memory for future intentions: Investigating pulses and steps. In M. M. Gruneberg, P. E. Morris, & R. N. Sykes (Eds.), *Practical aspects of memory: Current research and issues* (Vol. 1, pp. 371–376). Chichester, England: Wiley.

Ellis, J. A. (1996). Prospective memory or the realization of delayed intentions: A conceptual framework for research. In M. A. Brandimonte, G. O. Einstein, & M. A. McDaniel (Eds.), *Prospective memory: Theory and applications* (pp. 1–22). Mahwah, NJ: Lawrence Erlbaum Associates.

Elster, J. (1989). Social norms and economic theory. *Journal of Economic Perspectives, 3,* 99–117.

Engeström, Y. (1987). *Learning by expanding: An activity-theoretical approach to developmental research.* Helsinki, Finland: Orienta-Konsultit.

Engeström, Y. (2001). Expansive learning at work: Toward an activity theoretical reconceptualization. *Journal of Education and Work, 14,* 133–156.

Engeström, Y., Engeström, R., & Kerosuo, H. (2003). The discursive construction of collaborative care. *Applied Linguistics, 24,* 286–315.

Engeström, Y., Engeström, R., & Vähäaho, T. (1999). When the center does not hold: The importance of knotworking. In S. Chaiklin, M. Hedegaard, & U. J. Jensen (Eds.), *Activity theory and social practice* (pp. 345–374). Aarhus, Denmark: Aarhus University Press.

Engeström, Y., & Escalante, V. (1996). Mundane tool or object of affection? The rise and fall of the postal buddy. In B. Nardi (Ed.), *Context and consciousness: Activity theory and human–computer interaction.* Cambridge, MA: MIT Press.

Engeström, Y., Miettinen, R., Punamaki, & Heath, C. (Eds.). (1999). *Perspectives on activity theory.* Cambridge, England: Cambridge University Press.

Erasmus. (1974). *Praise of folly.* London: Folio.

Erickson, T. (2000a). Towards a pattern language for interaction design. In P. Luff, J. Hindmarsh, & C. Heath (Eds.), *Workplace studies: Recovering work practice and informing systems design* (pp. 252–261). Cambridge, England: Cambridge University Press.

Erickson, T. (2000b). Lingua francas for design: Sacred places and pattern languages. In D. Boyarski & W. A. Kellogg (Eds.), *Proceedings of the ACM Conference on Designing Interactive Systems* (pp. 357–368). New York: ACM Press.

Erickson, T., & Kellogg, W. A. (2003). Social translucence: Using minimalist visualizations of social activity to support collective interaction. In K. Höök, D. Benyon, & A. Munro (Eds.), *Designing information spaces: The social navigation approach* (pp. 17–42). London: Springer.

Erickson, T., Huang, W., Danis, C., & Kellogg, W. A. (2004). A social proxy for distributed tasks: The design and evaluation of a working prototype. In *Proceedings of the ACM 2004 Conference on Human Factors in Computing Systems, 559–566.*

Falcone, R., & Castelfranchi, C., (2000). Delegation-based conflicts between client and contractor. *Journal of Intelligent Automation and Soft Computing, 6,* 193–204.

Farr, R. (1996). *The roots of modern social psychology.* Oxford, England: Blackwell.

Fisher, P., Daniel, R., & Siva, K. V. (1990). Specification and design of input devices for teleoperation. *Proceedings of the IEEE International Conference on Robotics and Automation, Cincinnati, OH, 1,* 540–545.

Fisher, S. (1986). Telepresence master glove controller for dexterous robotic end-effectors. In D. P. Casasent (Ed.), *Advances in intelligent robotics systems, Pro-*

ceedings of SPIE 726: Symposium on Optical and Optoelectronic Engineering (pp. 396–401).

Fogg, B. J. (2002). *Persuasive technology: Using computers to change what we think and do.* San Francisco: Kaufmann.

Fong, T., Nourbakhsh, I., Dautenhahn, K. (2003). A survey of socially interactive robots. *Robotics and Autonomous Systems, 42,* 143–166.

Forlizzi, J., Lee, J., & Hudson, S. E. (2003). The Kinedit system: Affective messages using dynamic texts. *CHI letters. Proceedings of CHI 2003 Conference on Human Factors in Computing Systems,* 233–247).

Foucault, M. (1961). *L'Histoire de la folie à l'âge classique.* Paris: Gallimard. (Published in English as *Madness and civilization: A history of madness in the Age of Reason*)

Foucault, M. (1970). *The order of things: An archeology of the human sciences.* New York: Random House.

Friedman, B., Kahn, P. H., Jr., & Hagman, J. (2003). Hardware companions? What online AIBO discussion forums reveal about the human–robotic relationship. *Proceedings of the CHI 2003 Conference on Human Factors in Computing Systems,* 273–280.

Friedman, M. (1975). *There's no such thing as a free lunch.* LaSalle, IL: Open Court.

Frijda, N. (1997). *Commemorating.* In J. W. Pennebaker, D. Paez, & B. Rimé (Eds.), *Collective memory of political events: Social psychological perspectives* (pp. 103–127). Mahwah, NJ: Lawrence Erlbaum Associates.

Gadamer, H.-G. (1982). *Truth and method* (J. Weinsheimer & D. G. Marshall, Trans.). New York: Crossroad. (Original work published 1965)

Galati, D. (2002). *Prospettive sulle emozioni e teorie del soggetto* [Views on the emotions and theories of the subject]. Turin: Bollati Boringhieri.

Galison, P. (1997). *Image and logic: The material culture of microphysics.* Chicago: University of Chicago Press.

Garau, M., Slater, M., Bee S., & Sasse, M. A. (2001). The impact of eye gaze on communication using humanoid avatars. *CHI letters, Proceedings of the CHI 2003 Conference on Human Factors in Computing Systems,* 309–316.

Gargani, A. G. (1995). La figura del maestro: Esemplarità, autenticità, e inautenticità [The figure of the master: Exemplarity, authenticity and inauthenticity]. In G. Vattimo (Ed.), *Filosofia '94* (pp. 15–35). Bari, Italy: Laterza.

Gedenryd, H. (1998). How designers work: Making sense of authentic cognitive activities. *Lund University Cognitive Studies, 75.* Retrieved June 1st, 2005, from http://www.lucs.lu.se/People/Henrik.Gedenryd/HowDesignersWork/index.html

Geyer, W., Vogel, J., Cheng, L., & Muller, M. (2003). Supporting activity-centric collaboration through peer-to-peer shared objects. *Proceedings of the 2003 ACM SIGGROUP Conference on Supporting Group Work,* 115–124.

Giardini, F., & Castelfranchi, C. (2003). *Beyond language* (ISTC Technical Report No. ISTC–07–2003. Rome: Istituto di Scienze e Tecnologie della Cognizione.

Gibson, J. J. (1966). *The senses considered as perceptual systems.* Boston: Houghton Mifflin.

Gibson, J. J. (1979). *The ecological approach to visual perception.* Boston: Houghton Mifflin.

Givens, D. B. (2003). *Nonverbal dictionary of gestures, signs and body language cues.* Spokane, WA: Center for Nonverbal Studies Press.

Gmytrasiewicz, P. J., & Lisetti, C. L. (2001). Emotions and personality in agent design and modeling. *Proceedings of the 2nd Workshop on Attitude, Personality and Emotions in User Adapted Interaction.* Berlin: Springer-Verlag.

Goertz, R. C. (1964). Manipulator system development at ANL. *Proceedings of the 12th American Nuclear Society Conference on Remote Systems Technology, Argonne National Laboratory, La Grange Park, IL, 1,* 117–136.

Goffman, E. (1961). *Encounters: Two studies in the sociology of interaction.* Indianapolis, IN: Bobbs-Merrill.

Goffman, E. (1963). *Behavior in public places: Notes on the social organization of gatherings.* New York: Macmillan.

Goffman, E. (1981). Footing. In E. Goffman (Ed.), *Forms of talk* (pp. 124–159). Philadelphia: University of Pennsylvania Press.

Goldie, P. (2000). *The emotions.* Oxford, England: Oxford University Press.

Gollwitzer, P. M. (1999). Implementation intentions: Strong effects of simple plans. *American Psychologist, 54,* 493–503.

Gollwitzer, P. M., & Brandstaetter, V. (1997). Implementation intentions and effective goal pursuit. *Journal of Personality and Social Psychology, 73,* 186–199.

Gomes de Sa, A., & Zachmann, G. (1999). Virtual reality as a tool for verification of assembly and maintenance processes. *Computers & Graphics, 23,* 389–403.

Goodwin, C. (1994). Professional vision. *American Anthropologist, 96,* 606–633.

Goodwin, C. (1995). Seeing in depth. *Social Studies of Science, 25,* 237–274.

Goodwin, C. (1996). Transparent vision. In E. Ochs, E. A. Schegloff, & S. Thompson (Eds.), *Interaction and grammar* (pp. 370–404). Cambridge, England: Cambridge University Press.

Goodwin, C. (2000). Action and embodiment within situated human interaction. *Journal of Pragmatics, 32,* 1489–1522.

Goodwin, C. (2003a). Conversational frameworks for the accomplishment of meaning in aphasia. In C. Goodwin (Ed.), *Conversation and brain damage* (pp. 90–116). Oxford, England: Oxford University Press.

Goodwin, C. (2003b). *Il senso del vedere: Pratiche sociali della significazione* [The sense of sight: The social practices of meaning] (Vol. 20). Rome: Melterri.

Goodwin, C., Goodwin, M. H., & Olsher, D. (2002). Producing sense with nonsense syllables: Turn and sequence in the conversations of a man with severe aphasia. In B. Fox, C. Ford, & S. Thompson (Eds.), *The language of turn and sequence* (pp. 56–80). Oxford, England: Oxford University Press.

Goodwin, M. H. (1990). *He-said-she-said: Talk as social organization among black children.* Bloomington: Indiana University Press.

Goodwin, M. H., & Goodwin, C. (2001). Emotion within situated activity. In A. Duranti (Ed.), *Linguistic anthropology: A reader* (pp. 239–257). Malden, MA: Blackwell. (Reprinted from *Communication: An arena of development,* pp. 233–254, by N. Budwig, I. C. Uzgris, & J. V. Wertsch (Eds.). 2000). Stamford, CT: Ablex.

Grasso, F., Cawsey, A., & Jones, R. (2000). Dialectical argumentation to solve conflicts in advice giving: A case study in the promotion of healthy nutrition. *International Journal of Human–Computer Studies, 53,* 1077–1115.

Grice, P. (1957). Meaning. *Philosophical Review, 66,* 377–388.

Grimes, G. J. (1983). *Digital data entry glove interface device* (U.S. Patent No. 4,414,537, November 8). Murray Hill, NJ: Bell Telephone Laboratories.

Guerini, M., Stock, O., & Zancaro, M. (2003). Toward intelligent persuasive interfaces. *Proceedings of the IJCAI Workshop on Natural Argumentation* (International Joint Conference on Artificial Intelligence, Acapulco, Mexico).

Guerini, M., Stock, O., & Zancaro, M. (2004). Persuasive strategies and rhetorical relation selection. *Proceedings of the ECAI Workshop on Computational Models of Natural Argument, Valencia, Spain.*

Guignon, C. (2002). *Philosophy and authenticity: Heidegger's search for a ground for philosophizing.* In M. Wrathall & J. Malpas (Eds.), *Heidegger, authenticity and modernity: Essays in honor of Hubert L. Dreyfus, Vol. 1* (pp. 79–101). Cambridge, MA: MIT Press.

Guyau, J.-M. (1891). *Les problèmes de l'esthétique contemporaine* [The problems of contemporary esthetics]. Paris: Alcan.

Habermas, J. (1968). *Erkenntnis und Interesse* [Knowledge and Human Interests]. Frankfurt am Main: Suhrkamp.

Habermas, J. (1981) Urbanisierung der Heideggerschen Provinz [The urbanization of the Heideggerian province]. In J. Habermas (Ed.), *Philosophisch-politische profile*. Frankfurt am Main: Suhrkamp.

Habermas, J. (1985). *Theorie des Kommunikativen Handelns* [The Theory of Communicative Action]. Frankfurt am Main: Suhrkamp.

Halbwachs, M. (1925). *Les Cadres sociaux de la mémoire* [The Social Frameworks of Memory]. Paris: Alcan. (Original work published 1992)

Halbwachs, M. (1967). *La Mémoire collective* [*The collective memory*]. Paris: Presses Universitaires de France. (Original work published 1950)

Hall, E. T. (1983). *The dance of life: The other dimension of time*. New York: Anchor.

Haraway, D. J. (1991). *Situated knowledges: Simians, cyborgs, and women: The reinvention of nature*. London: Free Association Books.

Harrison, G. (1986). *Politica ecologica e ecologia politica* [Ecological politics and political ecology]. Padua: Francisci.

Harrison, S., & Dourish, P. (1996). Re-place-ing space: The roles of place and space in collaborative systems. In *Proceedings of CSCW'96 Conference on Computer Supported Cooperative Work* (pp. 67–76). New York: ACM Press.

Heath, C., & Luff, P. (2000). *Technology in action*. Cambridge, MA: Cambridge University Press.

Hedman, A. (2004). *Visitor orientation in context: The historically rooted production of soft places*. Stockholm: Royal Institute of Technology [KTH] Numerical Analysis and Computer Science.

Heidegger, M. (1962). *Being and time*. Oxford, England: Blackwell. (Original work published 1927)

Heidegger, M. (1977). *The question concerning technology and other essays*. New York: Harper & Row.

Heidegger, M. (1993). *Gesamtausgabe, Band 58: Grundprobleme der Phaenomenologie (1919/20)* [Collected works, Vol. 58: Basic problems of phenomenology]. Frankfurt am Main: V. Klostermann.

Hersh, N., & Treadgold, L. (1994). NeuroPage: The rehabilitation of memory dysfunction by prosthetic memory and cueing. *Neurorehabilitation, 4*, 187–197.

Heskett, J. (2002). *Toothpicks and logos: Design in everyday life*. Oxford, England: Oxford University Press.

Holland, O. E., & Beckers, R. (1996). The varieties of stigmergy: Indirect action as a control strategy for collective robotics. *Robotica*.

Holman, J., & Zaidi, F. (2004). *The economics of prospective memory*. Manuscript in preparation.

Hommel, B. (2003). Planning and representing intentional actions: *Scientific World Journal, 3*, 593–608.

Hong, T., & Tau, X. (1989). Calibrating a VPL dataglove for teleoperating the Utah/MIT hand. *Proceedings of the IEEE International Conference on Robotics and Automation, Scottsdale, AZ, 3*, 1752–1757.

Huizinga, J. (1950). *Homo ludens: A study of the play element in culture*. New York: Roy.

Hutchins, E. (1995). *Cognition in the wild*. Cambridge, MA: MIT Press.

Huysman, M., Wenger, E., & Wulf, V. (Eds.). (2003). *Communities and technologies*. Dordrecht, The Netherlands: Kluwer.

Interaction Design Institute Ivrea. (2003). *Telekatessen: Digital delicacies*. Retrieved March 10, 2005, from http://www.interaction-ivrea.it/en/gallery/telekatessen/index.asp

Ishida, T. (1998). *Community computing: Collaboration over global information networks.* New York: Wiley.

Ishida, T., & Isbister, K. (2000). *Digital cities: Technologies, experiences and future perspectives.* Berlin: Springer.

Jacobs, J. (1961). *The death and life of great American cities.* New York: Random House.

Jacobsen, J. C., Iversen, K. E., Knutti, D. F., Johnson, R. T., & Biggers, K. B. (1986). Design of the UTAH/MIT dexterous hand. *Proceedings of the IEEE International Conference on Robotics and Automation, San Francisco, 2,* 1520–1532.

Jeamsinkul, C. (2002). *Methodology for uncovering motion affordance in interactive media.* Unpublished doctoral dissertation, Illinois Institute of Technology, Chicago.

Jeamsinkul, C., & Poggenpohl, S. (2002). Methodology for uncovering motion affordance in interactive media. *Visible Language, 36,* 254–280.

Johnsen, E .G., & Corliss, W. R. (1995). Teleoperators and human augmentation: An AEC/NASA technology survey. In C. H. Gray (Ed.), *The Cyborg handbook* (pp. 83–93). New York: Routledge. (Original work published 1967)

Johnson, M. H., & Morton, J. (1991). *Biology and cognitive development: The case of face recognition.* Oxford, England: Blackwell.

Johnson, M. K., Hashtroudi, S., & Lindsay, D. S. (1993). Source monitoring. *Psychological Bulletin, 114,* 3–28.

Jones, J. C. (1970). *Design methods.* New York: Van Nostrand Reinhold.

Jones, J. C. (1977). How my thoughts about design methods have changed during the years. *Design Methods and Theories, 11,* 50–62.

Jones, K. S. (2003). What is an affordance? *Ecological Psychology, 15,* 107–114.

Kaptelinin, V. (2003). UMEA: Translating interaction histories into project contexts. *Proceedings of the CHI 2003 Conference on Human Factors in Computing Systems, New York,* 353–360.

Keller, C. M., & Keller, J. D. (1996). *Cognition and tool use: The blacksmith at work.* Cambridge, England: Cambridge University Press.

Kendon, A. (1990). *Conducting interaction: Patterns of behavior in focused encounters.* Cambridge, England: Cambridge University Press.

Kerosuo, H., & Engeström, Y. (2003). Boundary crossing and learning in creation of new work practice. *Journal of Workplace Learning, 15,* 345–351.

Kim, H. J., Burke, D. T., Dowds, M. M., Boone, K. A. R., & Park, G. J. (2000). Electronic memory aids for outpatient brain injury: Follow-up findings. *Brain Injury, 14,* 187–196.

Kindred, J. B. (1999). "8/18/97 bite me": Resistance in learning and work. *Mind, Culture, and Activity, 6,* 196–221.

Kintsch, W. (1995). Introduction to F. C. Bartlett (1932; reissued 1995). In F. C. Bartlett (Ed.), *Remembering: A study in experimental and social psychology* (pp. xi–xv). London: Cambridge University Press.

Kliegel, M., Martin, M., McDaniel, M. A., & Einstein, G. O. (2001). Varying the importance of a prospective memory task: Differential effects across time- and event-based prospective memory. *Memory, 9,* 1–11.

Kobsa, A. (2002). Personalized hypermedia and international privacy. *Communications of the ACM, 45,* 64–67.

Koriat, A., & Pearlman-Avnion, S. (2003). Memory organization of action events and its relationship to memory performance. *Journal of Experimental Psychology: General, 3,* 435–454.

Koszegi, B. (2000). *Ego utility and information acquisition,* Mimeo.

Kramer, J., & Leifer, L. (1990). *The talking glove: A speaking aid for non vocal deaf and deaf-blind individuals.* Palo Alto, CA: Palo Alto Veterans Administration.

Kreutzer, M. A., Leonard, C., & Flavell, J. H. (1975). An interview study of children's knowledge about memory. *Monographs of the Society for Research in Child Development, 40*(1), Serial No. 159.

Kukai [Kobodaishi] (1983). *Himitsu Mandara Jujushinron* [The ten abiding stages of the secret mandalas], *Book 1.* Tokyo: Tsukuma Shobo.

Latour, B. (1992). Where are the missing masses?: The sociology of a few mundane objects. In W. E. Bijker & J. Law (Eds.), *Shaping technology/building society: Studies in sociotechnical change* (pp. 225–258). Cambridge, MA: MIT Press.

Latour, B. (1999). *Pandora's hope: Essays on the reality of science studies.* Cambridge, MA: Harvard University Press.

Lave, J., & Wenger, E. (1991). *Situated learning: Legitimate peripheral participation.* Cambridge, England: Cambridge University Press.

Lazarus, R. S. (1966). *Psychological stress and the coping process.* New York: McGraw-Hill.

Lazarus, R. S. (1991). *Emotion and adaptation.* New York: Oxford University Press.

Lederman, S. J., & Klatzky, R. (1990). Haptic exploration and object representation. In M. A. Goodale (Ed.), *Vision and action: The control of grasping* (p. 367). Norwood, NJ: Ablex.

Legrenzi, P. (2005). *Creatività e innovazione: Come nascono le nuove idée* [Creativity and innovation: How new ideas are born]. Bologna: Il Mulino.

Lehnert, H., & Blauert, J. (1991). Virtual auditory environment. *Proceedings of the 91 ICAR: 5th International Conference on Advanced Robotics, Pisa, Italy, 1*, 211–216.

Leiva-Lobos, E. (1999). *From social complexity to cooperative awareness support.* Unpublished doctoral dissertation, University of Milano, Milan.

Leone, G. (1998). *I confini della memoria: I ricordi come risorsa sociale nascosta* [The boundaries of memory: Memories as a hidden social resource]. Soveria Mannelli, Italy: Rubbettino.

Leone, G. (2001). *Cosa è* sociale *nella memoria?* [What is *social* in memory?]. In G. Bellelli, D. Bakhurst, & A. Rosa (Eds.), *Tracce memoria collettiva e identità sociali* (pp. 49–69). Naples: Liguori.

Leontiev, A. N. (1975). Lo sviluppo della memoria [The development of memory]. In *Problemi dello sviluppo psichico* (pp. 355–88). Rome: Editori riuniti. (Original work published 1931)

Leontiev, A. N. (1978). *Activity, consciousness, and personality.* Englewood Cliffs, NJ: Prentice Hall.

Lerner, J., & Tirole, J. (2002). Some simple economics of open source. *Journal of Industrial Economics, 50*, 197–234.

Leroy-Gourhan, A. (1964). *Le geste et la parole* [The gesture and the word] (Vols. 1–2). Paris: Michel. (Original work published 1977)

Lévi-Strauss, C. (1962). *La pensée sauvage.* Paris: Plon.

Liestøl, G. (2003). "Gameplay": From synthesis to analysis (and vice versa). In G. Liestøl, A. Morrison, & T. Rasmussen (Eds.), *Digital media revisited: Theoretical and conceptual innovations in digital domains* (p. 402). Cambridge, MA: MIT Press.

Loftus, E. F. (1979). *Eyewitness testimony.* Cambridge, MA: Harvard University Press.

Long, N. (2001). *Development sociology: Actor perspectives.* London: Routledge.

Loomis, J. M., & Lederman, S. J. (1986). Tactual perception. In K. Boff, L. Kaufman, & J. Thomas (Eds.), *Handbook of perception and human performance* (31-1–31-41). New York: Wiley.

Löwgren, J., & Stolterman, E. (2004). *Thoughtful interaction: A design perspective on information technology.* Cambridge, MA: MIT Press.

Macintyre, A. (1981). *After virtue: A study in moral theory.* London: Duckworth.

MacKay, D. (1969). *Information, mechanism, and meaning.* Cambridge, MA: MIT Press.

Magri, T. (Ed.). (1999). *Filosofia ed emozioni* [Philosophy and emotions]. Milan: Feltrinelli.

Mandler, G. (1984). *Mind and body: Psychology of emotion and stress.* New York: Norton.

Mann, W. C., & Thompson, S. (1987). Rhetorical structure theory: A theory of text organization. In L. Polanyi (Ed.), *The structure of discourse.* Norwood, NJ: Ablex.

Manovich, L. (2001). *The language of new media.* Cambridge, MA: MIT Press.

Marcus, B. A., Lucas, W., & Churchill, P. J. (1989). Human hand sensing for robotics and teleoperations. *Sensors, 6,* 26, 28–31.

Marti, P., Palma, V., Pollini, A., Rullo, A., & Shibata, T. (2005). My gym robot. *Proceedings of the AISB'05 International Symposium on Robot Companions: Hard Problems and Open Challenges in Human–Robot Interaction,* 64–73.

Marti, P., Pollini, A., Rullo, A., & Shibata, T. (2005, September). Engaging with artificial pets. In N. Marmarmas, T. Kontogiannis, & D. Nathanael (Eds.), *Proceedings of the Annual Conference of the European Association of Cognitive Ergonomics, EACE 2005* (pp. 99–106). Chania, Crete.

Martin, M. (1986). Ageing and patterns of change in everyday memory and cognition. *Human Learning, 5,* 63–74.

Mason, M. T., & Salisbury, J. K. (1985). *Robot hands and the mechanics of manipulation.* Cambridge, MA: MIT Press.

Matsunaga, M. (2000). *The birth of I-mode: An analogue account of the mobile internet.* Singapore: Chuang Yi.

Mattioda, M., Norlen, L., & Tabor, P. (Eds.). (2005). *Interaction design almanacco 2004.* Ivrea, Italy: Interaction Design Institute Ivrea.

Mazzara, B. M., & Leone, G. (2001). Collective memory and intergroup relations. *Revista de Psicologia Social, 16,* 349–367.

Mazzoni, G., & Vannucci, M. (1998). Ricordo o conosco: Quando gli errori di memoria sono considerati ricordi veri [I remember or I know? When mistaken memories are considered to be true]. *Giornale Italiano di Psicologia, 25,* 79–99.

McCloud, S. (1993). *Understanding comics: The invisible art.* New York: HarperCollins.

McDaniel, M. A., & Einstein, G. O. (1993). The importance of cue familiarity and cue distinctiveness in prospective memory. *Memory, 1,* 23–41.

McDaniel, M. A., & Einstein, G. O. (2000). Strategic and automatic processes in prospective memory retrieval: A multiprocess framework. *Applied Cognitive Psychology, 14,* S127–S144.

McQuail, D., & Windahl, S. (1993). *Communication models.* New York: Longman.

Mecacci, L. (1990). Introduction to L. S. Vygotskij. In *Pensiero e linguaggio: Ricerche psicologiche* (pp. v–x). Rome: Laterza.

Meijers, A. W. M. (2002). Dialogue, understanding and collective intentionality. In G. Meggle (Ed.), *Social facts and intentionality* (pp. 225–254). Frankfurt: Hänsel-Hohenhausen.

Miller, P. J. (1994). Narrative practices: Their role in socialization and self-construction. In U. Neisser & R. Fivush (Eds.), *The Remembering self: Construction and accuracy in the self-narrative* (pp. 158–179). Cambridge, England: Cambridge University Press.

Mizoguchi, H., Sato, T., & Tagaki, K. (1997) Realization of expressive mobile robot. *Proceedings of the 1997 IEEE International Conference on Robotics and Automation, 1,* 581–586.

Moggridge, B. (in press). *Designing interactions.* Cambridge, MA: MIT Press.

Moles, A. (1976). *Micropsychologie de la vie quotidienné* [Micropsychology of everyday life]. Paris: Denoël Gonthier.

Moles, A. (1986). The legibility of the world: A project of graphic design. *Design Issues, 3.1,* 43–53.

Moran, T. P., & Carroll, J. M. (1996). *Design rationale: Concepts, techniques, and use.* Mahwah, NJ: Lawrence Erlbaum Associates.

Mori, M. (1970). Bukimi no tani [The uncanny valley]. *Energy, 7,* 33–35.

Morin, E. (1992). *Method: Towards a study of humankind, Vol. 1: The nature of nature.* New York: Peter Lang. (Original work published 1977)

Morris, W. (1992). *The collected works of William Morris.* London: Routledge/ Thoemmes Press.

Nancy, J.-L. (1990). *La communauté désoeuvrée.* Paris: Bourgois.

Napier, J. R. (1956). The prehensile movements of the human hand. *Journal of Bone and Joint Surgery, 38B,* 912–913.

Nardi, B. A., & Miller, J. R. (1991). Twinkling lights and nested loops: Distributed problem-solving and spreadsheet development. In S. Greenberg (Ed.), *Computer supported cooperative work and groupware* (pp. 29–52). London: Academic.

Nardi, B. A., & O'Day, V. (1999). *Information ecologies: Using technology with heart.* Cambridge, MA: MIT Press.

Negroponte, N. (1998). Beyond digital. *Wired, 6.12,* 288.

Neisser, U. (Ed.). (1982). *Memory observed: Remembering in natural contexts.* San Francisco: Freeman.

Nelson, P. (1970). Information and consumer behavior. *Journal of Political Economy, 78,* 311–329.

Nonaka, I., & Takeuchi, H. (1995). *The knowledge creating company.* New York: Oxford University Press.

Norman, D. A. (1986). Cognitive engineering. In D. A. Norman & S. W. Draper (Eds.), *User centered system design: New perspectives on human–computer interaction* (pp. 31–61). Hillsdale, NJ: Lawrence Erlbaum Associates.

Norman, D. A. (1993). *Things that make us smart: Defending human attributes in the age of the machine.* Reading, MA: Addison-Wesley.

Norman, D. A. (1994). How might people interact with agents. *Communications of the ACM, 3*(7), 68–71.

Norman, D. A. (1997). *The psychology of everyday things.* New York: Basic Books.

Norman, D. A. (1998). *The invisible computer.* Cambridge, MA: MIT Press.

Norman, D. A. (2002). *The design of everyday things.* New York: Basic Books.

Norman, D. A. (2004). *Emotional design: Why we love (or hate) everyday things.* New York: Basic Books.

Normann, R. (2001). *Reframing business: When the map changes the landscape.* Chichester, England: Wiley.

Nussbaum, M. (2001). *Upheavals of thought: The intelligence of emotions.* Cambridge, England: Cambridge University Press.

Okano, M. Y. (1990). *Depth psychology of Buddhism.* Tokyo: Aotosha.

Oksenberg Rorty, A. (1980). *Explaining emotions.* Berkeley: University of California Press.

Omicini, A., Ricci, A., Viroli, M.,m Castlefranchi, C., & Tummolini, L. (2004, July). Coordination artifacts: Environment-based coordination for intellect agents. Paper presented at Autonomous Agents and Multi-Agent Systems, 3rd International Joint Conference, New York.

Orr, J. (1996). *Talking about machines: An ethnography of a modern job.* Ithaca, NY: Cornell University Press.

Ortony, A., Clore, G. L., & Collins, A. (1988). *The cognitive structure of emotions.* Cambridge, England: Cambridge University Press.

Ortony, A., Norman, D. A., & Revelle, W. (2005). The role of affect and proto-affect in effective functioning. In J.-M. Fellous & M. A. Arbib (Eds.), *Who needs emotions? The brain meets the machine* (pp. 173–202). New York: Oxford University Press.

Owen, C. (2001). Structured planning and design: Information-age tools for product development. *Design Issues, 17.1*, 27–43.

OXO International Ltd. (n.d.). *Description of the OXO Good Grips Uplift Tea Kettle.* Retrieved January, 2004, from https://www.oxo.com/catalog/index.asp?getcategory=00001070

Papandreou, A. (1994). *Externality and institutions.* Oxford, England: Clarendon Press.

Parisi, D., & Castelfranchi, C. (1981). A goal analysis of some pragmatic aspects of language. In H. Parret, M. Sbisà & J. Verschueren (Eds.), *Possibilities and limitations of pragmatics* (Studies in Language Companion Series, Vol. 7, pp. 47–60). Amsterdam: Benjamins.

Parmentier, R. (1994). *Signs in society: Studies in semiotic anthropology.* Bloomington: Indiana University Press.

Pearce, W. B. (1989). *Communication and the human condition.* Carbondale: Southern Illinois University Press.

Peirce, C. S. (1998). "Harvard Lectures on Pragmatism." In The Peirce Edition Project (Ed.), *The Essential Peirce: Selected Philosophical Writings, Vol. 2, 1893–1913.* Bloomington, IN: Indiana University Press.

Perelman, C., & Olbrechts-Tyteca, L. (1969). *The new rhetoric: A treatise on argumentation.* Notre Dame, IN: Notre Dame Press.

Petty, R. E., & Cacioppo, J. T. (1986). *Communication and persuasion: Central and peripheral routes to attitude change.* New York: Springer.

Poggenpohl, S., Chayutsahkij, P., & Jeamsinkul, C. (2004). Language definition and its role in developing a design discourse. *Design Studies, 25*, 579–605.

Poggi, I., Castelfranchi, C., & Parisi, D. (1981). Answers, replies and reactions. In H. Parret, M. Sbisà, & J. Verschueren (Eds.), *Possibilities and limitations of pragmatics* (Studies in Language Companion Series, Vol. 7, pp. 61–73). Amsterdam: Benjamins.

Poggi, I., Pelachaud, C., & De Carolis, B. (2001). To display or not to display? Towards the architecture of a reflexive agent. *Proceedings of the 2nd Workshop on Attitude, Personality and Emotions in User Adapted Interaction,* Berlin, Germany.

Porter, O. W. (1969). Non verbal communication. *Training and Development Journal,* 3–8.

Preece, J., Rogers, Y., Sharp, H., Benyon, D., Holland, S., & Carey, T. (1994). *Human–computer interaction.* Reading, MA: Addison-Wesley.

Prendinger, H., & Ishizuka, M. (2004). Introducing the cast for social computing: Life-like characters. In H. Prendinger & M. Ishizuka (Eds.), *Life-like characters: Tools, affective functions and applications* (pp. 3–16). Berlin: Springer.

Pressman, J. L., & Wildavsky, A. (1984). *Implementation* (3rd ed.). Berkeley: University of California Press.

Prigogine, I., & Stengers, I. (1986). *Order out of chaos.* New York: Bantam. (Original work published 1984)

Putnam, R. D. (2000). *Bowling alone: The collapse and revival of American community.* New York: Simon & Schuster.

Radin, P., Jung, C. E., & Kerènyi, C. (1956). *The trickster: A study in American Indian mythology.* New York: Schocken.

Rakison, D. H., & Poulin-Dubois, D. (2001). The developmental origin of the animate–inanimate distinction. *Psychological Bulletin, 2*, 209–228.

Rao, P. K. (2002). *The economics of transaction costs: Theory, methods, and applications.* London: Palgrave Macmillan.

Reason, J. (1990). *Human error.* Cambridge, England: Cambridge University Press.

Redström, J. (2001). *Designing everyday computational things* (PhD thesis). *Gothenburg Studies in Informatics, 20,* Gothenburg, Sweden.

Reisberg, D. (1987). External representations and the advantages of externalizing one's thoughts. In E. Hunt (Ed.), *Proceedings of the Ninth Annual Conference of the Cognitive Science Society* (pp. 281–293). Hillsdale, NJ: Lawrence Erlbaum Associates.

Reiter, E., Sripada, S., & Robertson, R. (2003). Acquiring correct knowledge for natural language generation. *Journal of Artificial Intelligence Research, 18,* 491–516.

Ritella, A. (2004). *L'influenzabilità sociale della memoria: dal conformismo pubblico alla distorsione dei ricordi* [How memory can be socially influenced: From public conformity to memory distortions]. Unpublished doctoral dissertation, Università degli Studi Bari, Italia, A. A.

Robertson, T. (2002). The public availability of actors and artefacts. *Computer Supported Cooperative Work, 11,* 299–316.

Rogers, E. (1994). *A history of communication study.* New York: Free Press.

Rosella, F., Slocombe, D., Sunesson, L., & Torstensson, M. (2003). Telekatessen. In G. Crampton Smith, M. Mattioda, & P. Tabor (Eds.), *Interaction design almanacco 2003* (p. 62).

Ross, M. (1997). Validating memories. In N. L. Stein, P. A. Ornstein, B. Tversky, & C. Brainerd (Eds.), *Memory for everyday and emotional events* (pp. 49–81). Mahwah, NJ: Lawrence Erlbaum Associates.

Roth, A. E. (2002). The economist as engineer: Game theory, experimentation, and computation as tools for design economics. *Econometrica, 70,* 1341–1378.

Saito, T., Shibata, T., Wada, K., & Tanie, K. (2002). Examination of change of stress reaction by urinary tests of elderly before and after introduction of mental commit robot to an elderly institution. *Proceedings of the 7th AROB International Symposium on Artificial Life and Robotics, 1,* 316–319.

Salen, K., & Zimmerman, E. (2004). *Rules of play: Game design fundamentals.* Cambridge, MA: MIT Press.

Salisbury, J. K. (1984). Design and control of an articulated hand. *Proceedings of the 1st International Symposium on Design and Synthesis, Tokyo.*

Sarason, S. B. (1974). *The psychological sense of community: Prospects for a community psychology.* San Francisco: Jossey-Bass.

Sastry, L., & Boyd, D. R. S. (1998). Virtual environments for engineering applications. *Virtual Reality, 3,* 235–244.

Sawasdichai, N. (2004). *User goal-based approach to information search and structure on web site.* Unpublished doctoral dissertation, Illinois Institute of Technology, Chicago.

Sawasdichai, N., & Poggenpohl, S. (2003). User analysis framework. *Visible Language, 37*(1), 59–92.

Scheler, M. (1957). Ordo amoris [The order of love]. In L. Boella (Ed.), *Il valore della vita emotive.* Milan: Guerini.

Scherer, K. R. (1995). Expression of emotion in music. *Journal of Voice, 9,* 235–248.

Scholl, B. J., & Tremoulet, P. D. (2000). Perceptual causality and animacy. *Trends in Cognitive Science, 4,* 299–309.

Schön, D. (1987). *Educating the reflective practitioner: Towards a new design for teaching and learning in the professions.* San Francisco: Jossey-Bass.

Schön, D. A. (1983). *The reflective practitioner: How professionals think in action.* New York: Basic Books.

Schuler, D. (1996). Community networks: Building a new participatory medium. *Communications of the ACM, 39,* 52–63.

Schultze, U., & Leidner, D. E. (2002). Studying knowledge management in information systems research: Discourses and theoretical assumptions. *MIS Quarterly, 26,* 213–242.

Shapiro, C., & Varian, H. R. (1999). *Information rules: A strategic guide to the network economy.* Cambridge, MA: Harvard Business School Press.

Shedroff, N. (2002). *Experience design: A manifesto for the creation of experiences.* Indianapolis, IN: New Riders.

Sheridan, T. B. (1992a). Musings on telepresence and virtual presence. *Presence: Teleoperators and Virtual Environments, 1,* 120–126.

Sheridan, T. B. (1992b). *Telerobotics and human supervisory control.* Cambridge, MA: MIT Press.

Shibata, T., & Tanie, K. (2001). Emergence of affective behaviors through physical interaction between human and mental commit robot. *Journal of Robotics and Mechatronics, 13*(5), 505–516.

Sias, R. (Ed.). (1987). Ergonomia nell'ufficio computerizzato: Isao Hosoe intervista E. Grandjean e S. Bagnara [Ergonomics in the computerized office: Isao Hosoe interviews E. Grandjean and S. Bagnara]. [Special Issue]. *Ufficiostile, 6/7.*

Sillince, J. A. A., & Minors, R. H. (1991). What makes a strong argument? Emotions, highly-placed values and role playing. *Communication and Cognition, 24,* 281–298.

Simon, Herbert (1972). *The sciences of the artificial.* Cambridge, MA: MIT Press.

Slater, M. (1999). Measuring presence: A response to the Witmer and Singer questionnaire. *Presence: Teleoperators and Virtual Environments, 8,* 560–566.

Slater, M., Usoh, M., & Steed, A. (1994). Depth of presence in virtual environments. *Presence: Teleoperators and Virtual Environments, 3,* 130–144.

Smith, A. (1904). *An inquiry into the nature and causes of the wealth of nations* (5th ed.). London: Methuen. (Original work published 1776)

Smith, D. C., Armogida, F. P., & Mummery, H. L. (n.d.). *Green man.* Kaneohe, HI: Hawaian Laboratory of Naval Ocean System Center.

Smith, L. B., & Heise, D. (1992). Perceptual similarity and conceptual structure. In B. Burn (Ed.), *Percepts, concepts and categories.* Amsterdam: Elsevier.

Snow, C. P. (1959). *The two cultures and the scientific revolution.* Cambridge, England: Cambridge University Press.

Spinosa, C., Flores, F., & Dreyfus, H. L. (1997). *Disclosing new worlds: Enterpreneurship, democratic action, and the cultivation of solidarity.* Cambridge, MA: MIT Press.

Srinivasan, M. A. (1991). Tactual interfaces: The human perceiver. *Proceedings of the Engineering Foundation and NASA Conference on Human–Machine Interfaces for Teleoperators and Virtual Environments, 1,* 9–10.

Stock, O., Zancanaro, M., & Not, E. (2005). Intelligent interactive information presentation for cultural tourism. In O. Stock, & M. Zancanaro (Eds.), *Intelligent multimodal information presentation.* New York: Springer.

Suchman, L. A. (1987). *Plans and situated actions: The problem of human machine communication.* Cambridge, England: Cambridge University Press.

Sudnow, D. (2001). *Ways of the hand: A rewritten account.* Cambridge, MA: MIT Press.

Sutter, P. H., Iatridis, J. C., & Thakor, N. V. (1989). Response to reflected-force feedback to fingers in teleoperations. In G. Rodriguez & H. Seraji (Eds.), *Transactions of the NASA Conference on Space Telerobotics* (pp. 65–74). Pasadena, CA: NASA Jet Propulsion Laboratory.

Svanaes, D. (n.d.). *Understanding interactivity: Steps to a phenomenology of human–computer interaction.* Retrieved September 14, 2004, from www.idi.ntnu.no~dags/interactivity.pdf

Tappolet, C. (2000). *Emotions et valeurs* [Emotions and values]. Paris: Pressess Universitaires de France.

Thöne-Otto, A. I. T., Schulze, H., Irmscher, K., & von Cramon, D. Y. (2001). MEMOS: Interaktive elektronische Gedächtnishilfe für hirngeschädigte Patienten [MEMOS: Interactive electronic memory aids for brain injury patients]. *Deutsches Ärzteblatt, Ausgabe B, 11,* B598–B600.

Thöne-Otto, A. I. T., & Walther, K. (2003). How to design an electronic memory aid for brain-injured patients: Considerations on the basis of a model of prospective memory. *International Journal of Psychology, 38,* 1–9.

Toulmin, S. (1958). *The use of argument.* Cambridge, England: Cambridge University Press.

Toulmin, S. (1978, September 28). The Mozart of psychology. *New York Review of Books, 25*(14), 51–57.

Trope, Y., & Liberman, N. (2003). Temporal construal. *Psychological Review, 3,* 403–421.

Tulving, E. (2002). Episodic memory: From mind to brain. *Annual Review of Psychology, 53,* 1–25.

Van den Broek, M. D., Downes, J., Johnson, Z., Dayus, B., & Hilton, H. (2000). Evaluation of an electronic memory aid in the neuropsychological rehabilitation of prospective memory deficits. *Brain Injury, 14,* 455–462.

Varian, H. (2002, August 29). Avoiding the pitfalls when economics shifts from science to engineering. *The New York Times,* B–2.

Vera, A. H., & Simon, H. A. (1993). Situated action: A symbolic interpretation. *Cognitive Science, 17,* 7–48.

Vertut, J., & Coiffet, P. (1985). *Les robots: Téléopération, Vol. 3A & 3B.* Paris: Hermès.

Vertut, J., & Coiffet, P. (1985). *Teleoperation and Robotics, Vol. 3B.* London: Kogan Page.

Victor, B., & Boynton, A. (1998). *Invented here: Maximizing your organization's internal growth and profitability.* Boston: Harvard Business School Press.

Vygotsky, L. S. (1962). *Thinking and speaking* (E. Hanfmann, & G. Vakar, Eds., Trans.). Cambridge, MA: MIT Press. (Original work published 1934)

Vygotsky, L. S. (1980). *Mind in society: The development of higher psychological processes.* Cambridge, MA: Harvard University Press.

Vygotsky, L. S. (1983). The history of the development of the higher mental functions. In L. S. Vygotsky (Ed.), *Collected works, Vol. 3: Problems of the development of mind* (pp. 84–91). Moscow: Pedagogica. (Original work published 1931)

Wade, T. K., & Troy, J. C. (2001). Mobile phones as a new memory aid: A preliminary investigation using case studies. *Brain Injury, 15,* 305–320.

Walton, D. N. (2000). The place of dialogue theory in logic, computer science and communication studies. *Synthesis: An International Journal for Epistemology, logic, and philosophy of Science, 123,* 327–346.

Warschauer, M. (2003). *Technology and social inclusion: Rethinking the digital divide.* Cambridge, MA: MIT Press.

Watzlawick, P., Beavin, J., & Jackson, D. D. (1967). *Pragmatics of human communication: A study of interactional patterns, pathologies, and paradoxes.* New York: Norton.

Weiser, M. (1991, September). The computer for the 21st century. *Scientific American, 265*(3), 94–110.

Weiser, M. (1993). Some computer science problems in ubiquitous computing. *Communications of the ACM, 36*(7), 75–84.

Wenger, E. (1998). *Communities of practice: Learning, meaning and identity.* Cambridge, England: Cambridge University Press.

Whyte, W. H. (1988). *City: Return to the center.* New York: Anchor.

Wilkins, A. J., & Baddeley, A. D. (1978). Remembering to recall in everyday life: An approach to absentmindedness. In M. M. Gruneberg, P. E. Morris, & R. N. Sykes (Eds.), *Practical aspects of memory: Current research and issues* (pp. 27–34). Chichester, England: Wiley.

Williamson, O. E. (1975). *Market and hierarchies: Analysis and antitrust implications.* New York: Free Press.

Williamson, O. E. (1979). Transaction-cost economics: The governance of contractual relations. *Journal of Law and Economics, 22,* 233–261.

Williamson, O. E., & Masten, S. E. (Eds.). (1999). *The economics of transaction costs.* Cheltenham, England: Edward Elgar.

Wilson, B. A. (2002). Memory rehabilitation. *Zeitschrift für Neuropsychologie, 13,* 252.

Winograd, E. (1988). Some observations on prospective remembering. In M. M. Gruneberg, P. E. Morris, & R. N. Sykes (Eds.), *Practical aspects of memory: Current research and issues* (Vol. 1, pp. 348–353). Chichester, England: Wiley.

Winograd, T. (Ed.). (1996). *Bringing design to software.* New York: ACM Press.

Winograd, T., & Flores, T. (1986). *Understanding computers and cognition.* Norwood, NJ: Ablex.

Wittgenstein, L. (1953). *Philosophical investigations.* Oxford, England: Blackwell.

Wittgenstein, L. (1961). *Tractatus logico-philosophicus.* London: Routledge & Kegan Paul.

Wright, P., Rogers, N., Hall, C., Wilson, B., Evans, J., Emslie, H., et al. (2001). Comparison of pocket-computer memory aids for people with brain injury. *Brain Injury, 15,* 878–900.

Yamaguchi, M. (1975). *Trickster's world.* Tokyo: Tsukuma Shobo.

Yonemitsu, S., Higashi, Y., Fujimoto, T., & Tamura, T. (2002). Research for practical use of rehabilitation support equipment for severe dementia. *Gerontechnology 2002, 2,* 91.

Zimmer, H. D., Cohen, R. L., Guynn, M. J., Engelkamp, J., Kormi-Nouri, R., & Foley, M. N. (Eds.). (2001). *Memory for action: A distinct form of episodic memory?* New York: Oxford University Press.

Zuboff, S. (1988). *In the age of the smart machine: The future of work and power.* New York: Basic Books.

Zuckerman, I., Jinah, N., McConachy, R., & George, S. (2001). *Recognizing intentions from rejoinders in a bayesian interactive argumentation system. PRICAI2000.* Melbourne, Australia.

Author Index

Subject Index

Printed and bound by CPI Group (UK) Ltd, Croydon, CR0 4YY

17/10/2024

01775656-0008